U1AMR

TELEPEN

D1806472

ERRATUM

GROUP PLANNING AND PROBLEM-SOLVING METHODS IN ENGINEERING, *Shirley A. Olsen, editor*

The text on page 119, line 29, should read:

2. Optimization is discarded for a safe-fail strategy.

GROUP PLANNING AND PROBLEM-SOLVING METHODS IN ENGINEERING MANAGEMENT

CONSTRUCTION MANAGEMENT AND ENGINEERING
Edited by John F. Peel Brahtz

CONSTRUCTION PERFORMANCE CONTROL BY NETWORKS
H. N. Ahuja

CONSTRUCTION MANAGEMENT: Principles and Practices
Stanley Goldhaber, Chandra K. Jha,
and Manuel C. Macedo, Jr.

VALUE MANAGEMENT FOR CONSTRUCTION
Manuel C. Macedo, Jr., Paul V. Dobrow,
and Joseph J. O'Rourke

PLANNING, ENGINEERING, AND CONSTRUCTION OF ELECTRIC
POWER GENERATION FACILITIES
Jack H. Willenbrock and H. Randolph Thomas

SUCCESSFUL CONSTRUCTION COST CONTROL
H. N. Ajuha

HUMAN FACTORS/ERGONOMICS FOR BUILDING AND
CONSTRUCTION
Martin Helander, Editor

MANAGEMENT, MACHINES, AND METHODS IN CIVIL
ENGINEERING
John Christian

ACCOUNTING AND FINANCIAL MANAGEMENT FOR
CONSTRUCTION
Charles H. Mott

GROUP PLANNING AND PROBLEM-SOLVING METHODS IN
ENGINEERING MANAGEMENT
Shirley A. Olsen, Editor

GROUP PLANNING AND PROBLEM-SOLVING METHODS IN ENGINEERING MANAGEMENT

Edited by
SHIRLEY A. OLSEN
Ohio State University

A Wiley-Interscience Publication
JOHN WILEY & SONS
New York · Chichester · Brisbane · Toronto · Singapore

The University of
M...........................of
Science and Technology

13 JAN

LIBRARY

658.403

016

UML WITHDRAWN

Copyright © 1982 by John Wiley & Sons, Inc.

All rights reserved. Published simultaneously in Canada.

Reproduction or translation of any part of this work
beyond that permitted by Section 107 or 108 of the
1976 United States Copyright Act without the permission
of the copyright owner is unlawful. Requests for
permission or further information should be addressed to
the Permissions Department, John Wiley & Sons, Inc.

Library of Congress Cataloging in Publication Data:

Main entry under title:

Group planning and problem-solving methods in
   engineering management.

   (Construction management and engineering, ISSN
0193-9750)
   "A Wiley-Interscience publication."
   Include indexes.
   1. Problem solving, Group.   2. Engineering—
Management.   I. Olsen, Shirley A., 1939–
II. Series.
HD30.29.G76   658.4'036   81-19675
ISBN 0-471-08311-9   AACR2
Printed in the United States of America
10 9 8 7 6 5 4 3 2 1

CONTRIBUTORS

Dr. Charles Burnette

Charles Burnette & Associates
234 South Third Street
Philadelphia, Pennsylvania

Dr. André Delbecq

Graduate School of Business
University of Santa Clara
Santa Clara, California

Mr. Paul Fitzgerald

Consulting International Associates
1108 Main Street
Wakefield, Rhode Island

Ms. Sandra Gill

868 College Avenue
Santa Clara, California

Dr. Donald P. Grant

Architecture Department
California Polytechnic State University
San Luis Obispo, California

Dr. Julius Kane

P. O. Box 100
Bellingham, Washington

Dr. Harold Linstone

Director, Futures Research Institute
Portland State University
Portland, Oregon

Ms. Shirley A. Olsen

Associate Professor
Department of Industrial Design
The Ohio State University
Columbus, Ohio

Mr. George M. Prince

Synectics, Inc.
26 Church Street
Cambridge, Massachusetts

Dr. Tudor Rickards

The INCA Research Programme
Manchester Business School
Manchester, England

Dr. Thomas J. Snodgrass

Department of Engineering
University of Wisconsin
Madison, Wisconsin

Dr. John N. Warfield

P. O. Box 6189
Charlottesville, Virginia

Dr. Bernadine Young

Division of Academic Scheduling
University of Massachusetts
Amherst, Massachusetts

SERIES PREFACE

Industry observers agree that most construction practitioners do not fully exploit the state of the art. We concur in this general observation. Further, we have acted by directing this series of works on Construction Management and Engineering to the continuing education and reference needs of today's practitioners.

Our design is inspired by the burgeoning technologies of systems engineering, modern management, information systems, and industrial engineering. We believe that the latest developments in these areas will serve to close the state of the art gap if they are astutely considered by management and knowledgeably applied in operations with personnel, equipment, and materials.

When considering the pressures and constraints of the world economic environment, we recognize an increasing trend toward large-scale operations and greater complexity in the construction product. To improve productivity and maintain acceptable performance standards, today's construction practitioner must broaden his concept of innovation and seek to achieve excellence through knowledgeable utilization of the resources. Therefore our focus is on skills and disciplines that support productivity, quality, and optimization in all aspects of the total facility acquisition process and at all levels of the management hierarchy.

We distinctly believe our perspective to be aligned with current trends and changes that portend the future of the construction industry. The books in this series should serve particularly well as textbooks at the graduate and senior undergraduate levels in a university construction curriculum or continuing education program.

JOHN F. PEEL BRAHTZ

La Jolla, California
February 1977

PREFACE

This book is about methods and strategies that may be useful in team planning and problem-solving. It is directed to students, educators, and practicing professionals involved in policy making, planning, and problem-solving who see the need for a more cooperative attitude and participatory approach to designing. Such an approach incorporates the knowledge, values, and experiences of those parties affected by the outcome of a problem solution and those responsible for its implementation. This is an emergent attitude that has grown out of a dissatisfaction with mechanistic models and purely quantitative methods, which have been unsuccessful in dealing with social issues, product systems, communications, and environments having far-reaching effects on large segments of society.

In order to solve these types of problems, which are complex and probabilistic, methods are needed that provide for the following:

1. The collection and interpretation of qualitative data.
2. Flexibility over time in the processing of data and its subsequent impact on the outcome.
3. The incorporation of opinions and values held by team participants.
4. The processing and storing of large quantities of qualitative and quantitative data impinging on the problem area.
5. Effective idea generation and idea structuring for creative problem-solving.
6. A structure to deal with interpersonal obstacles to communication.
7. The systematization, substantiation, and documentation of decisions.

Many useful methods having these characteristics are already being used and have been described in the literature of such disciplines as architecture, education, engineering, environmental design, management science, policy science, the social and behavioral sciences, and urban planning. There is, however, no single publication that brings these methods together for comparison, nor is there an existing framework that may be used in the selection and implementation of a method or combination of methods.

The objective of this book is to bring together in one volume a collection of methods that may be useful in team planning and problem-solving and to provide a guide for the selection and implementation of one or more of these methods.

The book is structured in the following manner.

An overview of the history of policy making, planning, and problem-solving is provided to indicate the context in which the new methods have evolved.

Principles pertaining to group dynamics, social technology, group processes, human communication, and group problem-solving have been collected to guide the policy maker, planner, or problem-solver in the selection and implementation of a method or methods.

Each method is described in depth by one or more persons, each of whom has expertise in the subject through involvement in its development and/or its application. This provides the reader with first-hand knowledge, experiences, and indications of future development.

An abstract of each method is presented for quick reference and to facilitate cross-comparisons.

In summary, I believe that this book provides a comprehensive overview of current group process methods that should be useful to the newcomer or student wanting an introduction to new approaches to group problem-solving. The principles outlined at the beginning of the book provide a theoretical base to guide the user in selecting the most appropriate method, and the detailed accounts provide the necessary information for putting the method into practice.

SHIRLEY A. OLSEN

Columbus, Ohio
January 1982

CONTENTS

GROUP PLANNING AND PROBLEM-SOLVING METHODS IN ENGINEERING MANAGEMENT

CHAPTER 1

BACKGROUND AND STATE
OF THE ART

SHIRLEY A. OLSEN

As problems are realized to be more complex and difficult, as problem situations are seen to require knowledge and information from
many disciplines, and as teams of people from diverse backgrounds,
values, and perspectives are brought together to assist in obtaining
creative solutions to problems, it becomes increasingly clear that
new ways to structure and facilitate such group efforts are needed.*

In response to this statement of need and others like it, this book has
been written. It is a collection of methods that have been found useful
by groups involved in policy making, planning, and problem-solving. The
methods have been brought together in one volume to fulfill the need for
methods that facilitate the design process requiring team effort and the
need for a more participatory, democratic approach to the activity of
designing. Design, as it is used here, is broadly defined as an activity
directed toward the development of a policy, plan, or product, that, if
implemented, will result in a desirable situation in which there are little
or no unforeseen or undesired side effects and aftereffects.[1] Design is
seen as being synonymous with policy making, planning, and problem-
solving.

* Reprinted by permission of the publisher from *Environmental Design Research* by Wolfgang F. E. Preiser, Dowden, Hutchinson, and Ross, Stroudsburg, Pa. © 1973.

In order to begin our discussion of the group methods and the scope of this inquiry it is appropriate that we first review the circumstances warranting a team or group approach.

First of all, given a complex problem, there usually are no experts knowledgeable about the total problem area. The problem is most often a unique one for which there are no precedents. Also, the problem area spans many disciplines and professions. The collective members of a group or team are able to bring to the problem a larger quantity of data than would an individual. Through interaction one member of the group can build on another member's input, resulting in the production of a larger number of ideas and alternative solutions. Also, the collective members of a group have a variety of approaches to solving a problem. In problem-solving situations in which the team consists of such participants as policy makers, planners, problem-solvers, implementors, consumers or users, and those affected by the outcome, there is usually a greater understanding of the solution and the means for implementation, which reduces the chances of communication failures. The likelihood of solution acceptance is also greater, resulting in more effective implementation and maintenance.

Many designers working in teams have found the experience to be a negative one and not very fruitful in terms of outcome and meeting individual needs. There are probably a number of reasons for this. First, there may have been a lack of knowledge about group processes in general. Second, there may have been little or no knowledge about those methods that foster creative group problem-solving. Most likely the format of the meetings was that of the conventional interactive committee meeting or routine meeting. Such meetings are appropriate for coordination and information exchange.[2] They are not appropriate for creative group problem-solving involving idea generation, idea structuring, and/ or idea evaluation. Some of the problems that might arise from using the conventional interactive meeting format are group pressure, domination by one individual, a conflicting secondary goal such as "winning the argument," and disagreement, which results in hard feelings,[3] all of which interfere with the task at hand.

The methods presented in this book are the result of an inventory of methods already being used by task-oriented groups in a number of disciplines, these being architecture, business administration, engineering, environmental design, marketing, military science, and urban planning. They were selected on the basis of certain characteristics that they have in common. These appear in the list of criteria in the following section, Scope of Inquiry.

SCOPE OF INQUIRY

The focus of this inquiry is on methods that may be used by task-oriented groups or teams in creative problem-solving situations. The types of problems that benefit most by the application of such methods may be classified as complex-probabilistic[4] or, as Rittel and Webber have termed them, "wicked" problems.[5] Such problems may be viewed from several perspectives, which must be reconciled. Depending on the point of view and the interests of those involved in solving the problem and those affected by the problem solution, a number of strategies and solutions exist; sometimes these are seen as being diametrically opposed. One of the difficulties in dealing with a complex probabilistic problem lies in the definition of its boundaries. Often these problems are enmeshed within other problems in such a way that in order to solve the original problem several others must be solved as well. Often, too, it is difficult to know at what level to attack the problem. One strategy may be to attack it at the local level, whereas another may be to attack it at the national or international level. Complex problems are also characterized by large amounts of qualitative and quantitative data and a large number of dependent variables affecting each other simultaneously. This makes it very difficult to keep track of the "total picture" or the effect of one or several changes on one component or subsystem of the total problem system. Also, after the problem solution has become operational the effect of some modification or malfunction within the system is often unpredictable. As mentioned earlier, these complex probabilistic problems affect a large segment of the population, and thus within the participatory democratic group required to solve such problems there are vested interests and differences in values and world views. And finally, it should be mentioned that once the solution is implemented, it is difficult to "take it back" or alter it as a whole. Some examples of these types of problems are transportation systems, mass communications involving social issues, international sign systems, housing for the elderly, community revitalization and renewal, prefabricated housing components such as kitchens and bathrooms, mass-produced housing, and the planning and design of new cities.

Those methods found useful in solving these types of problems and presented in this book have a majority of the following criteria in common. These criteria are based upon the needs of conventional interactive groups or committees plus the needs of task-oriented groups involved in solving complex problems. The methods presented provide for most or all of the following:

1. Establishment of an agenda.
2. Goal definition.
3. Role definition.
4. Rotating leadership.
5. Isolation of issues.
6. Distinction between facts and opinions.
7. Definition of terms.
8. Specification of level of approach to the problem.
9. Recording of ideas in order to keep track of progress.
10. A structure for interaction.
11. Fulfillment of personal needs.[6]
12. Separation of the information processing subtasks—idea generation, idea structuring, and idea evaluation.[7]
13. Separation of problem. definition from generation of solution strategies.[8]
14. Collection, interpretation, and storage of quantitative and qualitative data.
15. Recognition and incorporation of opinions and values held by team members.
16. Multilogue (multiperson simultaneous dialogue).[9]
17. Provision for use of holistic languages, for example, simulation, maps, graphs, diagrams, and iconic models (two and three dimensional).
18. Provision for both individual effort and group effort depending on the nature of the subtask and the desired outcome.
19. A relaxed nonstressful environment.[10]
20. Ego-supportive interaction in which there is an open give and take situation among participants.[11]
21. Absence of penalties attached to an espoused idea or position.[12]
22. Mechanisms that support changes in attitude, rethinking, and mediation.

DEFINITIONS

The following terms are considered to be key words in this introduction and are subject to varied interpretations. Therefore, in order to clarify this discussion, they will be defined.

We begin with the definition of *design* in order to further define the audience for this book.

Design. There are two definitions of design that are preferred for the purposes of this book. One was written by Bruce Archer and the other by Horst Rittel. Archer states that "design is a goal-directed problem-solving activity."[13] Rittel further states that "design is an activity aiming at the production of a plan, which if carried out, will result in a situation with desired characteristics and without undesired or unforseen side and after effects."[14] Both of these definitions are broad in scope so that they cover the activities of many disciplines. Design is meant to refer to the processes of problem-solving, policy making, and planning found in such disciplines as architecture, business administration, city and regional planning, education, engineering, industrial design, communication design, policy science, and social work.

The next group of definitions relate to the title of the book. They are presented in the order in which they appear in the title. Some terms that are synonymous with or similar to words used in the title are also included.

Group. Patton and Giffin define *group* as a "body of individuals in interdependent role relations, having a set of values (norms) that regulate the behavior of members in matters of concern to the group."[15] A noteworthy distinction is made between group, organization, and society by Steiner in his book *Group Process and Productivity*. Steiner sees the group as a collection of mutually responsive individuals, the organization as sets of mutually responsive groups, and societies as clusters of mutually responsive organizations.[16] There are two major types of groups based on orientation or purpose. They are the task-oriented group and the social group.

Task-Oriented Group. The task-oriented group focuses on group goals and generating a group product.[17]

Social Group. The social or interpersonally oriented group focuses on satisfying social-emotional needs of individual group members.[18]

Team. *Team* is used interchangeably with *task-oriented group*. Teams are groups consolidated around a common goal or task.[19]

Process. A process, in the context of a group, consists of all the actions taken by the members of the group, individually and collectively, in the performance of a given task. It includes all those intra- and interpersonal productive and nonproductive actions by which people transform information and matter-energy into a product.[20]

Group Dynamics. *Group dynamics* is defined by Lippitt as "unique patterns of forces" within a group,[21] for example, patterns of interaction between members of the group, other groups, and larger institutions; interpersonal relationships; communication problems; stages of development; group structure; and ways in which the members make decisions.

Planning. Planning is the activity of anticipating and specifying how an objective can be achieved.[22]

Problem. A problem is "a discrepancy between two types of knowledge; factual knowledge, or knowledge of what is and deontic knowledge, or the collection of one's images about what ought to be" (definition by D. Grant).[23]

Problem-Solving. Problem-solving "can be regarded as the search for and discovery of a means to achieve or prevent transformation from one state of affairs to another, where the affairs may be abstract or concrete" (definition by S. A. Gregory).[24]

Decision-Making. Decision-making is the reaching of a conclusion on the basis of reasoning from premises by connected thought.[25] A distinction is made in this book between problem-solving and decision-making. Whereas problem-solving involves a number of phases and idea action, decision-making involves primarily one idea action or mental operation— that of evaluation.

Creativity. Creativity consists of "bringing something new into being, with emphasis on the lack of previous existence of the product or idea."[26] In the context of design, creativity is the discovery of combinations of principles, materials, or components that are especially suitable as solutions to the problem at hand.[27]

Group Creative Problem-Solving. Group creative problem-solving consists of "group interactions leading to an original solution to a problem or to a more effective recombining of given elements into new forms" (definition by L. E. Fiedler).[28]

Method (Design). The design method is a pattern of behavior and procedures employed in developing or inventing things of value that do not yet exist (definition by S. A. Gregory).[29]

CONTEXT

In light of the evolution of design methodology the methods that have been selected for inclusion in this book may be termed "political" or "participatory" methods. They reflect a new attitude toward the process of designing, which began to emerge around the beginning of the 1970s. They are, as Horst Rittel calls them, "second generation" design methods.[30]

Before characterizing these second generation design methods let us first look at "first generation" methods as they evolved in response to the needs of the design professions.

First generation design methods are based on the following assumptions[31]:

1. That there is professional expertise that can be applied to other people's problems.
2. That the design process is a process wherein the professional informs himself about a client's problem and then formulates a solution on the basis of his professional expertise.
3. That any "publicizing" or exposure of the means by which decisions are reached is unnecessary because the professional is guided by his code of ethics.
4. That quantified, objective measures obviate any need for "objectification" or making understandable.
5. That the development of increasingly complex techniques and procedures leads to better solutions, albeit at the cost of making the professional designer increasingly indispensable.

The design process in which these methods are used may be characterized as one consisting of stages or phases of design activity such as getting the brief, research and review, analysis, synthesis, preliminary design, detail design, implementation, and evaluation. Models of the design process containing these stages can take a number of forms, such as linear, circular, feedback, and branching, as presented by Koberg and Bagnall in their book *The Universal Traveler*[32] and shown in Figure 1. The first analyses of the design process and resultant identification of stages of design activity were done by Morris Asimow and L. Bruce Archer. Asimow's model of the morphology of design was presented in his book *Introduction to Design,* published in 1962,[33] and Archer's model was presented in a series of articles entitled "Systematic Method for Designers" published in *Design* magazine in 1963 and 1964.[34] These models of the design process connote a primarily linear or sequential phasing of design activities that eventually lead to a problem solution. In such cases the solution or outcome is fairly well defined at the outset.

The methods utilized in solving these problems may be thought of as scientific or systematic methods. They are largely quantitative and objective, examples being cost–benefit analysis and operations research.

The modes of communication utilized in these methods may be characterized as "advanced." Such a mode, as described by Duke,[35] is basically sequential, suitable for in-depth inquiry with a limited range of

linear

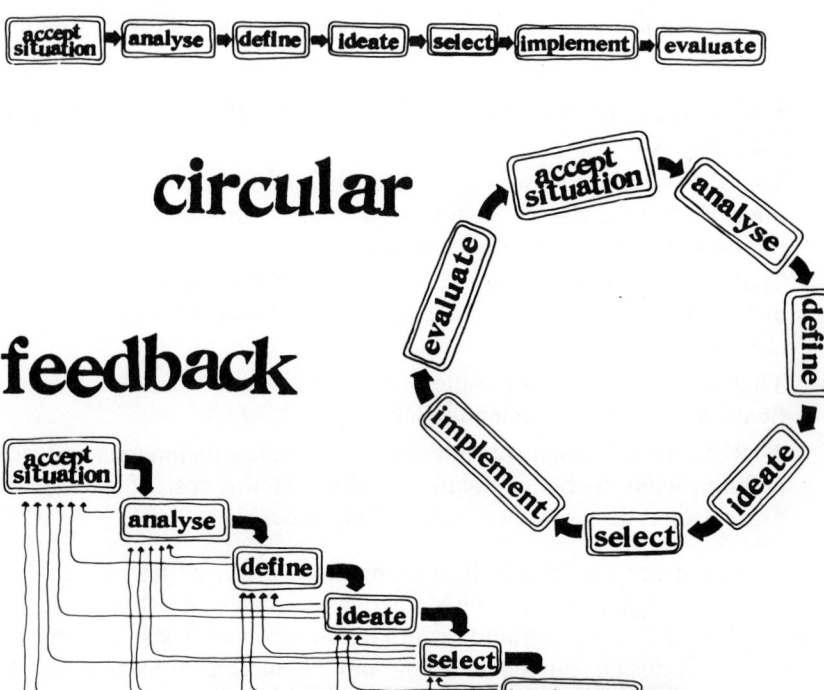

circular

feedback

branching

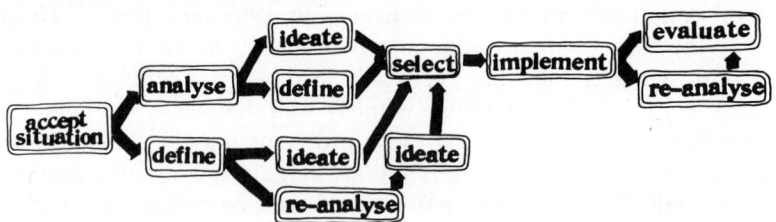

Figure 1 Models of the design process. (Reprinted by permission of the publisher from *The Universal Traveler* by Don Koberg and Jim Bagnall, William Kaufmann, Los Altos, Calif. © 1972, 1973, 1974, 1976).

applications, limited to those skilled in the particular mode, and capable of conveying sophisticated messages.

Second generation design methods on the other hand, are based on the following assumptions[36]:

1. Expertise does not reside solely in the professional, but in all those whose interests are affected by a design or planning problem (especially in the case of deontic knowledge, images of what ought to be).
2. Planning and design should be viewed as an argumentative process or as a network of issues to be argued and decided.
3. Any given issue can always be viewed as a symptom of some more fundamental one.
4. An ideal of "transparence" of argument.
5. The principle of "objectification" (making understandable) as a means toward forgetting less and stimulating doubt.
6. A client who delegates judgment to a professional must be able to maintain control over the delegated judgment.
7. The designer/planner conspires with his client to develop a solution, thus eliminating the problem of getting one's proposals implemented by his participation in producing the proposal.

Second generation problems may be characterized by a design process that is iterative and consisting of design activities that are carried out simultaneously. Solving the problem and defining the problem are the same. This is because there is no clear-cut image of the solution or of what ought to be.

The process is primarily a learning process that, after new data is collected and interpreted, requires change in a previous perception. The process, if it were to be modeled, would be a combination of the feedback and branching models described by Koberg and Bagnal (Figure 1). This approach to problem-solving first appeared in print in an interview with Horst Rittel conducted by Donald Grant and Jean-Pierre Protzen in the *DMG 5th Anniversary Report,* January 1972.

The methods utilized in solving these types of problems may be thought of as systems methods and "political" or participatory methods. This is because they require a holistic approach and participation by a heterogeneous group of people. The team or task-oriented group usually consists of from two to eight participants and may consist of, at one time or another, the client, the problem-solvers, those responsible for implementing the solution, those using or consuming the solution, and those affected by the solution. If the team consists of more than eight members, subgroups of two to eight members are usually formed. This type of approach requires a communication mode capable of conveying gestalt, which Duke terms an "integrated" mode.[37] Such a mode provides for multilogue, or multiperson simultaneous dialogue, in addition to one-way, two-way, and multiperson sequential dialogue (see Figure 2). When carefully constructed, "integrated" languages are suitable for diverse clientele and can convey a sophisticated message in a gestalt context.

It should be mentioned at this point that second generation methods are seen as *not* taking the place of first generation methods but as *additions* to the repertoire of the policy maker, planner, and problem-solver. They are meant to compliment the designer's skills and to be used in harmony with the first generation methods. The challenge to the designer is knowing which method or combination of methods is appropriate for the problem area.

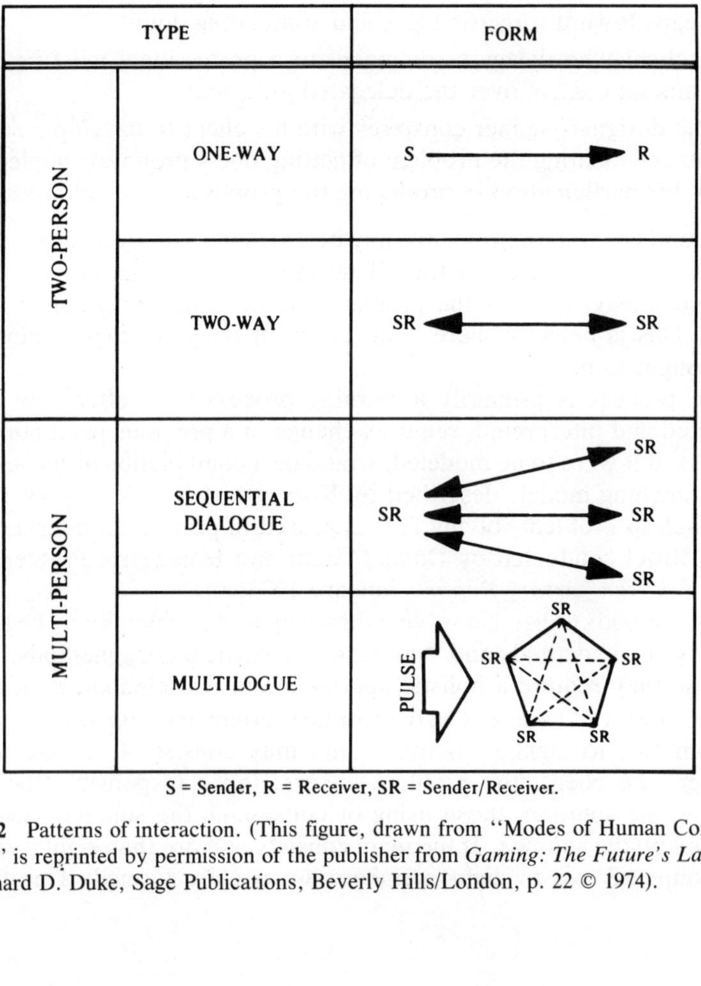

TYPE		FORM
TWO-PERSON	ONE-WAY	S ——————▶ R
	TWO-WAY	SR ◀——————▶ SR
MULTI-PERSON	SEQUENTIAL DIALOGUE	SR ◀——— SR / SR / SR
	MULTILOGUE	PULSE ⟹ (pentagon) SR SR SR SR SR

S = Sender, R = Receiver, SR = Sender/Receiver.

Figure 2 Patterns of interaction. (This figure, drawn from "Modes of Human Communication," is reprinted by permission of the publisher from *Gaming: The Future's Language* by Richard D. Duke, Sage Publications, Beverly Hills/London, p. 22 © 1974).

Second generation methods by necessity require a cooperative mode of action on the part of designers, a knowledge of group processes, and a knowledge of those methods that will support the previously stated assumptions.

For some designers the first requirement may be met by learning how to perform in a cooperative situation. This may be difficult for some who have been taught in environments fostering ego-centeredness and competitiveness. It may even require a "paradigm shift" or change in attitude and reasoning structure. Few designers are taught as students in design classes how to perform as members of a team. There may be team projects, but it may be left to chance as to whether the team members are able to work effectively with one another with productive results. It would help to know what kind of attitudes and behavior foster interaction, cooperation, and group creativity. To gain further insight into this subject the reader is referred to Magoroh Maruyama's article, "Paradigms and Communication,"[38] and Richard Duke's introductory remarks in his book *Gaming: the Future's Language.*[39]

In a team problem-solving situation it would also be beneficial to the designer to know something about group processes. Group processes include such topics as communication patterns, interpersonal relations, group development phases, group structure, and role functions. Most of the research on groups has been done within the disciplines of psychology and sociology. A recent book on the subject of task-oriented groups, entitled *Group Process and Productivity,* by Ivan D. Steiner (1972) is highly recommended. For an excellent history of research on task-oriented groups the reader is referred to the introductory chapter, by Nagao, Vollrath, and Davis, in the book *Dynamics of Group Decisions* (1978).[40] This chapter has been summarized for the reader in order to provide an overview of the primary contributors to task-oriented group problem-solving processes research (see Table 1). Of these contributors, those having a direct effect on the field of knowledge on which the second generation design methods are based are Shaw, Hackman, Steiner, and Moscovici. In 1931 Margaret Shaw[41] "established for years to come, the primacy of studying intellective performance (group problem-solving) in predominantly cooperative interaction settings." J. Richard Hackman and his associates[42] have shown that "direct interventions into the group process (such as instructions to discuss performance strategies) can lead to more efficient group performance, and hence engineer process gains rather than process losses."[43] Ivan Steiner[44] "developed a partial typology of tasks based upon his analysis of the interrelations among task demands, member responses, and group process" and also the formula Actual Productivity = Potential Productivity − Process Losses.[45] Serge

YEAR	CONTRIBUTOR(S)	RESEARCH AREA	CONTRIBUTION	REFERENCE
1898	Triplett	Social behavior	The first controlled study of social behavior	N. Triplett, "The Dynamogenic Factors in Pacemaking and Competition," American Journal of Psychology, 9 (1898), 507–533.
1902	Cooley	Primary groups and their importance in socialization, the military and organizations.	Cooley's interest in the social-emotional relations between interacting members set the direction for much of the subsequent work in sociology.	C. H. Cooley, Human Nature and the Social Order, Scribner, New York, 1902.
1920, 1924	Allport	Performance facilitation and decrement in a social context		F. H. Allport, Social Psychology, Houghton Mifflin, Boston, 1924.
1931	Shaw	Communication networks. Compared problem-solving (word puzzles) results of individuals working co-operatively in groups with individuals working independently.	Established a precedent for the study of group problem-solving in predominantly cooperative interaction settings for years to come.	M. E. Shaw, "A Comparison of Individuals and Small Groups in the Rational Solution of Complex Problems," American Journal of Psychology, 44 (1931), 491–504.
1935	Mayo	Primary groups and their importance in socialization, the military and organizations.		E. Mayo, The Human Problems of an Industrial Civilization, Macmillan, New York, 1933.

1935	Dashiell	Performance facilitation and decrement in a social context	J. F. Dashiell, "Experimental Studies of the Influence of Social Situations on the Behavior of Individual Human Adults," A Handbook of Social Psychology (C. Murchison, ed.), Clark University Press, Worcester, Mass., 1935.
1935	Sherif	Norms and the manner in which the individual subject is influenced by others nearby.	M. Sherif, "A Study of Some Social Factors in Perception," Archives of Psychology, Vol. 27, No. 187, 1935.
1948	Bavelas	The effect of communication, influence, and group cohesiveness, on the individual group member.	A. Bavelas, "A Mathematical Model for Group Structures," Applied Anthropology, 7 (1948), 16–30.

Table 1. Summary of Primary Contributors to Task-Oriented Group Problem-Solving Processes Research (This table, compiled from the "Introduction" by Dennis Nagao, David Vollrath, and James Davis, is reprinted by permission of the publisher from Dynamics of Group Decisions, (Sage Focus Ed., Vol. 5), Hermann Brandstatter, James H. Davis, and Heinz Schuler, Eds., Sage Publications Berverly Hills/London, pp. 11–22. © 1978).

YEAR	CONTRIBUTOR(S)	RESEARCH AREA	CONTRIBUTION	REFERENCE
1949	Deutsch	Mixed-motive interaction		M. Deutsch, "A Theory of Cooperation and Competition," Human Relations, 2 (1949), 129-152.
1950	Bales	Concerned with how variables such as group size affect the dynamics of the group process. Especially concerned with how to establish a balance between the time spent on the task and the time spent on the social-emotional problems of maintaining group structure.	Interaction Process Analysis, a 12 category checklist of behaviors and an instrument upon which each occurrence of the behavior over time can be coded.	R. F. Bales, Interaction Process Analysis: A Method for the Study of Small Groups, Addison-Wesley, Cambridge, Mass., 1950.
1951	Festinger	The effect of communication, influence, and group cohesiveness, on the individual group member.	Theory of cognitive dissonance (1957)	L. A. Festinger, "Informal Communications in Small Groups," Groups, Leadership and Men: Research in Human Relations (H. Guetzkow, ed.), Carnegie, Pittsburgh, 1951.

1953	Moreno	Group dynamics. Role playing and interpersonal processes conceived as patterned structure. Small group leadership styles—autocratic, democratic, laissez-faire. The individual and how the group experience effects changes in individual members.	Developed sociometry as a means of examining the emotional relations among group members. Moreno and Lewin's work mark the beginning of the sensitivity movement.	J. L. Moreno, Who Shall Survive? Beacon House, New York, Rev. ed. Beacon, 1953.
1958	Lewin	Group dynamics. The group as a vehicle for social and personal change.	Founded the Research Center for Group Dynamics in 1954. Workshop (1946) to investigate possible uses of the group as a vehicle for social and personal change. This was a precursor to the National Training Laboratories at Bethel, Maine.	K. Lewin, "Group Decision and Social Change," in E. E. Maccoby, T. M. Newcomb, and R. L. Hartley, eds., Readings in Social Psychology, Holt, Rinehart and Winston, New York, 1958.
1958	Zander	The effect of communication influence, group cohesiveness, on the individual group member.		A. Zander, "Group Membership and Individual Security," Human Relations, 11 (1958), 99–111.

Table 1 *(Continued)*

YEAR	CONTRIBUTOR(S)	RESEARCH AREA	CONTRIBUTION	REFERENCE
1960	Siegel and Fouraker	Mixed-motive interaction	Bargaining formulations useful as prescriptions for international conflict resolution.	S. Siegel and L. E. Fouraker, Bargaining and Group Decision Making: Experiments in Bilateral Monopoly, McGraw-Hill, New York, 1960.
1962	Wallach, Kogan and Bem	Risk taking in group decision-making under uncertainty	Marked the beginning of research activity in risk taking in groups.	M. A. Wallach, N, Kogan, and D. J. Bem, "Group Influence on Individual Risk Taking," Journal of Abnormal and Social Psychology, 25 (1973), 75-86.
1965	Zajonc	Individual performance facilitated by the presence of others on tasks requiring well-learned behaviors but inhibited on those requiring the acquisition of new information.		R. B. Zajonc, "Social Facilitation," Science, 149 (1965), 269-274.

1966, 1972	Steiner	Group productivity	Developed formula: Actual Productivity = Potential Productivity - Process Losses; and developed a partial typology of tasks based upon his analysis of the interrelations among task demands, member resources, and group process.	I. D. Steiner, "Models for Inferring Relationships Between Group Size and Potential Group Productivity," _Behavioral Science_, 11 (1966), 273–283. I. D. Steiner, _Group Process and Productivity_, Academic Press, New York, 1972.
1968	Cartwright and	Group dynamics	History of Group dynamics--"Origins of Group Dynamics" in Group Dynamics by Cartwright and Zander.	D. Cartwright and A. Zander, eds., Group Dynamics, Harper & Row, New York, 1968.
1974	Moscovici and Nemeth	Mixed-motive interaction. Nonmonetary motives of small group members; majority-minority relations and influence.		S. Moscovici and C. Nemeth, "Social Influence II: Minority Influence," in C. Nemeth, ed., Social Psychology: Classic and Contemporary Integrations, Rand McNally, Chicago, 1974.

Table 1 (*Continued*)

YEAR	CONTRIBUTOR(S)	RESEARCH AREA	CONTRIBUTION	REFERENCE
1974	Hackman, Weiss, and Brousseau	Group productivity	Hackman and colleagues have shown that direct interventions into the group process (such as instructions to discuss performance strategies) can lead to more efficient group performance, and hence engineer process gains rather than process losses.	J. R. Hackman, J. A. Weiss, and K. Brousseau, Effects of Task Performance Strategies on Group Performance Effectiveness, Technical Report No. 5, Department of Administrative Sciences, Yale University, New Haven, Conn., 1974.
1975	Hackman and Morris			J. R. Hackman and C. G. Morris, "Group Tasks, Group Interaction Process, and Group Performance Effectivens: A Review and Proposed Integration," in L. Berkowitz, ed., Advances in Experimental Social Psychology, Vol. 8, Academic Press, New York, 1975.
1976	Moscovici		Group polarization effect which reflects concern with social mechanics of interaction among members disparately inclined at the outset.	S. Moscovici, Social Influence and Social Change, Academic Press, London, 1976.

Year	Author	Description	Reference
1976	Davis, Laughlin, and Komorita	Mixed-motive interaction. Gaming simulation which provided concise social situations, into which independent variables were introduced to gauge effects on a simple dependent variable.	J. H. Davis, P. R. Laughlin, and S. S. Komorita, "The Social Psychology of Small Groups: Cooperative and Mixed-Motive Interaction," Annual Review of Psychology, 27 (1976), 501-541.
1976	Myers and Lamm	Polarization research	Summary of literature on polarization research. Working definition of polarization—"The average postgroup response will tend to be more extreme in the same direction as the pregroup responses."	D. G. Myers and H. Lamm, "The Group Polarization Phenomenon," Psychological Bulletin, 83 (1976), 602-627.

Table 1 (*Continued*)

Moscovici's research focus is on mixed-motive interaction.[46] His contributions have been in the area of "social mechanics of interaction among members disparately inclined at the outset" and nonmonetary motives of small group members.[47]

The third requirement for effectively solving second generation design problems is a knowledge of those methods that are appropriate for application. The objective of this book is to help fulfill this requirement.

REFERENCES

1. Horst Rittel, "Some Principles for the Design of an Educational System for Design," *DMG Newsletter,* January 1971, pp. 2–10.

2. André Delbecq, Andrew Van de Ven, and David Gustafson, *Group Techniques for Program Planning: A Guide to Nominal Group and Delphi-Processes,* Scott Foresman, Chicago, Ill., 1975.

3. Norman R. F. Maier, "Assets and Liabilities in Group Problem-Solving: The Need for an Integrative Function," *Psychological Review,* Vol. 74, No. 4, July 1967, p. 239.

4. Stafford Beer, *Cybernetics and Management,* English Universities Press, London, 1959.

5. Horst Rittel and Melvin Webber, "Dilemmas in a General Theory of Planning," *DMG–DRS Journal,* Vol. 8, No. 1, Jan.–Mar. 1974, pp. 31–39.

6. Gordon L. Lippitt, *Organization Renewal,* Appleton-Century-Crofts, New York, 1969.

7. André Delbecq, "The Management of Decision-Making Within the Firm: Three Strategies for Three Types of Decision-Making," *Academy of Management Journal,* Vol. 10, 1967.

8. Ibid.

9. Richard Duke, *Gaming: the Future's Language,* Wiley, New York, 1974.

10. Delbecq, 1967, op. cit.

11. Ibid.

12. Ibid.

13. L. Bruce Archer, "Systematic Method for Designers," reprint from *Design,* 1965.

14. Rittel, op. cit.

15. Bobby R. Patton and Kim Giffin, *Decision-Making Group Interaction,* Harper and Row, New York, 1978.

16. Ivan D. Steiner, *Group Process and Productivity,* Academic Press, New York, 1972.

17. David W. Wright, *Small Group Communication,* Kendall/Hunt Publishing, Dubuque, Ia., 1975.

18. Ibid.

19. Kenneth V. Stevens, "Systematic Methods: The Implications for Project Teams," *DMG–DRS Journal,* Vol. 9, No. 2, pp. 145–152.

20. Steiner, op. cit.

21. Lippitt, op. cit.

22. L. Bruce Archer, "Technological Innovation: A Methodology," Royal College of Art, London, 1970.

23. Donald P. Grant, "A Compromise-Generating Planning Information System for Public Housing Location Conflicts," *The DMG 5th Anniversary Report,* DMG Occasional Paper, No. 1, Jan. 1972, pp. 23–28.

24. S. A. Gregory, Ed. *The Design Method,* Plenum Press, New York, 1966.

25. Archer, 1970, op. cit.

26. J. C. Flanagan, "The Definition and Measurement of Ingenuity," *Scientific Creativity: Its Recognition and Development* (D. W. Taylor and F. Barron, Ed.), Wiley, New York, 1963.

27. Morris Asimow, *Introduction to Design,* Prentice-Hall, Englewood Cliffs, N.J., 1962.

28. L. E. Fiedler, "Leader Attitudes, Group Climate, and Group Creativity," *Journal of Abnormal Psychology,* Vol. 65, No. 5, 1962, pp. 308–318.

29. Gregory, op. cit.

30. Horst Rittel, "Son of Rittelthink," *The DMG 5th Anniversary Report,* DMG Occasional Paper, No. 1, Jan. 1972, pp. 5–10.

31. Donald P. Grant, "The Second Generation and Beyond," *The DMG Bulletin,* No. 73-1, Jan. 1973, pp. 3–4.

32. Don Koberg and Jim Bagnall, *The Universal Traveler,* William Kaufmann, Los Altos, Calif., 1973.

33. Asimow, op. cit.

34. L. Bruce Archer, "Systematic Method for Designers," *Design,* Nos. 172, 174, 176, 179, 181, and 185, 1963, 1964.

35. Duke, op. cit.

36. Grant, 1973, op. cit.

37. Duke, op. cit.

38. Magoroh Maruyama, "Paradigms and Communication," *Technological Forecasting and Social Change,* Vol. 6, No. 1, 1974, pp. 3–32.

39. Duke, op. cit.

40. Dennis H. Nagao, David Volrath, and James H. Davis, "Introduction: Origins and Current Status of Group Decision Making," *Dynamics of Group Decisions* (Hermann Brandstatter, James H. Davis, and Heinz Schuler, Eds.), Sage Publications, London, 1978.

41. Margaret Shaw, "A Comparison of Individuals and Small Groups in the Rational Solution of Complex Problems," *American Journal of Psychology,* Vol. 44, 1931, pp. 491–504.

42. J. R. Hackman, J. A. Weiss, and K. Brousseau, *Effects of Task Performance Strategies on Group Performance Effectiveness,* Technical Report No. 5, Department of Administrative Sciences, Yale University, New Haven, Conn., 1974; and J. R. Hackman and C. G. Morris, "Group Tasks, Group Interaction Process, and Group Performance Effectiveness: A Review and Proposed Integration," *Advances in Experimental Social Psychology* (L. Berkowitz, Ed.), Vol. 8, Academic Press, New York, 1975.

43. Nagao, op. cit.

44. Ivan Steiner, op. cit.

45. Nagao, op. cit.

46. Serge Moscovici, *Social Influence and Social Change,* Academic Press, London, 1976.

47. Nagao, op. cit.

CHAPTER 2

FRAMEWORK FOR SELECTION
AND IMPLEMENTATION

SHIRLEY A. OLSEN

One of the biggest challenges to the designer involved in solving complex probabilistic problems, is that of selecting the most appropriate strategy and method(s) to be used throughout the problem-solving process. This is because the team or group required to solve complex probabilistic problems is a living, dynamic, complex, adaptive system. The participants of the group, as well as the group as a whole, change with time. Such systems are "open" internally as well as externally so that interaction between the participants may result in significant changes in the nature of the participants themselves, with important consequences for the group as a whole.[1] Furthermore, all of the activities carried out by the participants of the group are "context sensitive,"[2] that is, based on the participants' state and relationship to each other at a particular point in time and in a particular environment. As a result it is very difficult to predict group activity and productivity. Ivan Steiner, in describing the central theme of his book *Group Process and Productivity*, states that, "our limited ability to predict group outcomes stems from uncertainty about the course [interpersonal] influences will take. It is usually much easier to deduce what a group *can* do than to predict what it *will* do."[3] Under such circumstances it is easy to see why some team efforts and team products fail. In order to increase the probability of success, information pertaining to group productivity and group maintenance is presented in this chapter to guide the designer in the selection and implementation of the appropriate design method(s) and strategies and related group processes.

23

The framework that is used to structure this information is that of General Systems Theory. General Systems Theory (GST) is defined by Miller as "a set of related definitions, assumptions, and propositions which deal with reality as an integrated hierarchy of organizations of matter and energy."[4] GST provides us with a perspective for perceiving the group as a system, as one element or subsystem in a hierarchy of systems. This perspective enables us to see more clearly the relational aspects of groups and also to perceive the group as a synergistic whole definable as a structure and process. In addition, GST serves as metatheory,[5] as a means of relating and integrating diverse concepts and theories. It serves as a kind of checklist that one can use as a guide to identify gaps and deficiencies in the research pertaining to task-oriented groups. It provides us with hypotheses that represent isomorphisms within and between various hierarchies of systems. These hypotheses serve to direct future research; they also provide insight as to possible outcomes of particular interactions between systems and system components.

Principles and conjectures pertaining to task-oriented groups have been selected from the literature of the behavioral and social sciences, management science, and design and organized within this framework of GST. A conscious effort was made to include only those findings related to group problem-solving. Thus, for example, conclusions drawn from experiments involving therapy groups and T groups have been excluded.

The principles are based on the conclusions of experiments that verify the results of earlier experiments under identical or nearly identical conditions. These principles, however, should be considered as tentative and equivocal.* This is because they are situation specific and have not yet been applied in problem-solving situations involving complex, probabilistic problems. Most, if not all, of the problems presented in the research experiments have been "tame" problems, that is, problems for which there is a definite right or wrong, true or false answer. Also, the groups have met under controlled laboratory conditions—conditions in which certain factors are operating that would not necessarily be found operating in everyday working situations.

The conjectures are based on projections made by social and behavioral scientists on the basis of research findings and on projections made by design practitioners and educators following the application of a technique or procedure in a team design situation with favorable results.

---

* "A statement of a principle or conjecture may be equivocal in that it has multiple interpretations or it may be equivocal in application in the sense that the validity may be variously perceived, depending on the situation" (John Warfield, "TOTOS: Improving Group Problem-Solving," *Approaches to Problem-Solving,* No. 3, sponsored jointly by the Battelle Institute and the Academy for Contemporary Problems, Columbus, Ohio, 1975, p. 2).

These principles and conjectures are meant to serve as guidelines for the selection and implementation of procedures that support planning, policy making, and problem-solving by groups. It will be through their use in complex, probabilistic problem situations that their applicability and usefulness will be demonstrated.* The next step in developing the body of knowledge related to team problem-solving will be the documentation of the design process and resulting outcome(s) in which the principles and conjectures have been used so that their applicability or nonapplicability can be determined.

In the literature pertaining to task-oriented group problem-solving and design methods certain factors stand out as being of primary importance in the maintenance of group stability and productivity; they are listed in Table 1. Their corresponding classification in a GST taxonomy is included to facilitate inquiry into the factor's position and relationship to other factors within the GST framework. In referring to its GST classification the factor is seen in a new light, in a broader context, and perhaps with greater clarity.

In the following presentation of factors and related principles and conjectures the design or behavioral science term will appear first, followed by its GST classification in parentheses. Each GST classification will be defined, the definitions coming from James Miller's taxonomy of GST concepts and hypotheses in his recent book *Living Systems*.[6]

1. GROUP TASK (PROCESS)

"All changes over time of matter-energy or information in a system is process."[7] Group task in a design sense and in terms of group productivity refers to the process(es) of altering matter-energy and/or information into other forms of matter-energy and/or information to achieve a particular goal.

As mentioned earlier, Ivan Steiner has made a significant contribution toward the development of a framework for task-oriented group research with his partial typology of tasks based on task demands, member resources, and group process. It is significant because it provides a means for reviewing past research and classifying it with respect to some standard. The classification system also helps identify gaps in the research, which serves to indicate possible future directions for inquiry.

---

* The following statement summarizes very succinctly the rationale for the approach taken in this chapter. It was made by John Warfield in the introduction to his publication, "TOTOS: Improving Group Problem-Solving," sponsored jointly by the Battelle Institute and the Academy for Contemporary Problems, 1975. "It may be that the *user* of social science should be responsible for formulating tentative principles based partly on social science research which the *user* believed to be at least worth testing in everyday life" (p. 2).

BEHAVIORAL SCIENCE AND DESIGN TERMINOLOGY	GENERAL SYSTEMS THEORY CLASSIFICATION
Group task	Process
Group structure	Structure
Roles	Subsystems
Group communication	Subsystems which process information
Communication network	Channel Net
Group cohesiveness Morale	Integration Morale
Group size	Number of members
Mental operations in problem-solving	Associating Memory Deciding
Problem complexity	Input

Table 1. Factors of Primary Importance in the Selection and Implementation of Design Methods and Meta-processes That Affect Group Stability and Productivity.

The factor of "group task" and a summary of Steiner's typology of tasks have been placed first in this discussion because many of the principles and conjectures that follow under other factor classifications contain terminology or make reference to concepts presented in his typology.

In Steiner's typology there are two major task classifications—*unitary* tasks and *divisible* tasks.

Unitary Tasks. "Unitary tasks cannot profitably be divided into subtasks and performed in piecemeal fashion by two or more individuals. Consequently, the group product must either be the outcome generated by a single member or some combination of the outcomes produced by several members."*

Divisible Tasks. Divisible tasks "can be profitably broken into specialized subtasks, each of which may be performed by a different person."*

The classifications that follow are all forms of unitary tasks, but they are also used in discussing divisible tasks. This is because divisible tasks can be broken down into subtasks, which ultimately become unitary or individual tasks.

There are seven types of unitary tasks, two of which help determine the "permitted process" or ways that members can combine their individual products.[8] These two types of tasks are called *maximizing* tasks and *optimizing* tasks.

Maximizing Tasks. "Maximizing tasks make success depend upon how many units of an outcome are produced, or how rapidly they are produced."*

Optimizing Tasks. Optimizing tasks make success a function of the degree to which the obtained outcome approximates a preferred or "best" value.[9]

The other five types of unitary tasks, which are in part determined by the above process tasks, are single-member, disjunctive, conjunctive, additive, and discretionary. Of these five types, there are two types of single-member tasks in which the product of the group is based on the product of one person. They are termed *disjunctive tasks* and *conjunctive tasks.*

Disjunctive Tasks. Disjunctive tasks require that total weight be assigned to the most productive member. "The group product is really an individual product that is sanctioned by the group. One member receives

* Reprinted by permission of the publisher and author from *Group Process and Productivity* by Ivan D. Steiner, Academic Press, New York. © 1972.

total 'weight' in determining the group product and others are accorded no weight at all. The task may be said to be disjunctive because it requires an 'either–or' decision; the group can accept only one of the available individual contributions as its own, and must reject all others.''*

Conjunctive Tasks. "Conjunctive tasks require that total weight be assigned to the least productive member. If a single member fails to contribute the unique information he or she has been given, the group cannot solve the problem. In this respect, the task is conjunctive, as are many real-life tasks that permit or require a division of labor.''*

Additive Tasks. "Additive tasks require that members' contributions be weighted equally and that they be summed.''* Additive tasks usually involve maximization.

Discretionary Tasks. Discretionary tasks "permit group members to select their own weighting and combinatorial rules. Within broad limits, members are free to employ whatever processes they prefer. Discretionary tasks seem invariably to require optimization. When an optimizing task is discretionary, the quality of the group product depends heavily upon the process the group actually employs.''*

All of the principles and conjectures pertaining to group tasks and the matching of group members to tasks refer to *potential productivity* and *actual productivity*. There are important distinctions between the two. They should therefore be defined before going further.

Potential Productivity. Steiner defines potential productivity as "the maximum level of productivity that can occur when an individual or group employs its resources to meet the task demands of a work situation. It is the level of productivity that will be attained if the individual or group uses its resources in the most advantageous way permitted by the rules under which it must operate.''* If the individual or group does *not* utilize its resources in the most advantageous way permitted and/or the rules are inappropriate, then there will be process losses and the resulting productivity is termed "actual productivity."

Actual Productivity. Actual productivity is "what an individual or group does, in fact accomplish.''* Steiner has developed the following formula to express the relationship between potential productivity and actual productivity:

Actual Productivity = Potential Productivity − Process Losses

---

* Reprinted by permission of the publisher and author from *Group Process and Productivity* by Ivan D. Steiner, Academic Press, New York. © 1972.

The following principles and conjectures are based on conclusions formulated by Steiner in summing up his review of the literature pertaining to the productivity of task-oriented groups and also on the conclusions drawn from the findings of specific experiments.

Potential and Actual Productivity

CONJECTURE 1.1

"How well a group will perform a task depends upon the relevant resources of its members, on task demands that determine which processes are permitted and prescribed, and on the process that members of the group actually employ."[10]

PRINCIPLE 1.1

"When actual process corresponds to prescribed process, actual productivity will equal potential productivity."[11]

Some of the factors that may prevent this from happening, as cited by Steiner, are the following[12]:

1. "Failure of status differences to parallel the quality of the contributions offered by participating members."
2. "The low level of confidence proficient members sometimes have in their own ability to perform the task."
3. "Social pressures that an incompetent majority may exert on a competent minority."
4. "The fact that the quality of individual contributions is often very difficult to evaluate."

Disjunctive Tasks

PRINCIPLE 1.2

"When a task is *disjunctive,* a group's *potential* productivity is determined by the resources of its most competent member. If at least one person has the ability to perform the task perfectly, the group has the potential ability to perform it perfectly."[13]

PRINCIPLE 1.3

When a task is *disjunctive,* "*actual* productivity of the group may fall below its *potential* productivity because the processes that transpire in the group deviate from the prescribed pattern."[14]

CONJECTURE 1.2

"Evidence suggests that process in *disjunctive* tasks is likely to be faulty when

1. a majority of the group members initially favor an outcome other than that generated by the most competent person,
2. the most competent person has low status in the group,
3. the most competent person is not very confident of his own ability to perform the task, or
4. the most competent person does not present his contribution very aggressively and does not evoke supportive reactions from others.''*

Principles 1.2 and 1.3 and Conjecture 1.2 are based on the research findings of Maier and Solem, 1952,[15] Torrance, 1954,[16] Thomas and Fink, 1961,[17] and Johnson and Torcivia, 1967.[18]

Conjunctive Tasks

At this time no principles or conjectures have been formulated pertaining to conjunctive tasks and productivity.[19]

Additive Tasks

PRINCIPLE 1.4

"When tasks are additive, the group product is the sum of the individual contributors, but there is no guarantee that members will contribute as much as they are capable of doing or that they will perform the summation process with better precision.''[20]

This principle is based on the research findings of Ringlemann in the 1920's as described in the chapter by D. J. F. Dashiell in the *Handbook of Social Psychology*, 1935.[21]

Discretionary Tasks

CONJECTURE 1.3

In *discretionary* tasks, which usually involve optimization, "the nature of the prescribed process will depend upon the distribution of

---

* Reprinted by permission of the publisher and author from *Group Process and Productivity* by Ivan D. Steiner, Academic Press, New York. © 1972.

individual contributions that are combined into a single product and upon the criterion value that is regarded as being correct. (Unfortunately, members will not ordinarily know which of many permitted processes is prescribed, and the group's success will depend upon the one they actually employ.)''*

Divisible Tasks

CONJECTURE 1.4

"Vertical divisions of a task into subtasks usually means that persons who work on different subtasks are conjunctively independent. This is the case whenever all subtasks must be completed in order to generate a single unit of the group product. Conjunctivity also exists if some members cannot perform their specialized functions until others have performed theirs, or if the several subtasks must be performed simultaneously, each being synchronized with the other.''*

Most team design tasks are divisible and discretionary/optimizing. As we learned from Conjecture 1.2 pertaining to discretionary tasks, permitted processes are left to the discretion of the facilitator or leader of the group or to the group as a whole. Given a divisible, discretionary task, it would seem that one of the activities receiving first priority would be to identify subtasks on the basis of problem requirements. Design subtasks are usually individual, disjunctive, conjunctive, or discretionary. With disjunctive and conjunctive subtasks we can at least identify permitted processes, which is a start in understanding and implementing the metaprocesses needed to support performance of the task. Even though we cannot so readily determine permitted processes for the discretionary and individual tasks, we can, however, gain some insight from the principles and conjectures pertaining to the various types. They alert us to the problems that might be encountered, thereby helping us to plan group activities in order to avoid them and, also, how to analyze and resolve group problems when they do occur.

There are more principles and conjectures related to task and productivity that will follow according to their major factor classification. For example, a principle referring to group task and productivity that is a function of number of group members is placed under the heading Group Size (Number of Members).

* Reprinted by permission of the publisher and author from *Group Process and Productivity* by Ivan D. Steiner, Academic Press, New York. © 1972.

2. GROUP STRUCTURE (STRUCTURE)

"The structure of a system is the arrangement of its subsystems and components in three-dimensional space at a given moment in time."[22] Group structure, within the context of the behavioral sciences, has been defined by Tuckman as "the pattern of interpersonal relations at a point in time. . . ."[23]

As was mentioned earlier in this chapter, the group is seen as an open, complex, adaptive system. One of the abilities of an adaptive system is that of maintaining its stability and increasing its efficiency and effectiveness over time under changing conditions. This process is expressed in the following principle based on the experiments of Burstein and Zajonc, 1965,[24] and Bavelas and Hastorf, 1968.[25]

PRINCIPLE 2.1

"Members will alter the structure of their group(s) as the abilities of their colleagues change."[26]

If we consider the process of growth, which is an open system process, we learn that a system changes in the direction of increasing division into subsystems or differentiation of functions. Hall and Fagen note that this kind of differentiation seems to appear in systems involving some creative processes or in evolutionary and developmental processes.[27]

Differentiation may be structural or functional. If within the task-oriented group differentiation is functional, it is based on activities or tasks to be performed by the members of the group. The activities or tasks for which the member is responsible constitutes his or her "role." This phenomenon is expressed in the following conjecture.

CONJECTURE 2.1

"In any small group engaged in a process of interaction, an internal differentiation between persons as concrete units exists initially or tends to develop."[28]

This brings us to our next section, that of Roles (Subsystems), in which the concept of role is discussed.

3. ROLES (SUBSYSTEMS)

Miller defines *subsystem* as "the totality of all the structures in a system which carry out a particular *process*."[29] A structure may consist of one or more structural units called components, members, or parts. (One or more processes may be carried out by two or more structural units.) In the social and behavioral sciences the process carried out by a member is called "role." Note that there is a distinction between a role and a subtask. A role is used to refer to the type of process for which a member is responsible. A subtask is a job or action leading to a specified output or product. The following conjectures refer to both roles and subtasks.

According to Steiner, there are three primary ways in which a division of labor can be established[30]:

1. Specified matching with specified subtasks.
2. Unspecified matching with specified subtasks.
3. Unspecified matching with unspecified subtasks.

A fourth way, specified matching with unspecified subtasks, is left out because it is hypothetical; it cannot be accomplished unless the subtasks are also specified.

The following conjectures pertain to the matching of members to subtasks or division of labor.

Specified Matching With Specified Subtasks

CONJECTURE 3.1

"When individuals are advantageously matched with [specified] subtasks, groups perform better than when members are permitted to evolve their own division of labor. (However, when individuals are inappropriately matched with specified subtasks, performance is notably worse than that of groups in which neither a division nor matching is imposed.)"*

Unspecified or Self-Matching With Specified Subtasks

CONJECTURE 3.2

"In divisible tasks, consisting of subtasks that are *unitary* and *disjunctive*, groups are more productive than individuals working alone and are usually quite successful in matching members with subtasks, i.e., assigning total weight to the particular person who can best perform a specific subtask."[31]

This conjecture is based on the research findings of Gurnee, 1937,[32] Barnlund, 1959,[33] Faust, 1959,[34] and Ryack, 1965.[35]

---

* Reprinted by permission of the publisher and author from *Group Process and Productivity* by Ivan D. Steiner, Academic Press, New York. © 1972.

Conjectures 3.1 and 3.2 support an approach to problem-solving developed by Charles Burnette, an architect practicing in Philadelphia. It is called "A Role-Oriented Approach to Problem-Solving"[36] and is one of the methods presented in this book. Role definition or "specified subtasks" are central to this method. Roles are either assigned to or chosen by the members of the group. There are seven roles that are based on those tasks found to be germane to the design process:

1. Problem designator/client.
2. Resource specifier.
3. Resource organizer.
4. Form giver.
5. Actualizer–doer.
6. User.
7. Evaluator.

Unspecified or Self-Matching With Unspecified Subtasks

CONJECTURE 3.3

"When both division into subtasks and the matching of individuals with subtasks are left to the discretion of the group, the lower is the probability that the best possible division of labor will be chosen."[37]

This conjecture is based on the research findings of Leavitt, 1951[38] and Guetzkow, 1960.[39]

The "lower probability that the best division of labor will be chosen" may be due to the fact that the group must expend additional effort in the tasks of determining not only what subtasks should be performed but who should perform them. In a design situation the identification of the subtasks should be taken care of prior to the scheduled problem-solving session. The procedure for matching members to tasks will depend on the problem-solving situation. For example, some groups are created to perform a specific task or achieve a certain goal. The members are selected on the basis of their expertise at performing certain tasks. In this particular circumstance "specified matching with specified tasks" is an obvious procedure. However, in situations in which the group already exists, the leader, facilitator, or "decider" (a systems term) will have to choose between the matching of members to subtasks prior to the session or allow the members to choose for themselves those subtasks that they feel they will do best. There may be a number of factors that will help determine how this is done. For example, group morale may have a lot to do with how division of labor is decided.

George Miller, in his book *Living Systems*, has defined 19 critical subsystems of living systems.[40] They are listed in Table 2 and should be of help to the designer as a checklist of roles representing subsystems that process both matter-energy and information, subsystems that process matter-energy, and subsystems that process information. Use of this list can help the planner or facilitator of problem-solving sessions and activities to be aware of all the processes characteristic of functioning, living systems. The group or team, being a living system, manifests all or most of the subsystems in the lists, depending on its purpose and goals.

There is another list of functional roles cited frequently in the social and behavioral sciences literature pertaining to task-oriented groups. This list consists of two major types of functions that are essential for efficient decision-making in groups.[41] It was developed in connection with the First National Training Laboratory in Group Development in 1947.[42] The two major types of functions (or roles) are "group task functions" and "group building and maintenance functions." Task functions facilitate and coordinate group effort in the selection and definition of a common problem and the solution of a problem.[43] Maintenance functions are necessary to alter or maintain the way in which members of a group work together, developing loyalty to one another and to the group as a whole.[44] The specific roles supporting each function[45] are shown in Table 3.

As a result of the analysis of functional member roles, it has been concluded that the various roles should be shared and performed by any or all of the members of the group or team in order to achieve maximum efficiency and maintain group stability. The following two conjectures reflect these conclusions. The first pertains to the leadership function in particular, and the second to all of the functions.

CONJECTURE 3.4

"Leadership can and should be exercised by any and all participants with the consent and approval of the other participants, whenever such behavior helps the group to achieve a mutually desired goal."[46]

CONJECTURE 3.5

"Groups are likely to operate at maximum efficiency when members perform both task and maintenance functions, and when these functions become the responsibility of all members rather than of the designated leader alone."[47]

1. Subsystems Which Process Both Matter-Energy and Information

    1.1 Reproducer   "the subsystem which is capable of giving rise to other systems similar to the one it is in."

    1.2 Boundary   "the subsystem at the perimeter of a system that holds together the components which make up the system, protects them from environmental stresses, and excludes or permits entry to various sorts of matter-energy or information."

2. Subsystems Which Process Matter-Energy

    2.1 Ingestor   "the subsystem which brings matter-energy across the system boundary from the environment."

    2.2 Distributor   "the subsystem which carries inputs from outside the system or outputs from its subsystems around the system to each component."

    2.3 Converter   "the subsystem which changes certain inputs to the system into forms more useful for the special processes of that particular system."

    2.4 Producer   "the subsystem which forms stable associations that endure for significant periods among matter-energy inputs to the system or outputs from the converter, the materials synthesized being for growth, damage repair, or replacement of components of the system, or for providing energy for moving or constituting the system's outputs of products or information markers to its suprasystem."

    2.5 Matter-   "the subsystem which retains in the system,
         energy   for different periods of time, deposits of
         storage   various sorts of matter-energy."

Table 2. The 19 Critical Subsystems of a Living System (Based on Table 3.1 in *Living Systems* by James Miller. Reprinted by permission of the publisher, McGraw-Hill, New York. © 1978.)

2.6 Extruder	"the subsystem which transmits matter-energy out of the system in the forms of products or wastes. If the system has a purpose or goal of transmitting beyond its boundary a certain form of matter-energy which its suprasystem lacks, the particular sort of matter-energy is its product. Its wastes are sorts of matter-energy which are excess to the system and thus not useful in achieving its purposes or goals, or which hamper the accomplishment of these purposes or goals. What are wastes to a system, however, may be useful inputs lacked by its suprasystem."
2.7 Motor	"the subsystem which moves the system or parts of it in relation to part or all of its environment or moves components of its environment in relation to each other."
2.8 Supporter	"the subsystem which maintains the proper spatial relationships among components of the system, so that they can interact without weighting each other down or crowding each other."

3. Subsystems Which Process Information

3.1 Input transducer	"the sensory subsystem which brings markers bearing information into the system, changing them to other matter-energy forms suitable for transmission within it."
3.2 Internal transducer	"the sensory subsystem which receives, from subsystems or components within the system, markers bearing information about significant alterations in those subsystems or components, changing them to other matter-energy forms of a sort which can be transmitted within it."
3.3 Channel and net	"the subsystem composed of a single route in physical space, or multiple interconnected routes, by which markers bearing

Table 2 (*Continued*)

information are transmitted to all parts
of the system. Channels may intersect at
points called <u>nodes</u> and may thus be interconnected to form a <u>net</u>. This net is similar to the distributor in that the former
conveys markers bearing information and the
latter conveys matter-energy to all parts
of the system."

3.4 Decoder   "the subsystem which alters the code of
information input to it through the input
transducer or internal transducer into a
'private' code that can be used internally
by the system."

3.5 Associator   "the subsystem which carries out the first
stage of the learning process, forming
enduring associations among items of
information in the system."

3.6 Memory   "the subsystem which carries out the second
stage of the learning process, storing
various sorts of information in the system
for different periods of time."

3.7 Decider   "the executive subsystem which receives
information inputs from all other subsystems
and transmits to them information outputs
that control the entire system."

3.8 Encoder   "the subsystem which alters the code of
information input to it from other information processing subsystems, from a
'private' code used internally by the
system into a 'public' code which can be
interpreted by other systems in its
environment."

3.9 Output   "the subsystem which puts out markers
    transducer   bearing information from the system,
changing markers within the system into
other matter-energy forms which can be
transmitted over channels in the system's
environment."

Table 2 *(Continued)*

TASK ROLES	MAINTENANCE ROLES
Initiator	Encourager
Information or opinion	Harmonizer
seeker	Compromiser
Information or opinion	Standard setter
giver	Group observer and commentator
Elaborator	Expediter and gate keeper
Coordinator	Follower
Orienter	
Evaluator-critic	
Energizer	
Procedural technician	
Recorder	

Table 3.   Group Task and Maintenance Role Functions.

Another conjecture resulting from the analysis of functional member roles pertains to the role requirements of the group.

Conjecture 3.6

"Groups in different stages of an act of problem selection and solution will have different role requirements."[48]

For example, at the beginning of the problem-solving process there may be a greater need for those processes carried out by the information and opinion seeker and orienter than for the evaluator-critic and energizer. To quote from Benne and Sheats, "the combination and balance of task role requirements is a function of the group's stage of progress with respect to its task."[49]

Within the framework of GST the task functions and maintenance functions are concerned primarily with information processing. If this list is used as a guide, the user should realize that those processes pertaining to matter-energy have not been included. James Miller, in his chapter "The Group" in *Living Systems*, discusses which subsystems the task functions and maintenance functions parallel.[50] These parallels are shown in Table 4.

In conclusion, the following conjecture seems appropriate.

Conjecture 3.7

Role identification and clarification can be used as a technique to improve team effectiveness. "Used on a regular basis rather than in

GROUP TASK ROLES	SUBSYSTEM (GST)
Initiator Information and opinion seeker Information and opinion giver	Input transducing
Elaborator Coordinator Orienter Evaluator-critic Energizer Procedural technician	Deciding
Recorder	Memory

GROUP MAINTENANCE ROLES	SUBSYSTEM (GST)
Encourager Harmonizer Compromiser Standard setter	Deciding
Group observer and commentator	Internal transducing
Expediter and gate keeper	Channel and net
Follower	Motor or output transducer

Table 4. Correspondence Between Group Task and Group Maintenance Roles to a Living System's Subsystems.

response to a crisis, the procedure can clear up misunderstandings before they become heated confrontations or simmering resentments. It can also help a team to redirect its efforts by revealing inefficiencies and duplication and might lead a team to adopt new task approaches by providing new insights into the team's use of its resources."[51]

This conjecture was formulated by Kenneth Stevens, who is a design practitioner and educator. He is currently in the process of developing and testing team training exercises and team building procedures at the Illinois Institute of Technology.

4. GROUP COMMUNICATION (SUBSYSTEMS THAT PROCESS INFORMATION)

According to Miller, "all matter-energy and information subsystems are coordinated by information transmissions [or communication]. Living systems also maintain their relationships with their environments or suprasystems by inputs and outputs of information."[52] The group, being a living system, adapts and maintains its stability and productivity through the functioning of its various subsystems and information transmissions. In order to ensure that adequate communication takes place there needs to be a minimum of distortion, noise, and ambiguity. Two factors that help to ensure that messages are received and interpreted relatively accurately are feedback and redundancy. Feedback is essentially the response by the receiver(s) to a communication and is a homeostatic process. It involves the receipt of stimuli elicited by the receiver and received by the source. This data, when interpreted, may constitute information and may be used by the subsystem (source) to produce new and additional communications or to modify itself in some way. The data may be interpreted as information if it is "novel," that is, it does not duplicate information already in the receiver. Redundancy implies some kind of repetition of signals or excess of signals. If a part of a communication is missing, distorted, or masked by noise, redundant signals provide clues or the added data needed to decode the communication.

Feedback and redundancy are essential for team integration and productivity. Adequate provision should therefore be made in every group endeavor for processes providing for feedback and redundancy. Special attention should be given to the subsystems that are responsible for the major portion of the group processes at a given stage in the problem-solving process. The following conjecture comes from an article by Kenneth Stevens, entitled "Systematic Methods: The Implications for Project Teams," (1975).[53] It is essentially a procedure for obtaining feedback.

CONJECTURE 4.1

One of the ways that a team may "sharpen its internal functioning is to incorporate 'reflective' procedures into each meeting as a method for closing the session and summing up issues. In the process the team should consider questions like the following:

How well was the agenda and the procedure for working planned?

How well did the team stick to its meeting goals?

How clearly did the team understand its problems?

How successful was the team in getting information?

Was the form of the information appropriate?

How well did the team test for agreement?

How well did the team accommodate for differences among members?"

Group communication represents a broad range of research topics. As the reader can see, from its parallel GST classification the process is accomplished by means of a number of subsystems, specifically those that process information. These are the input transducer, internal transducer, channel and net, decoder, associator, memory, decider, encoder, and output transducer. Of these nine subsystems, four will be discussed individually under their own factor headings. The other subsystems will be discussed in the context of other factors. For example, the internal transducer plays a major role in group integration, so those principles and conjectures relating to internal transducer will be presented under the factor classification Group Cohesiveness and Morale.

5. COMMUNICATION NETWORK (CHANNEL AND NET)

A channel is an information processing subsystem composed of a single route in physical space, or multiple interconnected routes, by which signals are transmitted to all parts of a system or from a system to the environment. "Channels may intersect at points called *nodes*. The interconnections between the nodes form a *net*."[54]

On the basis of his review of the literature pertaining to group subsystems, James Miller has found that group communication nets have been studied experimentally more than any other group subsystem. Much of the research in this area was stimulated by Alex Bavelas in 1948[55] when he suggested a procedure for examining in the laboratory the complex structures found in large organizations.[56] Much of the research that followed, utilizing this procedure, was designed to determine how accurately and rapidly groups could solve problems when their communication activities were guided by one specific pattern of available link-ages.[57] Examples of some possible alternative linkages in five-person groups[58] are shown in Figure 1.

The value of these experiments, however, has been questioned by Glanzer and Glaser[59] for these reasons:

1. "The findings are more applicable to the level of the organization where constraints on communication are the rule,
2. many relationships are not face-to-face,
3. inadequate information may be available about components [members] which are not nearby, and
4. all components [members] are forced to participate."[60]

Even so, there are some principles and conjectures that the author feels are valid and may be useful to the designer. They are of a general nature and refer to effects of predetermined networks in general or major types of networks, such as "centralized" networks or "decentralized" networks.

CONJECTURE 5.1

"Group performance, whether measured in terms of time required to reach a solution, number of messages sent, number of errors made, or

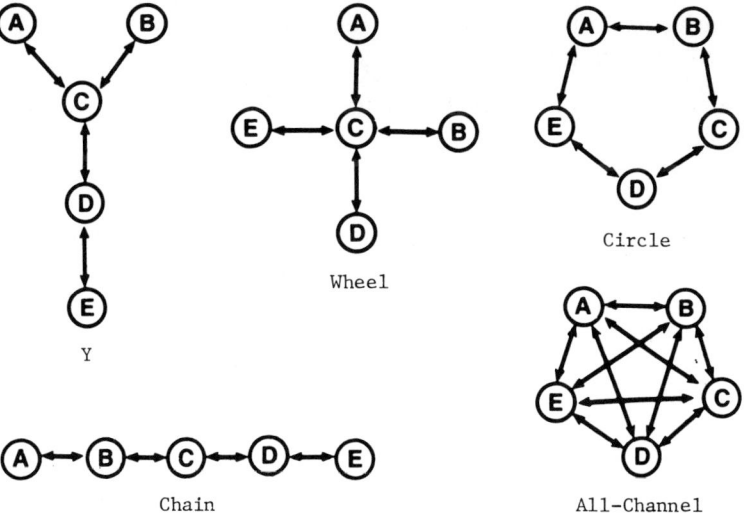

Figure 1 Five-person communication networks. (Reprinted by permission of the publisher and author from *Group Process and Productivity* by Ivan D. Steiner, Academic Press, New York. © 1972.)

rate of group learning, is affected by the communication structure imposed on the group."[61]

CONJECTURE 5.2

"The communication structure appears to affect performance by the relative difficulty it creates for the group in establishing an effective work organization.[62]

CONJECTURE 5.3

The "communication network structure in a group is a determinant not only of who transmits and receives information but also of who does the deciding."[63]

Conjectures 5.2 and 5.3 are based on the findings of Guetzkow and Simon, 1955.[64]

CONJECTURE 5.4

"A central member in a group's decision network is especially likely to be its decider or to participate in its decision making."[65]

CONJECTURE 5.5

"Increasing or decreasing the amount of information available to a member [of a group] has effects like those of increasing or decreasing his 'relative centrality' or ease of access for purposes of communication to the other members."[66]

Conjecture 5.5 is based on the findings of Shaw, 1954,[67] and Shaw and Rothschild, 1956.[68]

CONJECTURE 5.6

"The productivity of centralized networks will be low if the person who is most advantageously located to perform the 'keyman' role is disliked by his associates. Under these circumstances, the individual who is 'selected' by the network to function as the dominant member of the group may be reluctant to make decisions for others, and his colleagues may refuse to accept his decisions even if he is willing to make them."[69]

Conjecture 5.6 is based on the research of Mohanna and Argyle, 1960,[70] who found that wheel networks with unpopular central members were less productive than those in which a popular person was assigned to a central position.[71]

CONJECTURE 5.7

"Centralized groups appear to do better on a task requiring little use of logic or deduction, such as in the accumulation of information at one point in the group."[72]

CONJECTURE 5.8

"Decentralized groups are more effective when the task requires logical operations to be performed upon the information after it is accumulated in one place."[73]

Conjectures 5.7 and 5.8 are based on a review by Shaw, 1964,[74] of experiments on communication networks.

These principles and conjectures pertaining to channel and net should be of some help to the designer by first of all heightening his or her awareness about how communication networks effect group productivity. Given a particular problem-solving task that has been broken down into subtasks and analyzed according to the type of process involved and the nature of output, a network that supports the process can then be identified and implemented.

6. GROUP COHESIVENESS AND MORALE (INTEGRATION AND MORALE)

Integration exists "when multiple, simultaneous, separate processes of a system work under control of centralized decision-making, toward a common purpose or goal."[75] Integration in GST parallels cohesion in the social and behavioral science literature. In GST the term *cohesion* is also used, but to refer to "the tendency of systems to maintain sufficient closeness in space-time among subsystems and components, or between them and the channels in physical space which convey information among them to enable them to interact, resisting forces that would disrupt such relationships."[76] Morale plays a large part in the integration of components of a system. The following definition is based on factor analytic studies by Cattell, 1948.[77] "Morale includes both the amount of energy the group puts into its activities and its resistance to dispersion."[78] Cattell uses the word *synergy* to denote the energy available to the group. The energy is expended in performing the group task and integrating the group. This GST definition of *morale* parallels that found in the social and behavioral science literature.

Within a group or team the role of the internal transducer is to receive from subsystems or components within the group signals conveying potential information about changes in those subsystems or components, and modifying them or transforming them into other matter-energy forms that can be transmitted within it.[79] Since this function is essential for good group performance, it should be recognized and encouraged to function through feedback channels made available for this purpose. The internal transducer consists of one member or a subgroup of members of the task-oriented group or team, who receive and convey to the decider(s) information about group task, group integration, and individual internal states such as feelings about group processes.[80]

The following principles and conjectures should help facilitate the process of internal transducing and give some insight into those factors affecting group morale and group integration.

The first principle helps to explain group members' behavior with regard to their own performance and participation in the group task. It involves personal boundaries. Slater, 1966,[81] suggests that group interaction involves a continuing tension between the maintenance and loss of personal boundaries.[82] Both conditions can be equally attractive and threatening. According to Slater, "boundary maintenance implies autonomy and self possession, yet isolation; boundary loss implies intimacy and communication, yet a potentially frightening loss of the sense of individual self."[83] The following principle is supported by Bion, 1959,[84] Mills, 1964,[85] Slater, 1966,[86] and Hartman and Gibbard, 1974.[87]

PRINCIPLE 6.1

"Throughout the life of a group, there is a recurring or cyclical tension between the member's ambivalent desires for fusion in the group and the assertion of autonomy and individuality."[88]

This tension or shifting equilibrium of group boundaries may lead to what Hartman and Gibbard term "ego-state distress,"[89] which in turn must be dealt with by means of structural change, which brings us to the next conjecture.

CONJECTURE 6.1

"Anxiety and depression are signs or 'symptoms' of strain or shift in social equilibrium but also trigger defensive and adaptive maneuvers which lead to social change."[90]

Some self-oriented behaviors and defense mechanisms that disrupt group performance are the following:[91]

1. Fighting and controlling.
2. Withdrawing.

3. Depending and counterdepending.
4. Fixation (rigidity).
5. Projection.
6. Alienation.

The next set of principles and conjectures pertain to competitive and cooperative behavior.

CONJECTURE 6.2

"Groups are more likely to continue to hold their members if they have cooperative rather than competitive internal processes."[92]

Conjecture 6.2 is based on the research of Rosenthal and Cofer, 1948.[93]

PRINCIPLE 6.2

"Group task performance, as well as interpersonal relations among group members are better under conditions of *facilitative interdependence*, where the motivational conditions tend to foster cooperative behavior, than under contrient interdependence, where the motivational conditions tend to foster competition among group members."[94]

Principle 6.2 is based on the work of Deutsch, 1949,[95] Mintz, 1951,[96] and Thomas, 1957.[97]

CONJECTURE 6.3

"Group associative learning can enable the members to increase their cooperation, advance the common good, and discover the advantages of resisting temptations to substitute personal gain. (They are much more likely to develop the mutual trust necessary to do this if they can intercommunicate than if they cannot.)"[98]

Conjecture 6.3 is based on experimental games involving information-acquisition tasks by Rapoport, 1962,[99] and associative learning by Lanzetta and Roby, 1957.[100]

CONJECTURE 6.4

"Depending upon the way the payoffs are arranged, cooperation or competition may develop between players."[101]

CONJECTURE 6.5

"In general, group members cooperate when cooperation is rewarded and compete when it is punished."[102]

The following set of conjectures pertain to payoffs and rewards and the types of tasks for which they are appropriate. All of these were formulated by Steiner. Task types explained under factor 1, Group Task, are used in these conjectures and may have to be reviewed. Steiner makes a distinction between three types of payoffs—promotive, contrient, and independent. They are explained as follows:

1. *Promotive.* In a promotive payoff situation "everyone is encouraged to win equally. Payoffs are allocated equally within the group, but unequally among groups."[103]

2. *Contrient.** In a contrient payoff situation the disparity is great between payoffs given to the winner and loser. Competition is involved and "participants exert contrary effects upon one another's payoffs. To the extent that A succeeds, B fails."[104]

3. *Independent.* In an independent payoff situation each member's payoff depends solely upon how well his or her own performance satisfies previously established criteria; consequently, any number of members can obtain high payoffs if they perform well.[105]

CONJECTURE 6.6

"When the quality of individual outputs cannot be readily evaluated, a payoff system that promises a reward for contributing a product (regardless of its quality) should have the effect of increasing the number of alternatives among which the group may choose."[106] (Such a payoff system will create conditions resembling those characteristic of brainstorming, which is one of the methods presented in this book.)

CONJECTURE 6.7

"When neither the preferred pattern of subtasks nor the best matching of persons with subtasks is obvious, a promotive payoff system has the merit of at least permitting an appropriate pattern of specialization to emerge."[107]

CONJECTURE 6.8

"If a person's payoff depends entirely upon his personally meeting an established standard of task performance, he or she is likely to be strongly motivated to do what is required to meet it."[108]

---

* Steiner has adopted the term, "contrient" from M. Deutsch's article, "The Effects of Cooperation and Competition upon Group Process," *Human Relations,* Vol. 2, 1949, pp. 129–152 and 199–231, in which Deutsch used the term to denote the type of payoff system described here.

CONJECTURE 6.9

"In performing an *additive* task, a promotive payoff system may be advantageous when the task requires interpersonal coordination that is not readily achievable through a system of individual payoffs; people are rewarded for coordinating as well as performing."[109]

CONJECTURE 6.10

"When success depends upon the manner in which the contributions of individual members are weighted and assembled, a promotive payoff system has the merit of making self-interest parallel group interest."[110]

CONJECTURE 6.11

In conjunctive groups "no system of payoffs is likely to be entirely satisfactory, but a promotive arrangement probably has the greatest chance of maintaining tolerable internal relationships. It offers the least able person rewards that are contingent upon the goodness of his own performance and it provides others with an incentive for encouraging him to do his best."*

Another factor that has a strong effect on motivation besides payoffs is that of "evaluation apprehension." The following conjecture is related to this factor.

CONJECTURE 6.12

"When people work in the presence of an audience or co-workers who are doing the same task, their performance may be either facilitated or inhibited. Evaluation apprehension, competition, and modeling tend to intensify arousal and steer behavior along specific channels. Facilitative effects may be anticipated when task behaviors have been well learned and are expected to evoke favorable appraisals. Inhibiting effects may be anticipated when task behaviors have been poorly learned or are likely to elicit adverse appraisals."*

The rest of the principles and conjectures are quite varied, so no specific heading can be used to group them. They are all self-expanatory.

CONJECTURE 6.13

"The sort of direction which a group is given affects the satisfaction of its members."[111]

* Reprinted by permission of the publisher and author from *Group Process and Productivity* by Ivan D. Steiner, Academic Press, New York. © 1972.

In an experiment conducted in 1960 White and Lippitt[112] compared subjects' performance and productivity based on three leadership structures—autocratic, democratic, and laissez-faire. In the autocratic situation the subjects displayed hostility, aggression, and discontent, limiting their conversation to the immediate task. Subjects in the democratic situation produced less work than in the autocratic situation, but it was more original; also, the subjects liked their leader, were friendly, and praised others more than the subjects in the other groups. They were more willing to share. Subjects in the laissez-faire group produced less work than those under autocratic and democratic leadership and it was not of good quality. Play often replaced work.[113]

CONJECTURE 6.14

"If a system has multiple purposes and goals, and they are not placed in clear priority, and commonly known by all components or subsystems, conflict among them will ensue."[114]

Conjecture 6.14 is supported by research observations of Durkheim, 1947.[115]

CONJECTURE 6.15

"The more access a member has to task-relevant information the greater are his satisfactions."[116]

This conjecture is supported by the research findings of Bavelas, 1948.[117] Miller summarizes two examples from the research as follows. "A group with the circle net [shown in Figure 1], tends to transmit numerous messages and to be leaderless, unorganized and erratic, but pleasing to its members. A [wheel] group [also shown in Figure 1], on the other hand, sends many messages, has a leader, and is organized, but gives low average satisfaction."[118]

CONJECTURE 6.16

"No group can become fully productive until its members are willing to assume responsibility for the way the group acts."[119]

CONJECTURE 6.17

"The degree to which the members share the same norms as to how one should behave or what one should believe has sometimes been taken as an indicator of cohesiveness [integration]."[120]

Conjecture 6.17 is summarized from conjectures by and research findings of French, 1941,[121] Libo, 1953,[122] Coch and French, 1962,[123] and Schachter, 1962.[124]

CONJECTURE 6.18

"A group which participates in planning a task is more likely to have high morale and productivity than one which has not been involved in such planning."[125]

Conjecture 6.18 is based on the research of Bass and Leavitt, 1962–1963.[126]

CONJECTURE 6.19

"Confusion concerning the results of collective endeavor, resulting in part from the poor communication of what goes on in a group in the normal developmental sequence, suggests a very high probability that whatever results are developed in the task area will be abrogated by later action. This may not only change the group's results, but may devastate the group morale and induce hostile actions toward those who make such changes unilaterally."*

7. GROUP SIZE (NUMBER OF MEMBERS)

The term *group size* has been traditionally used in the social and behavioral sciences to refer to the number of members in a group. Miller, in his book *Living Systems*, however, uses the term to denote "spatial extent,"[127] or the physical space occupied by the members of the group when in "session." *Number of members* is used to refer to the number of members in the group. Similarly, for the purposes of this discussion *number of members* will be used.

A great amount of research has been conducted on the effect of the number of members in a group on group productivity and group information processes. But Miller has found that there are conflicting findings about the relationship of the number of members to the amount of influence they have over their associates and to the efficiency of decision-making. A comparison of results is difficult because the types of problems and conditions of decision-making in the groups differ.

There are a few general principles and conjectures, however, based on reviews of the literature. Steiner, by making a distinction between

---

* Reprinted by permission of Ralph Widner, President, Academy for Contemporary Problems, Columbus, Ohio, from "TOTOS: Improving Group Problem-Solving" by John Warfield, in *Approaches to Problem-Solving*, No. 3, 1975.

potential productivity and actual productivity, has been able to bring some order to the research findings pertaining to the effect of the number of members on group productivity. Consequently, most of the principles and conjectures under this factor heading come from his book *Group Process and Productivity*.[128]

Some general principles and conjectures from the literature will be presented first and these will be followed by those based on Steiner's review of the literature.

PRINCIPLE 7.1

"As groups increase in size, a smaller and smaller proportion of persons become central to the organization, make decisions for it, and communicate to the total membership." (Complaints about poor communication are thus more often heard in larger organizations than in smaller ones.)[129]

Principle 7.1 is based on research findings and summaries of research findings by Indik, 1963,[130] Thomas and Fink, 1963,[131] Willems, 1964,[132] and Porter and Lawler, 1965.[133]

CONJECTURE 7.1

"The size of a group should never exceed eight persons, except when it is understood that the principal role to be played by almost everyone in the group is that of listener."[134]

CONJECTURE 7.2

"A group numbering five is optimal for information processing tasks."[135] "A five person group appears to be large enough to permit members to express their feelings freely, make active efforts to solve problems even at the risk of antagonizing one another, and tolerate the loss of a member. . . . Members of groups with fewer than five members are too tense, passive, tactful, and restrained to work together efficiently and happily."*

The following principles and conjectures have been selected from Steiner's review of research findings and other summaries of research findings in his book *Group Process and Productivity*.

CONJECTURE 7.3

"The critical size that marks the upper limit of polarized [or centralized] groups appears to be a function of the extent and nature of

---

* Reprinted by permission of the publisher *Living Systems* by James G. Miller, McGraw-Hill, New York. © 1978.

the interaction that must occur between the central individual and each of his associates."[136]

CONJECTURE 7.4

"Motivation will ordinarily be lower in large groups [over eight members] than in small ones [three–eight members]. If other conditions remain constant, task motivation will tend to fall as new members are added to a group."[137]

CONJECTURE 7.5

"Costs and complications entailed in each-to-each communication become unacceptable when more than 8 or 10 persons are involved. (Above that size, groups either divide into subgroups or adopt a polarized pattern of interaction. Depending on the nature of the task, one or another of these changes may occur before size 8 is attained.)"†

CONJECTURE 7.6

"As a group increases in size, its organizational problems become more difficult to solve in the best possible manner."†

7.6.1. "If the task is *unitary*, large groups must discover and establish a procedure for coordinating and combining the efforts of many persons, whereas for small groups, the required integrative processes are generally less complex."†

7.6.2. "If the task is *divisible*, the number of subtasks into which it must be divided in order that a large group can make full use of its available resources is likely to be great, but even if the number of subtasks is constant, an increase in group size will compound the complexity of the problem of matching individuals with subtasks."†

CONJECTURE 7.7

"When tasks are *divisible*, groups can ordinarily achieve their full potential only if they are successful in establishing rather complex patterns of interpersonal coordination."†

7.7.1. "Prescribed process generally involves a division of labor. The quality of the collective effort depends heavily upon who does what when, and upon how various subtask performances are synchronized

† Reprinted by permission of the publisher and author from *Group Process and Productivity* by Ivan D. Steiner, Academic Press, New York. © 1972.

and combined. (Moreover, there is a strong element of conjunctivity because poor performance of any subtask may jeopardize group success.)''*

CONJECTURE 7.8

"The effect that a change in group size will have on *actual* productivity depends upon the impact of the change on potential productivity and process losses.''[138]

CONJECTURE 7.9

"A group is most successful when it contains the number of members that maximizes the positive discrepancy between potential productivity and process losses.''[139]

Conjecture 7.9 is based on a proposal by Thelen, 1949.[140]

CONJECTURE 7.10

If a task is *additive, disjunctive,* or *divisible*, an increase in the number of members of the group generally leads to accelerating increases in *potential* productivity. However, this increase is counteracted by a parallel accelerating increase in process losses.[141]

There are many other conjectures in addition to Conjecture 7.10 pertaining to potential productivity and they are mentioned by Steiner in his book *Group Process and Productivity* (pp. 68–78). They have not been included, however, because most complex, probabilistic problem-solving situations involve divisible and discretionary tasks. Under such circumstances permitted processes are left to the discretion of the team and/or decider(s). Among the possible alternatives there is no guarantee that the best process will be or was chosen and no way to determine or measure potential productivity. Principles and conjectures pertaining to potential productivity are not of much use for our purposes. Therefore only those principles and conjectures related to actual productivity have been presented.

In concluding this discussion about numbers of members the following list of phenomena that occur as the number of members in a group increases is presented. It is based on an evaluative review of research findings by Thomas and Fink, 1963.[142] The list was formulated by Miller and is quoted from his book *Living Systems*.†

---

* Reprinted by permission of the publisher and author from *Group Process and Productivity* by Ivan D. Steiner, Academic Press, New York. © 1972.
† Reprinted by permission of the publisher *Living Systems* by James G. Miller, McGraw-Hill, New York. © 1978.

"As the number of members in a group increases, the following things take place:

1. under some conditions both quality of group performance and group productivity tend to increase; less numerous groups are never superior in these aspects;
2. expressions of disagreements and dissatisfactions occur more frequently;
3. individual members can interact less;
4. individual members have less opportunity to lead;
5. the average amount of participation per member decreases;
6. group cohesion, measured chiefly by sociometric choices, decreases;
7. the least active member contributes less;
8. members vary more among themselves in interaction;
9. a wider range of ways to solve problems is suggested and adopted by the group;
10. each member suggests a wider range of ways to solve problems;
11. more votes are required to reach decisions;
12. more suggestions are passed on to others by those who did not originate them;
13. there is more specialization in ways to solve problems;
14. there is more organization and division of labor in the group;
15. there are more cliques in the group;
16. member satisfaction in dicussion groups trying to solve problems decreases;
17. the group becomes more representative of the population from which it comes;
18. the group becomes more heterogeneous; and
19. the group increases geometrically in the number of possible relations which can exist among members.''

8. MENTAL OPERATIONS IN PROBLEM-SOLVING (ASSOCIATING AND DECIDING)

The associator subsystem "carries out the first stage of the learning process, forming enduring associations among items of information in the group."[143] The decider subsystem "receives information inputs from all the subsystems and transmits to them information outputs that control the entire group."[144] These two subsystems are responsible for those processes that have been referred to in the literature on creativity as "mental operations." Knowing what mental operations will be involved in the processing of information in the performance of a particular problem-solving task can help the designer select a design method.

CONJECTURE 8.1

"Greater productivity can be achieved in problem-solving tasks by conscious separation of mental activity into distinct idea actions, each being carried out with methodology most appropriate to it." [145]

This conjecture, termed by Warfield, "The Separation Conjecture," supports a view of problem-solving in which information processing tasks are thought of as "idea actions" or mental operations. The literature regarding design methodology has traditionally emphasized a "phase" or "stage" design process approach to selecting a design method. However, there are certain information processing activities that occur repeatedly throughout the problem-solving process. This is especially the case with complex-probabilistic or "wicked" problems, which involve iterative processes.

Some methods may be useful in more than one stage, depending on the task and mental operation involved. By selecting a method on the basis of the information processing task *and* primary mental operations required to perform the task, rather than problem-solving phases, we are released from a view of the design process as a linear process and given a new viewpoint that supports the new strategies and methods required for solving complex probabilistic problems.

J. P. Guilford, in his general theory of intelligence, cites five basic mental operations that the organism performs in the processing of information:

1. Cognition.
2. Memory.
3. Divergent production.
4. Convergent production.
5. Evaluation.

He has incorporated these in a model of the intellect in which there are also products of the intellectual process and content or mental notation systems. Guilford's model is reproduced in Figure 2 to indicate the interrelationships between these three parameters.

In this dicussion we shall be concerned primarily with the operations. They are defined by Guilford as follows*:

---

* Reprinted by permission of the publishers from *The Nature of Human Intelligence* by J. P. Guilford, McGraw-Hill, New York. © 1967.

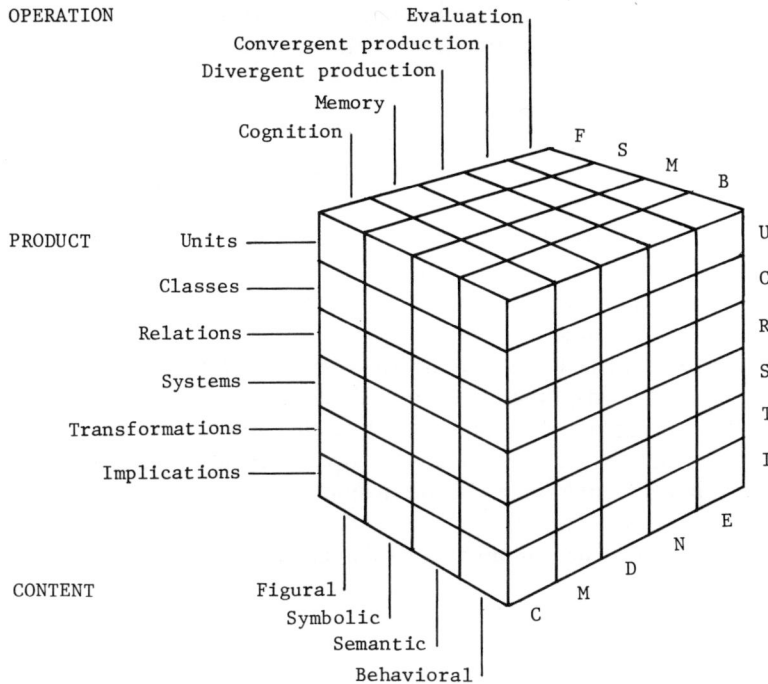

Figure 2 The structure-of-intellect model, with three parameters. (Reprinted by permission of the publisher from *The Nature of Human Intelligence* by J. P. Guilford, McGraw-Hill, New York. © 1967.)

Cognition.   "Awareness, immediate discovery or rediscovery, or recognition of information in its various forms; comprehension or understanding."

Memory   "Retention or storage, with some degree of availability, of information in the same form in which it was committed to storage and in connection with the same cues with which it was learned."

Divergent production.   "Generation of information from given information where the emphasis is upon variety and quantity of output from the same source; likely to involve transfer."

Convergent production.   "Generation of logical conclusions from given information, where emphasis is upon achieving unique or conventionally best outcomes. It is likely that the given (cue) information fully determines the outcome, as in mathematics and logic."

Evaluation.   "A process of comparing a product of information with known information according to logical criteria, reaching a decision concerning criteria satisfaction."

Within the framework of GST, cognition, divergent production, and convergent production are functions of the associator, memory is a function of memory, and evaluation is a function of the decider. Memory will not be included in the following discussion, since this function is well understood and either incorporated as part of a method or provided for throughout the total process as a means of keeping records and documenting tasks and decisions. In order to give the reader some idea as to which of the design methods presented in this book provide for the memory function, it will be included in a matrix of second generation design methods that have been correlated with the mental operations or information processing functions they support.

Before this matrix is presented, however, it should be noted that information processing functions have already been used as a basis for classifying design methods and modeling the problem-solving process. The information processing functions have been used to denote "idea actions," phases, and stages of the design process. The following overview is presented to familiarize the reader with other authors' interpretations and to provide a context for this author's approach.

Idea Actions

John Warfield, in his model of idea management, discusses three idea actions—idea generating, idea structuring, and idea communicating.[146] In this model the actions of "idea generating" and "idea structuring" are considered to be the two most basic operations that permit learning, evaluation, and memory to take place.[147] These terms encompass Guilford's operations of cognition, memory, divergent production, convergent production, and evaluation. The action of communication is included in the model because it is considered to be instrumental in carrying out the actions of idea generating and idea structuring.

Design Phases

Horst Rittel describes the design process as "an alternating sequence of two kinds of basic mental activities, followed by periods of unproblematic routine work (like converting a working drawing into a contract drawing)."[148] Each phase contains activities analogous to at least two mental operations in the Guilford model of the intellect. The first phase emphasizes divergent production and the second phase emphasizes convergent production. They are the following:

1. Initially, a phase of generating variety: the search for a set of relevant possibilities which might solve the problem at hand. (This is the

process of developing ideas. It ends with a set of alternatives which contains at least one element);"[149]

2. this is followed by a phase of reducing variety; the alternatives are evaluated for their feasibility and desirability, and a decision is made in favor of the most desirable, feasible alternative, which is incorporated into a plan until another of the same difficulty arises, which gives rise to another cycle."[150]

Design Stages

In his book on *Design Methods* J. Christopher Jones uses three major categories for the purpose of classifying design methods. These are divergence, transformation, and convergence. He uses these terms to denote a three-stage process of designing.[151] Divergence refers to "the act of extending the boundary of a design situation so as to have a large enough, and fruitful enough, search space in which to seek a solution."* Transformation represents "the stage of pattern-making, fun, high-level creativity, flashes of insight, changes of set, inspired guesswork. . . . This is the stage in which judgments of values, as well as of technicalities, are combined in decisions that should reflect the political, economic, and operational realities of the design situation. Out of all this comes the general character, or pattern, of what is being designed, a pattern that is perceived as appropriate but cannot be proved to be right."* Convergence "is the stage after the problem has been defined, the variables have been identified and the objectives have been agreed upon. The designers' aim becomes that of reducing the secondary uncertainties progressively until one of many possible alternative designs is left as the final solution to be launched into the world."* It should be noted that Jones includes methods of evaluation under this heading.

Classification of Group Process Methods

As mentioned earlier, the five basic mental operations specified by Guilford will be used as the means for classifying the group process methods presented in this book. Each operation is considered by the author to be basic, and each definition within a design context can be used to represent each information processing activity. In classifying the group process

---

* Reprinted by permission of the publisher from *Design Methods* by J. Christopher Jones, Wiley, New York. © 1970.

methods the definitions of the mental operations of memory, divergent production, and convergent production remain unchanged. The definition of cognition, however, is expanded to include insight, invention, and "creative leap"; and the definition of evaluation is expanded to refer explicitly to the process of decision-making. The matrix in Table 5 shows which mental operations are specifically supported by each design method.

MENTAL OPERATIONS → / METHODS ↓	Cognition	Memory	Divergent Production	Convergent Production	Evaluation
Brainstorming	x	x	x		
Delphi		x	x	x	
ISM		x		x	x
IBIS		x	x	x	x
KSIM	x	x	x	x	x
NGT		x	x	x	
PPM			x	x	x
Role-Oriented Approach ...	x		x	x	x
Synectics	x		x	x	
Value Engineering	x		x	x	x

Table 5. Classifications of Group Process Methods According to Mental Operations.

Problem-Solving and Decision-Making

In the literature regarding task-oriented groups the terms *problem-solving* and *decision-making* are often used interchangeably. The author of this chapter, however, makes a distinction between these two terms.

Problem-solving refers to the total process of "the search for and discovery of a means to achieve or prevent a transformation from one state of affairs to another, where the affairs may be abstract or concrete."[152] Decision-making is one aspect of problem-solving and involves the mental operation of evaluation.

There are a number of articles and publications regarding decision-making that may be of some help to the reader in identifying appropriate strategies for implementation, given a particular situation. The first source is a book entitled *Leadership and Decision-Making* by Vroom and Yetton. In Chapters 2 and 3 these authors provide the basis for "A Normative Model of Leadership Styles." In Chapter 2 there is a very useful decision process flow chart that may be used in selecting acceptable decision-making strategies.[153] It is reproduced in Figure 3. The capital letters are used to denote these strategies and are described in detail in Chapter 2 of *Leadership and Decision-Making*.

Another model of decision-making was developed by Thompson and Tuden.[154] It is a sociological model based primarily on problem situations. In this model the authors show the following:

1. There are several types of decisions to be made in and on behalf of collective enterprises.
2. Each type of decision calls for a different strategy or approach,
3. There are several varieties of organizational structures which facilitate these several strategies, and
4. The resulting behavior defines variations in decision processes.

The 'model, shown in Figure 4, by Thompson and Tuden, is based on a fourfold typology of decision issues. In essence, what the model helps us to do is first identify the type of problem situation and then the type of organization, in terms of composition, needed for organizational decision-making. Thus in a problem situation in which there is agreement about causation and preference, *computational* decision-making is required and the type of organization needed for this is one consisting of experts or *specialists*. In a problem situation in which there is uncertainty about causation and preferences are clearly known, then *judgmental* decision-making is required and the type of organization needed for this is what may be called a *collegium* or a "self-governing voluntary group,

A. If decision were accepted, would it make a difference which course of action were adopted?
B. Do I have sufficient information to make a high quality decision?
C. Do subordinates have sufficient additional information to result in high quality decision?
D. Do I know exactly what information is needed, who possesses it, and how to collect it?
* Is necessary additional information to be found within my entire set of subordinates?
† Is it feasible to collect additional information outside group prior to making decisions?
E. Is acceptance of decision by subordinates critical to effective implementation?
F. If I were to make the decision by myself, is it certain that it would be accepted by my subordinates?
G. Can subordinates be trusted to base solutions on organizational considerations?
H. Is conflict among subordinates likely in preferred solutions?

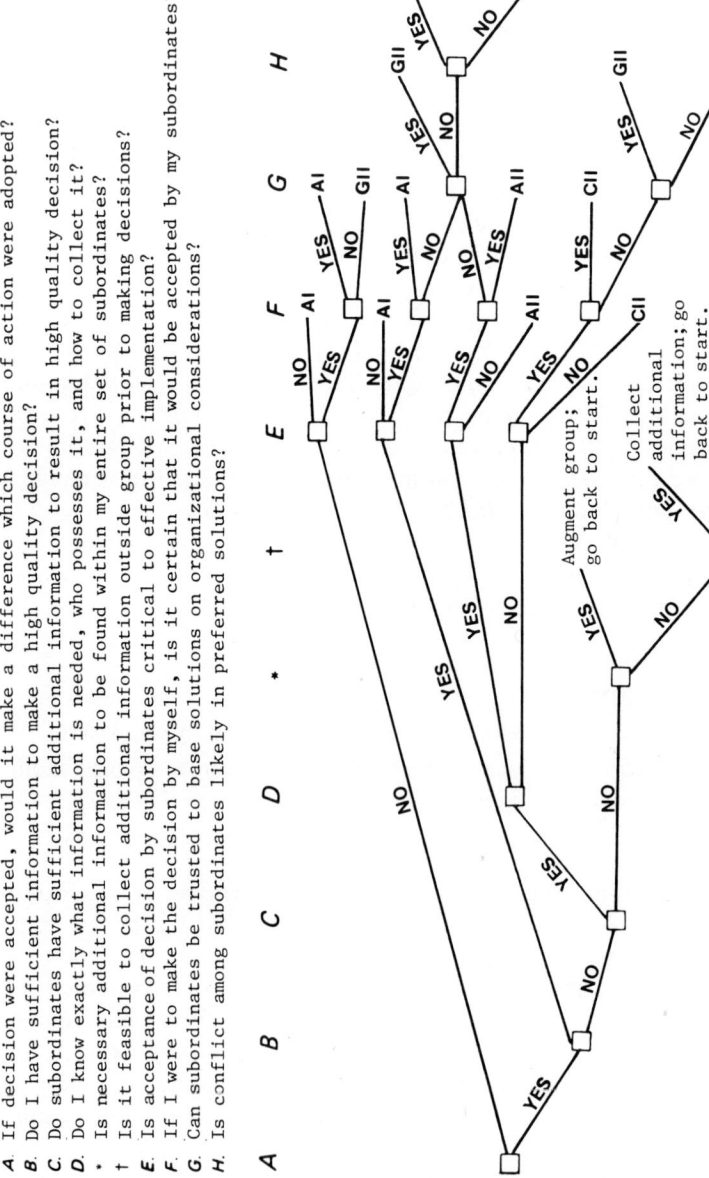

Figure 3 Decision process flow chart. (Reprinted by permission of the publisher from *Leadership and Decision-Making* by Victor H. Vroom and Philip W. Yetton, University of Pittsburgh Press, Pittsburgh, Pa. © 1973.)

PREFERENCES ABOUT POSSIBLE OUTCOMES

		Agreement	Disagreement
BELIEFS ABOUT CAUSATION	Agreement	Computation	Compromise
	Disagreement	Judgment	Inspiration

Figure 4 Fourfold typology of decision issues. (Reprinted by permission of the publisher from "Strategies, Structures, and Processes of Organizational Decision," by James D. Thompson and Arthur Tuden, in *Comparative Studies of Administration,* edited by the staff of the Administrative Science Center. Published in 1959 by the University of Pittsburgh Press).

with authority vested in the members."[155] Similarly, in a problem situation in which there is agreement about causes and uncertainty about preferences, decisions are made through *compromise* and the type of organizational structure needed is one that will facilitate bargaining, that is, a *representative body*. In problem situations in which there is no agreement about causes or preferences, an unstable situation usually exists and a *charismatic leader* is frequently the solution. Decisions are made by *inspiration.*

André Delbecq, in his article, "The Management of Decision-Making Within the Firm," develops this model further.[156] He brings greater detail to the first three classifications requiring group decision-making. He uses the term *strategy* to refer to types of decision-making and the structure of the group required to make the decisions. The three strategies are the following:

Strategy One. Routine decision-making (analagous to Simon's "programmed decision situation"[157] and Thompson and Tuden's "computational" decision).[158]

Strategy Two. Creative decision-making (analagous to Simon's "heuristic"[159] and Thompson and Tuden's "judgmental" decision-making).[160]

Strategy Three. Negotiated decision-making.

In his article Delbecq also characterizes each group in terms of group structure, group roles, group process, group style, and group norms.

9. PROBLEM COMPLEXITY (INPUT)

Input is matter-energy or information that is transmitted to a system from the suprasystem or environment through the system's boundaries. There can also be input from one subsystem to another subsystem within a system. One subsystem's output may be another subsystem's input.

▷ Problem complexity is one aspect of input that should be considered by designers when selecting a strategy or method for solving a problem. A problem as it is presented to or defined by a design team may be characterized according to certain attributes. For example, in some cases the end product or output of the design process is clearly defined, whereas in other cases the desired results are ambiguous. There may be a moderate amount of data or a very large amount of data to be collected and analyzed. There may be total agreement as to how to approach the problem or there may be several alternative approaches to the problem or total disagreement about the approach. The variables in one problem situation may be fairly independent, whereas in other problem situations the variables may be simultaneously interdependent. The effect of the outcome of one problem solution on the environment may be fairly predictable, whereas the effect of the outcome of another problem solution may be very unpredictable.

Knowledge of attributes such as these and the problem types they signify should help the designer identify which of two major categories of design methods—first or second generation—has the most potential for supporting the design tasks and producing the desired results.

There are three classification systems of problems according to type. One was developed by Herbert Simon and Allen Newell, another by Horst Rittel and Melvin Webber, and a third by Stafford Beer.

The classification system developed by Simon and Newell appeared in their article, "Heuristic Problem-Solving: The Next Advance in Operations Research."[161] They make a distinction between well-structured and ill-structured problems.

In general, well-structured problems benefit most from first generation design methods and ill-structured problems are best served by second generation methods. However, there are always exceptions, and there may be a combination of first and second generation methods, depending on the subproblems. The characteristics of first and second generation methods have been presented in Chapter 1.

Rittel and Webber, in their classification of problem types, also make a distinction between two types of problems. In their article, "Dilemmas in a General Theory of Planning,"[162] they cite the characteristics of "tame" and "wicked" problems. These attributes are summarized in Table 7. Tame problems benefit from first generation methods, whereas wicked problems benefit most from second generation methods.

Another approach to the classification of problem types can be based on Stafford Beer's classification of systems. A parallel can be drawn

TAME PROBLEMS	WICKED PROBLEMS
1. An exhaustive formulation can be stated, containing all the information the problem-solver needs for understanding and solving the problem--provided he or she knows the "art."	1. No definitive formulation. The formulation of a wicked problem is the problem.
2. There are criteria that tell when the or a solution has been found.	2. Have no stopping rule.
3. Solutions are true-or-false.	3. Solutions are not true-or-false, but good-or-bad.
4. On the spot determination of how good a solution-attempt has been.	4. No immediate and no ultimate test of a solution.
5. Problem-solver can try various runs in problem-solving without penalty.	5. Every solution is a "one-shot operation," because there is no opportunity to learn by trial-and-error; every attempt counts significantly.
6. Tool chest of operations or rules in reaching one's goal in solving the problem is more or less explicit.	6. Do not have enumerable set of potential solutions, nor is there a well-described set of permissable operations that may be incorporated into the plan.

Table 7. Some Characteristics of "Tame" and "Wicked" Problems (Reprinted in tabular form by permission of the publisher from "Dilemmas in a General Theory of Planning" by Horst W. J. Rittel and Melvin M. Webber, *Policy Sciences*, Vol. 4, No. 1 (March 1973), Elsevier Scientific Publishing, Amsterdam.)

TAME PROBLEMS	WICKED PROBLEMS
7. There are explicit characteristics of tame problems that define similarities among them.	7. Every wicked problem is essentially unique.
8. Not necessarily a problem in a hierarchical complex of problems.	8. Every wicked problem can be considered to be a symptom of another problem.
9. There are usually rules or procedures for determining the "correct" explanation of a discrepancy and subsequent solution to a problem.	9. The existence of a discrepancy representing a wicked problem can be explained in numerous ways. The choice of explanation determines the nature of the problem's resolution.
10. Immunity granted to members of the scientific community in the postulating of hypotheses that may be later refuted.	10. The planner/designer has no right to be wrong.

Table 7 (*Continued*)

between systems and problems based on the problem area or problem definition. Beer's classification refers to system complexity in which six distinctions are made.[163] They are simple deterministic, complex deterministic, exceedingly complex deterministic, simple probabilistic, complex probabilistic, and exceedingly complex probabilistic. The classifications are shown with accompanying examples in Figure 5.

The six distinctions are based on two distinct criteria, one being a system's complexity, that is, simple, complex, and exceedingly complex; the other is based on whether or not the system is deterministic or probabilistic. Under the classification of complexity Beer defines a simple system as one that is least complex, "simple but dynamic." "A system which is not simple, but which has become highly elaborate and is richly interconnected is called complex but describable." A system that has

SYSTEMS	SIMPLE	COMPLEX	EXCEEDINGLY COMPLEX
DETERMINISTIC	Window catch	Electronic digital computer	Empty
	Billiards	Planetary system	
	Machine-shop layout	Automation	
PROBABILISTIC	Penny tossing	Stockholding	The economy
	Jellyfish movements	Conditioned reflexes	The brain
	Statistical quality control	Industrial profitability	The company

Figure 5 Beer's model of system complexity. (Reprinted by permission of the author from *Cybernetics and Management* by Stafford Beer, Hodder and Stoughton, London, 1959).

"become so complicated that, while it may still be designated as complex, it cannot be described in a precise and detailed fashion" is called exceedingly complex. Note that the exceedingly complex deterministic set is empty, because any fully deterministic system can eventually be described in detail. It is just indescribable at this point in time owing to man's limited abilities to describe the system. An example of such a system is an astronomical system in the universe.[164]

A summary of characteristics of complex probabilistic problems or "wicked" problems appears in Chapter 1 under the heading Context.

Simple and complex deterministic problems and simple probabilistic problems benefit from first generation methods. Complex probabilistic and exceedingly complex probabilistic problems benefit from second generation methods. Again, as with the first two classification systems, there are exceptions, and there may be a combination of both first and second generation methods, depending on the subproblems of the total problem area.

SUMMARY

Nine factors have been presented that are considered to be of primary importance in the selection and implementation of design methods and metaprocesses that effect group stability and group productivity. They are the following:

1. Group task (process).
2. Group structure.
3. Roles (subsystems).
4. Communication (subsystems that process information).
5. Channel and net.
6. Cohesiveness (integration and morale).
7. Number of members.
8. Mental operations in problem-solving (associating and deciding).
9. Problem complexity (input).

Principles, conjectures, and theoretical constructs pertaining to each of these factors were selected primarily from the literature of the social and behavioral sciences, management science, and design.

The factors and related principles and conjectures were placed within the framework of GST. This was done in order to provide a context for the theory and research findings pertaining to group processes and to clarify the relationships between the elements and processes that characterize the group. It also provides us with a perspective for viewing the group as a structure and a process and as a living system within a hierarchy of living systems.

As a result of this presentation of factors, a pattern regarding procedure emerges. This pattern is expressed in the form of a checklist of procedures for identifying design methods that may be used throughout the design process. The checklist serves as a guide for the analysis of the problem area. The analysis is primarily a systems analysis in which the basic elements and processes of the problem "system" are identified. Those elements and processes considered to be primary are input (nature of problem), output (goal), tasks, subtasks, roles, the information processing subsystems of associator and decider (mental operations), the number of members, and the channel and net. The resulting analysis, along with the principles, conjectures, and theoretical constructs, provides a guide for the selection of design methods and meta-processes that support group stability and group productivity. The checklist is as follows:

1. Identify the problem type (and subproblems) based on attributes; for example, is the problem well-structured or ill-structured, tame or wicked, deterministic or probabilistic? Once the nature of the problem is identified, the designer can determine whether or not the

problem requires an approach involving first generation methods or second generation methods. If a second generation approach is indicated on the basis of the problem attributes, the following procedures may help the designer at the meta-process level, that is, with the selection and implementation of processes that support problem-solving activities.

2. Identify the goal and desired outcomes of the problem.

3. Based upon the goal and desired outcome, identify tasks and subtasks. This may be expressed in the form of a mission profile.

4. Identify the roles required to carry out the subtasks.

5. Match team participants with roles or tasks.

6. Analyze the tasks to determine the types of processes and subsystems involved; that is, are they subsystems that process matter-energy, information, or both matter-energy and information?

    6.1 If the tasks involve information processing by the associators or deciders, the mental operations required to perform these tasks may be identified; that is, we can ask, "Are the mental operations or idea actions cognition, memory, divergent production, convergent production, or evaluation?"

7. Analyze the tasks to determine the number of members needed to perform the task and the communication channel and net that would support the group tasks.

8. Analyze the problem in terms of the the nature of the data to be processed, the context in which the problem occurs, the desired results, the physical environment in which the problem is to be solved, the time allotment for solving the problem, and the budget. These factors are situation specific and are in addition to the nine factors presented in this chapter, which could be discussed in more general terms.

9. Select the appropriate method(s).

10. Select and implement meta-processes that will complement the methods and foster group stability and productivity.

This checklist, as well as the body of principles and conjectures presented in this chapter, are by no means complete. It is hoped that they will serve as a framework that will help designers implement a participatory democratic approach to problem-solving and to stimulate further contributions toward a more comprehensive body of knowledge pertaining to task-oriented group problem-solving.

REFERENCES

1. Walter Buckley, "Society as a Complex Adaptive System," *Modern Systems Research for the Behavioral Scientist*, (Walter Buckley, Ed.), Aldine, Chicago, Ill., 1968, pp. 490–513.

2. T. Baumgartner, T. R. Burns, P. DeVille, and L. D. Meeker, "A Systems Model of Conflict and Change in Planning Systems with Multi-Level, Multiple Objective Evaluation and Decision Making," *General Systems*, Vol. 20, 1976, pp. 167–183.

3. Ivan D. Steiner, *Group Process and Productivity*, Academic Press, New York, 1972.

4. James Grier Miller, *Living Systems*, McGraw-Hill, New York, 1978.

5. Ervin Laszlo, *The Relevance of General Systems Theory*, George Braziller, New York, 1972.

6. Miller, op. cit.

7. Miller, op. cit., p. 33.

8. Steiner, op. cit., p. 16.

9. Steiner, op. cit., p. 38.

10. Ibid.

11. Ibid.

12. Steiner, op. cit., pp. 38–39.

13. Steiner, op. cit., pp. 27, 28.

14. Steiner, op. cit., p. 28.

15. N. R. F. Maier and A. R. Solem, "The Contribution of a Discussion Leader to the Quality of Group Thinking: The Effective Use of Minority Opinions," *Human Relations*, Vol. 5, 1952, pp. 277–288.

16. E. P. Torrance "Some Consequences of Power Differences on Decision Making in Permanent and Temporary Three-Man Groups," *Research Studies, State College of Washington*, Vol. 22, 1954, pp. 130–140.

17. E. J. Thomas and C. F. Fink, "Models of Group Problem-Solving," *Journal of Abnormal and Social Psychology*, Vol. 63, 1961, pp. 53–63.

18. H. H. Johnson and J. M. Torcivia, "Group and Individual Performance in a Single-Stage Task as a Function of Distribution of Individual Performance," *Journal of Experimental Social Psychology*, Vol. 3, 1967, pp. 266–273.

19. Steiner, op. cit., p. 31.

20. Steiner, op. cit., p. 33.

21. J. F. Dashiell, "Experimental Studies of the Influence of Social Situations on the Behavior of Individual Human Adults, *Handbook of Social Psychology*, (C. Murchison, Ed.) Clark University Press, Worcester, Mass., 1935, pp. 1097–1158.

22. Miller, op. cit., p. 22.

23. B. W. Tuckman, "Developmental Sequence in Small Groups," *Psychological Bulletin*, Vol. 63, 1965, pp. 384–399.

24. E. Burnstein and R. Zajonc, "Individual Task Performance in a Changing Social Structure," *Sociometry*, Vol. 28, 1965, pp. 16–29.

25. Alex Bavelas and Albert A. Hastorf, "Experiments on the Alteration of Group Structure," *Group Dynamics* (Dorwin Cartwright and Alvin Zander, Eds.), Harper and Row, New York, 1968, pp. 527–537.

26. Dorwin Cartwright and Alvin Zander, Eds., *Group Dynamics*, Harper and Row, New York, 1968.

27. A. D. Hall and R. E. Fagen, "Definition of System" *Modern Systems Research for the Behavioral Scientist* (Walter Buckley, Ed.), Aldine, Chicago, Ill., 1968.

28. Robert F. Bales, *Interaction Process Analysis*, Addison-Wesley, Cambridge, Mass., 1950.

29. Miller, op. cit., p. 30.

30. Steiner, op. cit., p. 65.

31. Steiner, op. cit., p. 47.

32. H. Gurnee, "Maze Learning in the Collective Situation," *Journal of Psychology*, Vol. 3, 1937, pp. 437–443.

33. D. C. Barnlund, "A Comparative Study of Individual, Majority, and Group Judgment," *Journal of Abnormal and Social Psychology*, Vol. 58, 1959, pp. 55–60.

34. W. L. Faust, "Group Versus Individual Problem-Solving," *Journal of Abnormal and Social Psychology*, Vol. 59, 1959, pp. 68–72.

35. B. L. Ryack, "A Comparison of Individual and Group Learning of Nonsense Syllables," *Journal of Personality and Social Psychology*, Vol. 2, 1965, pp. 296–299.

36. Charles Burnette, Gary Moore, and Lynn Simek, "A Role Oriented Approach to Problem-Solving by Groups," *Environmental Design Research* (Wolfgang Preiser, Ed.), 4th International EDRA Conference, Vol. 1, 1973, Dowden, Hutchinson, and Ross, Stroudsburg, Pa., pp. 481–492.

37. Steiner, op. cit., p. 55.

38. H. J. Leavitt, "Some Effects of Certain Communication Patterns on Group Performance," *Journal of Abnormal Psychology*, Vol. 46, 1951, pp. 38–50.

39. H. Guetzkow, "Differentiation of Roles in Task-Oriented Groups," *Group Dynamics: Research and Theory* (Dorwin Cartwright and Alvin Zander, Eds.), Row, Peterson, Evanston, Ill., 1960, pp. 683–704.

40. Miller, op. cit., p. 54.

41. Gordon L. Lippitt, *Organization Renewal*, Appleton-Century-Crofts, New York, 1969.

42. Kenneth D. Benne and Paul Sheats, "Functional Roles of Group Members," *Small Group Communication: Selected Readings* (Victor D., Wall, Jr., Ed.), Collegiate Publishing, Columbus, Ohio, 1976, pp. 290–299.

43. Lippitt, op. cit., p. 104.

44. Lippitt, op. cit., p. 105.

45. Benne and Sheats, op. cit., pp. 292–294.

46. Bobby Patton and Kim Giffin, *Decision-Making Group Interaction*, Harper and Row, New York, 1978, p. ix.

47. Lippitt, op. cit., p. 106.

48. Benne and Sheats, op. cit., p. 296.

49. Ibid.

50. Miller, op. cit., p. 566.

51. Kenneth V. Stevens, "Systematic Methods: The Implications for Project Teams," *DMG–DRS Journal*, Vol. 9, No. 2, Apr.–Jun. 1975, pp. 145–152.

52. Miller, op. cit., p. 60.

53. Stevens, op. cit., p. 151.

54. Miller, op. cit., p. 63.

55. Alex Bavelas, "A Mathematical Model of Group Structures," *Applied Anthropology*, Vol. 7, No. 3, 1948, pp. 16–30.

56. Steiner, op. cit., p. 49.

57. Steiner, op. cit., p. 50.

58. Ibid.

59. M. Glanzer and R. Glaser, *Techniques for the Study of Team Structure and Behavior. II. Empirical Studies of the Effects of Structure*, American Institute for Research, Contract N70NR-37008, NR-154-079, Pittsburgh, Pa., June 1957.

60. Miller, op. cit., p. 538.

61. Cartwright and Zander, op. cit., p. 498.

62. Ibid.

63. H. Guetzkow and H. A. Simon, "The Impact of Certain Communication Nets upon Organization and Performance in Task-Oriented Groups," *Management Science*, Vol. 1, 1955, pp. 233–250.

64. Ibid.

65. Miller, op. cit., p. 587.

66. Miller, op. cit., p. 535.

67. M. E. Shaw, "Some Effects of Unequal Distribution of Information upon Group Performance in Various Communication Nets," *Journal of Abnormal and Social Psychology*, Vol. 49, 1954, pp. 547–553.

68. M. E. Shaw and G. H. Rothschild, "Some Effects of Prolonged Experience in Communication Nets," *Journal of Applied Psychology*, Vol. 40, 1956, p. 281.

69. Steiner, op. cit., p. 57.

70. A. Mohanna and M. Argyle, "A Cross-Cultural Study of Structural Groups with Unpopular Central Members," *Journal of Abnormal and Social Psychology*, Vol. 60, 1960, pp. 139–140.

71. Steiner, op. cit., p. 57.

72. Cartwright and Zander, op. cit., p. 498.

73. Ibid.

74. M. E. Shaw, "Communication Networks," *Advances in Experimental Social Psychology*, (L. Berkowitz, Ed.), Vol. 1, Academic Press, New York, 1964.

75. Miller, op. cit., pp. 80, 81.

76. Miller, op. cit., pp. 79, 80.

77. R. B. Cattell, "New Concepts for Measuring Leadership in Terms of Group Syntality," *Psychological Review*, Vol. 55, 1948, pp. 58–59.

78. Miller, op. cit., p. 577.

79. Miller, op. cit., p. 532.

80. Ibid.

81. P. E. Slater, *Microcosm: Structural, Psychological and Religious Evaluation in Groups*, Wiley, New York, 1966, pp. 169–181.

82. Thomas Dolgoff, "Small Groups and Organizations: Time, Task, and Sentient Boundaries," *General Systems*, Vol. 20, 1975, pp. 135–141.

83. Slater, op. cit.

84. W. R. Bion, *Experiences in Groups*, Basic Books, New York, 1959.

85. T. M. Mills, *Group Transformation: An Analysis of a Learning Group*, Prentice-Hall, Englewood Cliffs, N.J., 1964.

86. Slater, op. cit.

87. J. J. Hartman and G. S. Gibbard, "Anxiety, Boundary Evolution, and Social Change," *Analysis of Groups* (G. S. Gibbard et al., Eds.), Jossy-Bass, San Francisco, Calif., 1974, pp. 154–176.

88. Dolgoff, op. cit., p. 139.

89. Hartman and Gibbard, op. cit., p. 159.

90. Ibid.

91. Patton and Giffin, op. cit., pp. 37–38.

92. Miller, op. cit., pp. 577–578.

93. D. Rosenthal and C. N. Cofer, "The Effect on Group Performance of an Indifferent and Neglectful Attitude Shown by One Group Member," *Journal of Experimental Psychology*, Vol. 38, 1948, pp. 568–577.

94. Joseph E. McGrath, "Group Composition and Structure," *Foundations of Communication Theory* (Kenneth Sereno and C. David Mortensen, Eds.), Harper and Row, New York, 1970.

95. M. Deutsch, "A Theory of Cooperation and Competition," *Human Relations*, Vol. 2, 1949, pp. 129–152.

96. A. Mintz, "Non-Adaptive Group Behavior," *Journal of Abnormal and Social Psychology*, Vol. 46, 1951, pp. 150–159.

97. E. J. Thomas, "Effects of Facilitative Role Interdependence on Group Functioning," *Human Relations*, Vol. 10, 1957, pp. 347–366.

98. Miller, op. cit., p. 546.

99. Anatol Rapoport, "Some Self-Organization Parameters in Three Person Groups," *Principles of Self-Organization* (H. Von Foerster and G. W. Zapf, Eds.), Pergamon Press, New York, 1962, pp. 1–24.

100. J. T. Lanzetta and T. B. Roby, "Group Learning and Communication as a Function of Task and Structure Demands," *Journal of Abnormal and Social Psychology*, Vol. 55, 1957, pp. 121–131.

101. Miller, op. cit., p. 579.

102. Ibid.

103. Steiner, op. cit., p. 143.

104. Steiner, op. cit., pp. 141, 142.

105. Steiner, op. cit., p. 146.

106. Steiner, op. cit., p. 157.

107. Steiner, op. cit., p. 158.

108. Steiner, op. cit., p. 148.

109. Steiner, op. cit., p. 156.

110. Steiner, op. cit., p. 158.

111. Miller, op. cit., p. 568.

112. R. K. White and R. Lippitt, *Autocracy and Democracy*, Harper, New York, 1960.

113. Miller, op. cit., p. 568.
114. Miller, op. cit., p. 108.
115. E. Durkheim, *Division of Labor in Society*, translated from the French by George Simpson, Ph.D., The Free Press, Glencoe, Ill., 1947.
116. Miller, op. cit., p. 568.
117. Bavelas, op. cit.
118. Miller, op. cit., p. 568.
119. Lippitt, op. cit., p. 101.
120. Cartwright and Zander, op. cit., p. 70.
121. J. R. P. French, Jr., "The Description and Cohesion of Groups," *Journal of Abnormal and Social Psychology*, Vol. 36, 1941, pp. 361–377.
122. L. Libo, *Measuring Group Cohesiveness*, Institute of Social Research, Ann Arbor, Mich., 1953.
123. Lester Coch and John R. P. French, Jr., "Overcoming Resistance to Change," *Group Dynamics* (D. Cartwright and Alvin Zander, Eds.), Row, Peterson, Evanston, Ill., 1962, pp. 319–341.
124. Stanley Schachter, "Deviation, Rejection, and Communication," *Group Dynamics* (Dorwin Cartwright and Alvin Zander, Eds.), Harper and Row, New York, 1968, pp. 260–285.
125. Miller, op. cit., p. 568.
126. B. M. Bass and H. J. Leavitt, "Some Experiments in Planning and Operating," *Management Science*, Vol. 9, 1962–1963, pp. 574–585.
127. Miller, op. cit., p. 516.
128. Steiner, op. cit.
129. Cartwright and Zander, op. cit., p. 499.
130. B. Indik, "Some Effects of Organization Size on Member Attitudes and Behavior," *Human Relations*, Vol. 16, 1963, pp. 369–384.
131. E. J. Thomas and C. F. Fink, "Effects of Group Size," *Psychological Bulletin*, Vol. 60, 1963, pp. 371–384.
132. E. P. Willems, "Review of Research," *Big School, Small School* (R. B. Barker and P. V. Gump, Eds.), Stanford University Press, Stanford, Calif., 1964.
133. L. Porter and E. E. Lawler, "Properties of Organization Structure in Relation to Job Attitudes and Job Behavior," *Psychological Bulletin*, Vol. 64, 1965, pp. 23–51.
134. John Warfield, "TOTOS: Improving Group Problem-Solving," *Approaches to Problem-Solving*, No. 3, sponsored jointly by Battelle Institute and the Academy for Contemporary Problems, Columbus, Ohio, 1975, p. 5.
135. P. E. Slater, "Contrasting Correlates of Group Size," *Sociometry*, Vol. 21, 1958, pp. 129–139.
136. Steiner, op. cit., p. 102.
137. Steiner, op. cit., p. 87.
138. Ibid.
139. Steiner, op. cit., p. 95.
140. H. A. Thelen, "Group Dynamics in Instruction: Principles of Least Group Size," *School Review*, Vol. 57, 1949, pp. 139–148.
141. Steiner, op. cit., p. 95.

142. E. J. Thomas and C. F. Fink, "Effects of Group Size," *Psychological Bulletin*, Vol. 60, 1963, pp. 371–384.

143. Miller, op. cit., p. 543.

144. Miller, op. cit. p. 548.

145. Warfield, op. cit., p. 11.

146. John Warfield, *Societal Systems*, Wiley, New York, 1976.

147. Ibid., p. 58.

148. Horst Rittel, "Some Principles for the Design of an Educational System for Design," Part One, *DMG Newsletter*, Vol. 4, No. 12, Dec. 1970, 3–10.

149. Ibid., p. 8.

150. Ibid.

151. J. Christopher Jones, *Design Methods*, Wiley, London, 1970.

152. S. A. Gregory, Ed., *The Design Method*, Plenum Press, New York, 1966.

153. Victor H. Vroom and Philip Yetton, *Leadership and Decision-Making*, University of Pittsburgh Press, Pittsburgh, Pa., 1973, p. 39.

154. James D. Thompson and Arthur Tuden, "Strategies, Structures, and Processes of Organizational Decision," *Comparative Studies in Administration*, edited by the staff of Administrative Science Center, University of Pittsburgh Press, Pittsburgh, Pa., 1959, pp. 195–216.

155. Ibid., p. 200.

156. André L. Delbecq, "The Management of Decision-Making Within the Firm: Three Strategies for Three Types of Decision-Making," *Academy of Management Journal*, Vol. 10, 1967, pp. 329–339.

157. Herbert A. Simon and Allen Newell, "Heuristic Problem-Solving: The Next Advance in Operations Research," *Operations Research*, Vol. 6, Jan.–Feb. 1958, p. 6.

158. Thompson and Tuden, op. cit., pp. 198–199.

159. Simon and Newell, op. cit., p. 6.

160. Thompson and Tuden, op. cit., pp. 199–200.

161. Simon and Newell, op. cit., pp. 4, 5.

162. Horst W. J. Rittel and Melvin Webber, "Dilemmas in a General Theory of Planning," *Policy Sciences*, Vol. 4, No. 2, June 1973, pp. 155–167.

163. Stafford Beer, *Cybernetics and Management*, English Universities Press Ltd., London, 1959, pp. 9–19.

164. Ibid., p. 16.

CHAPTER 3

BRAINSTORMING

TUDOR RICKARDS

Brainstorming is a creative problem-solving method that promotes and supports idea generation and divergent thinking. Its use results in large numbers of ideas or possible solutions to a stated question or problem. The central principle of the technique is that of deferred judgment. Although large numbers of variations in operating procedures are known and applied, only when individuals or groups are undertaking to generate ideas under conditions of deferred judgment can a session be accurrately called brainstorming.

Brainstorming was popularized by the American advertising executive Alex Osborn in the 1930s being initially applied to commercial and educational problems. In his book *Applied Imagination* he described a range of techniques for stimulating the mind and generating new and useful ideas. These approaches were for use sometimes by individuals, sometimes by groups. The book became a worldwide success, but in popular memory Osborn is associated with one technique—brainstorming, and brainstorming is also popularly assumed to be only a group technique. Furthermore, the term has come to mean any unstructured or undisciplined meeting, bull session, or informal discussion, contrasted with "normal" meetings by lack of control, the absence of logic, and an output that is more crazy then creative. (One commentator coined the term "cerebral popcorn" in a scathing and misinformed attack on the technique.) As we

This chapter is an updated version of Chapter 5 and Case Study 14 from *Problem-solving through Creative Analysis* by Tudor Rickards, with permission of the author and Gower Press, Epping, Essex, 1974.

will see, there is a rationale and a discipline necessary to achieve successful brainstorming.

At its simplest, Osborn's approach was to overcome status in business meetings by having all suggestions written down without criticism until all ideas had been noted. It was from this starting point that the later versions, and misconceptions, grew.

As the approach was tested in increasingly varied fields, many claims were made for new products, millions of dollars saved through sessions on cost savings and marketing methods and other technical, aesthetic, and commercial outputs from meetings. Enthusiasm was at its peak in the 1950s in the United States, after which there was a more realistic appraisal of the technique, which was to become subsequently incorporated into a range of second generation technique systems such as value analysis (for finding economies in new product and process operations) and in the five-stage creative problem-solving approach developed by Osborn's colleagues at the State University of New York at Buffalo.

In Western Europe a similar process of experimentation took place in the 1960s, with a diffusion of knowledge and experience into commercial and educational fields. A recent survey of diffusion of creative problem-solving techniques revealed that in Europe and America the most frequently mentioned and applied technique was brainstorming. In addition, there were large numbers of modifications that had in common the essential principle of deferment of judgment.

DEFERMENT OF JUDGMENT PRINCIPLE

Although most textbooks on brainstorming emphasize the importance of deferring judgment in a brainstorming session, less is said about just what is meant by the expression. From a practical point of view we can distinguish between a conscious discipline in a meeting where participants may be highly evaluative of ideas, and yet prepared to suppress their evaluations until the ideas are produced, and, in contrast, a group that also generates a set of ideas but whose members have reached a state of nonevaluation.

In other words, deferred judgment is more than the conscious suppression of criticism. It is a state in which ideas are dreamed up without self-criticism and presented to other group members who accept the ideas without internal or external evaluation.

Any experienced brainstormer knows that the concept of absolute deferment of judgment is unattainable and perhaps undesirable (all outputs would be of equivalent worth—a random word generator could do as well). But practice can reduce the tendency to react evaluatively to an idea and can develop a channelizing of residual emotions and evaluations towards producing additional ideas (hitchhiking).

As one experienced trainer puts it, "If it's worth thinking it's worth writing . . . if it's worth writing it's worth sharing." The brainstorming climate helps to bring about such attitudes of mind.

SUITABLE TOPICS FOR BRAINSTORMING

In the overall scheme of open-ended problem-solving, brainstorming is suggested along with synectics, another method presented in this book, as the preferred techniques for idea generation and for situations requiring creative group insights. If the problem is particularly complex, it should first be restructured and only then tackled by brainstorming or synectics.

Some help in deciding which technique to adopt can be obtained through considering the importance of the client, the need for creative insights, and the sort of output desired. Brainstorming is preferable when

	FACTORS FAVORING BRAINSTORMING	FACTORS FAVORING SYNECTICS
PRESENCE OF THE CLIENT	Not a key factor for production of ideas (but to be recommended from motivational considerations)	Ideas not likely to be produced as easily in the absence of a client (regardless of motivational considerations)
SIGNIFICANCE OF NEW INSIGHTS	Usually a bonus but may not be necessary	Solutions may require a high level of new insights
OUTPUT FROM A PROBLEM-SOLVING SESSION	Simple "crystallized" concepts in large quantity needed quickly and covering as many aspects of the problem situation as possible	Smaller numbers of concepts developed in some detail and complexity often requiring subsequent tests before their value can be established

Figure 1 Factors favoring brainstorming rather than synectics as an idea-generation technique.

PROBLEM TYPES	SPECIFIC REASONS FOR SUITABILITY	EXAMPLES
New concepts for products	Large number of ideas needed from differ-	New uses for glass
New concepts for markets	ent people with different experiences	New markets for a commercial patent
		New food concepts for consumer testing
Trouble shooting and planning	Need for a rapid collection of views	Anticipating problems during a scale-up project
	Need to identify as many significant factors as possible	Reducing factory-chimney emissions
		Identifying divisional or company future needs
Managerial problems	Problem often requires production of a cross-section of views without undue inhibition by status	Improving safety performance
		Reducing warehouse losses
		Tangible rewards for outstanding work performance
Process improvements	Suggestions are additive	Value-analysis exercise (various)
		How to meter bulk materials better
		How to maintain a cold-room more cheaply

Figure 2 Problems suited to brainstorming exercises.

PROBLEM TYPES	SPECIFIC REASONS FOR UNSUITABILITY	EXAMPLES
Problems with one or a small number of correct answers; problems which appear to have only one sort of answer	Problem is not sufficiently open-ended	Who should be in charge of the company's diversification?
	Analytical thinking preferred	Which chemicals should we try next to produce the desired effect?
Extremely diffuse and complex problems	Initial sessions produce a wide range of vague solutions which essentially redefine the problem	What should the company do to save money?
	Restructuring becomes necessary; might have been better before brainstorming	How to reduce global pollution?
Problems in which the main obstacle to solution is the decision-making process	Analytical type of thinking preferred	Where should we relocate our research laboratories?
Problems demanding an extremely high level of specific technical exper- tise (e.g. mathematical or organic chemical problems) which seem to necessitate a group with rather homogeneous background experiences or simply individual activity	Heterogeneous groups are preferred for brainstorming	How to synthesize compound X by a novel patentable approach?
Problems requiring the manipulation or motivation of people who can never be involved in the exercise; problems without a client fully committed to the brainstorming	Incorrect evaluation of the spheres of influence of the participants	How to persuade the company board to give my group more money?

Figure 3 Problems not suited to brainstorming exercises.

the presence of an expert is not likely to be critical for a successful idea to be produced, when a high level of creativity is a bonus rather than a necessity, and when the output is needed as a quantity of ideas derived rapidly from a heterogeneous group of people. These factors are summarized in Figure 1. In practice, the decision tends to be tackled intuitively and for a significant proportion of problems no clear preference can be established. In these cases both methods could be tried. Figure 2 consists of a list of problem types that have been brainstormed satisfactorily in recent exercises. To reinforce the point further Figure 3 consists of a list of problem types that have been brainstormed but which, for various reasons, proved less suited to the technique.

PARTICIPANTS

Osborn recommended a group of approximately 6 to 10 people, half of whom were experienced "core" participants, that is, participants who had previously participated in a brainstorming session. The group should represent as many disciplines as possible that will be able to understand and contribute to the problem. The members should not, however, be selected for expertise in the problem; on the contrary, they should be disinterested but capable of contributing. For technical problems some level of scientific training might be desirable, but no more than one would expect an intelligent person to have picked up through high school studies.

OPERATIONAL PROCEDURES

WARM-UP SESSION

If the brainstorming team is experienced, there is a temptation to dispense with the warm-up stage. It should be remembered that no team slips immediately into a nonevaluative mode. The pressures and prejudices imposed on the team members in their everyday actions outside the room cannot be cast off at will. Therefore it is wiser to keep some form of warm-up, regardless of the experience of the team. To maintain interest the nature of the session may have to be modified by substituting a single longer exercise (e.g., inventing a new game) and perhaps reducing the total time spent on warming up to about half an hour. Some warm-up procedures are shown in Figure 4.

EXERCISE	NOTES
Introduction: discussion and presentation (5–15 minutes)	Ground rules for brainstorming outlined. Leader emphasizes principles rather than details. Deals with questions that members introduce.
Group building (5–30 minutes)	Some means of encouraging group members to interact and get to know each other. Can be brief introduction about jobs, etc.; or a topic such as "a good idea and how I got it," or "what makes people laugh?"
Individual warm-up brainstorming (5–10 minutes)	First practice with "postpone judgment" principle. Short (two minute) spells on items of interest to group members.
Non-verbal puzzles and exercises (5–10 minutes)	To encourage non-verbal fluency--set manipulative puzzles, jig saw puzzles, join dots with straight lines, etc. Or encourage doodles/drawing to stimulate the imagination.
Interactive warm-up (10–20 minutes)	A dry run with the Osborn rules to practice "postponement of judgment," e.g. "In what ways can we..." on a non-threatening but interesting topic (...reduce car park vandalism,...make a second income,... etc.). Afterwards the output is discussed for examples of hitchhiking and freewheeling.
Reinforcing warm-up (0–20 minutes)	A chance to introduce additional techniques that may be used during the session, or reinforce some aspect not well dealt with in the interactive warm-up (e.g. hitch-hiking).

According to the experience and mood of the group, the warm-up might take 30–90 minutes.

Figure 4   Some methods of starting a brainstorming ("warm-up" procedures).

PREPARING THE PROBLEM FOR THE BRAINSTORMING

Before the idea generation can start, the problem needs to be presented in a form suited to brainstorming. Practitioners of synectics will have techniques of problem definition at their disposal that are extremely suited to this purpose. For groups and leaders without synectics skills the problem redefinition can be agreed upon either at the premeeting, or during the actual meeting. The result should be a statement (or several statements) that is immediately evident as one from which a range of different types of ideas is obtainable. ("In what ways can we. . . .") Extremely broad statements can sometimes be usefully split into more easily manageable statements. (*In what ways can we improve our company's profitability* is very broad and might be turned into *in what ways can we make product X more cheaply, sell more of Y,* etc?)

Once the problem statement has been agreed upon, the brainstorming is allowed to flow. Participants can be reminded of the basic elements of the technique by displaying a summary of the key aspects as in Figure 5. Ideas are called out by the team, and the leader writes down all the ideas on large sheets of flip chart paper. When a sheet is filled, it is displayed prominently on one of the walls, where it can be seen, in order to increase the chances of additional "hitchhiked" ideas.

To the experienced leader the nature of the ideas and the speed of their production are indicators of the mood of the group. Crisp statements capturing the essence of ideas of varied types are hallmarks of a free-wheeling group. When there are long pauses and the ideas are haltingly expressed together with justifications and qualifications (or, even worse, verbal dissent from time to time), the signs are of a group having great difficulty postponing judgment. Such sessions are not true to the intended spirit of brainstorming. They tend to be the result of poor premeeting preparation, group selection, and warm-up.

Under conditions of postponed judgment the speed of idea production from a group of six to eight participants is just low enough for the leader to be able to write all ideas legibly. (It helps to write in a mixture of upper- and lower-case letters for maximum legibility, with abbreviations of words rather than interpretations and précis by the leader to avoid discussions.)

```
Try to relax and let the ideas come to you of their own
accord.

Don't evaluate ideas at this stage.

Listen and improve on other people's ideas.

The more ideas produced, the more good ideas will be
present.
```

Figure 5 Principles of "Osborn-type" brainstorming.

Over a period of 10 years the Manchester Business School creativity program has carried out hundreds of brainstorming sessions on many different types of problems. In general, provided that the premeeting conditions are followed, the actual idea generation proceeds rather easily. Even inexperienced groups can produce 50 to 100 ideas in less than half an hour. (The inexperienced group is often astonished at the output.)

For a proportion of the sessions, however, there can come a period of slump in the energy of the group. This may correspond to the period after the immediately obvious ideas have been called out and before the group begins to hitchhike and obtain more complex ideas. The leader should be prepared to encounter such a slump and either ride it out or try one of the other subroutines described below. (If the hiatus is assumed to result from the group's failing to hitchhike, rather than due to unexpected difficulties with the problem, the leader should persist, letting the silence introduce a little tension into the proceedings. Ideas then often emerge that are of high quality).

Groups tend to become excessively frivolous or bored after about half an hour and a change of procedure is recommended. At this stage the meeting should be particularly relaxed and evaluation should be postponed until some of the other idea-generation variations have been worked through.

Group interaction is a tremendous spur to creativity in itself. It is relatively rare that everyone in the group "runs dry." As one person articulates a fresh idea it provides a stimulus for all the other participants. However, an experienced leader who wishes to further stimulate the flow may switch to techniques such as the following: checklists, forced relationships, attribute listing, and a morphological approach.

1.   *Checklists.*   By means of a checklist, the brainstorming team can find stimuli to cover the various possibilities intrinsic to the problem. The rationale of a checklist is to avoid bias in the search process. Two checklists have been found particularly valuable during brainstorming: the first, a well-known operations research tool, sometimes known as the Kipling list (it was mentioned in one of Kipling's poems), and the second proposed by Alex Osborn himself.

The Kipling list is *who? what? when? how? where? and why?* Because problems have dimensions that can be associated with most or all of these words, the checklist serves its purpose—to help the user avoid overlooking aspects of the problem.

The Osborn list was recently used in conjunction with trigger sessions (see below) by a team of 10 U.K. brainstormers to generate 280 different ideas in eight minutes on "how to regenerate the British economy." The list, which was displayed in view of the members during the attempt, was a shortened version of the full Osborn list to be found in *Applied Imagination*. It read simply, "Make bigger? smaller? combine? take away from? reverse? take apart? multiply? modify?" The session was authenticated by national press observers.

2. *Forced Relationships*   One way of encouraging a brainstorming group to hitchhike and produce more complex, integrated ideas is to arrange for them to seek and force relationships between apparently unrelated objects or concepts. This skill has been said to lie at the heart of the creative process. As there is always a range of unconnected items around after a brainstorming, it is a simple procedure to pair off ideas at random (ignoring "obviously related" pairs) and challenge the group or individual to find an idea that draws on elements of the two early ones to suggest a more interesting idea. The leader has to guide the group between two extreme positions—on the one hand, there is the danger of becoming "stuck" and frustrated with a difficult force fit; on the other hand, the group should not rapidly jump from pairs of words until an "easy" fit emerges. Easy fits tend to be obvious and unimaginative.

3. *Attribute Lists.*   When a problem is concerned with new uses for a product or service, attribute listing has been found to be a valuable ingredient in a brainstorming session. In one company alone several new product opportunities have been discovered by using the technique created by industrial innovator John Carson.[1]

The technique, using individual or Osborn brainstorming, is applied by a team or an individual to list all the aspects or attributes of the product that can be dreamed up. While it should not be necessary to stress the point, there does seem to be a tendency for people to overlook obvious and unpleasant attributes. (Attributes that are unpleasant in one context may be highly valuable in another.)

An example of the technique from the project mentioned above listed about 100 attributes of an industrial chemical—chrome oxide—for which new uses were being sought. The attributes included the following: colored (green), stable, solid, powder, free flowing, abrasive, nongritty, smooth, etc. Afterwards a member of the team observed that it was a strange attribute for an abrasive to appear smooth to the touch. The upshot was the realization that the material, then sold as an abrasive, might be used as a special lubricant. This is a good example of brainstorming helping a group to overcome an assumption in their previous thinking—that an abrasive could not also be a lubricant.

4. *Morphological Analysis.*   This technique, like brainstorming itself, has found wide acceptance and modification in industrial and other

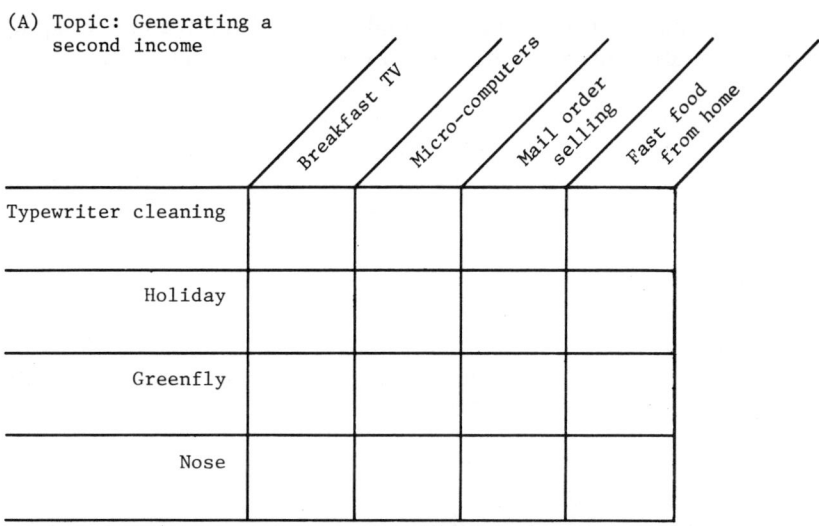

(A) Topic: Generating a
    second income

	Breakfast TV	Micro-computers	Mail order selling	Fast food from home
Typewriter cleaning				
Holiday				
Greenfly				
Nose				

Through forced relationships, the most outrageous of the 16 combinations are studied:
e.g. Idea linking Typewriter cleaning with Breakfast TV
            ... Greenfly with Micro-computer
            ... Nose with Mail order selling, etc.

(B) Topic: Improving
    our school

LOCATIONS

METHODS

	Class rooms	Playfields	Assembly Hall	Toilets
Keep cleaner				
Make safer				
Make more interesting				
Find new use for				

Figure 6   Two examples of a morphological matrix suited to brainstorming applications.

problem-solving contexts. The essential aspect of the technique is, like checklists, a structuring of the problem to help the user achieve a less biased idea search. In this technique the structuring involves setting up a matrix in which the main dimensions and elements within those dimensions can be explored. Many books have described the matrix, but few stress that there is a need for considerable skill and experience in setting up the appropriate dimensions prior to using the matrix to generate ideas.

In a simple version appropriate to a brainstorming session the set of ideas selected for forced relationships might be written along one axis of a matrix (Figure 6), with another set along the second axis. Each cell of the matrix is then examined in turn. Examples of more complex matrices can be found in the account of Carson's new product searches mentioned above.

BRAINSTORMING SUBROUTINES

TRIGGER SESSIONS

A trigger session is a group idea-generation process in which members work independently for a period, producing a list of ideas. At the end of a given time each person reads out his list, thus generating stimuli for the rest of the group to produce more ideas. Some practitioners prefer trigger sessions to the Osborn method and make them the central part of the brainstorming process, perhaps lasting up to an hour. A strict time schedule should be set, but the atmosphere should be relaxed rather than formal. For an inexperienced group 10 ideas in five minutes is about right. The most fluent members of the group will produce up to 60 ideas. If the trigger session is the first idea-generation procedure of the particular day's activities, a warm-up similar to that described in the previous sections is desirable, modified to illustrate the principles of trigger sessions. The parallel between creativity tests and individual idea generation can be pointed out: the importance of "building" on the individual lists of ideas can also be emphasized. If two short (two-minute) warm-up exercises are run, one without instructions to defer judgment and the second with the deferred judgment advice, most group members will have increased idea-generation rates in the second trial. This demonstrates the value of postponed judgment.

One advocate of trigger sessions uses a series of different colored cards for the different stages of the meeting. Ideas produced by each member independently are written on cards of one color; ideas developed subsequently by listening to the lists being read out are jotted down on cards of a different color. This can be extended to other modifications of the technique (such as further stages under different conditions imposed by the leader). The color coding cards can later help in evaluating the effectiveness of the modifications introduced.

Trigger sessions operate on the principle of introducing a certain amount of competitive presure into the meeting. Some people perform better under (controlled) levels of stress, but others are uncomfortable and become resentful. When the stress is taken off, ideas sometimes flood into the mind as it relaxes. I have never deliberately tried this on a real problem, but in training sessions the following device has also produced good results. A group of observers, watching a second group generate ideas, is asked to jot down observations and any ideas that occur. The second group sees the problem in a fresh and detached way and produces a variety of ideas triggered by the working group. The second group may at the same time, however, have an inhibiting effect on the first one.

In general, ideas from a trigger session are of a noncomplex nature but drawing on a wide range of associations, some close to and some distant from the immediately obvious ideas. Once again it seems expedient to work the group in short bursts, perhaps five minutes of individual thinking, followed by a period of building on the ideas being read out (which can take about two minutes and produces a dozen ideas for each member on the average). The individual nature of much of the work in a trigger session means that there is more flexibility regarding the size of the groups. Some experts have developed methodologies for very large groups, split into subgroups of 6 to 10 people. Given freedom of selection, the organizer might be advised to select 6 to 10 people from a range of backgrounds if the trigger session needs fresh views on a problem. (If the exercise is concerned with sharing perceptions of subgroups within an organization, the Nominal Group Technique discussed in Chapter 8 becomes more appropriate.)

RECORDED ROUND ROBIN

A recent variation of the trigger session has developed in Europe and is particularly popular in Germany. It is particularly suited for rather small groups, optimally six people (for reasons explained below). It is another subroutine often following an Osborn-type warm-up, problem redefinition, and even idea-generation process.

Each member of the group receives a subproblem and three blank filing cards. (Bank computer cards can also be used.) He or she is asked to write the problem on each card and to then add one idea to each card.

The cards are then passed to other members of the group according to some sort of system to ensure that one person is receiving different ideas and, as far as possible, from different people. Each time the cards are exchanged the last idea on each card is used as a stimulus to generate fresh or modified ideas. For six people it is possible to exchange the cards five times, according to the scheme outlined in Figure 7. In practice, this does lead to some confusion if the organizer is not well prepared, and even more complicated arrangements are necessary for groups with more than six people. Therefore the technique may be more conveniently operated in a truncated form, with only two or three passes with none predetermined.

The method has genuine merits: There is high group energy aided by the physical passage of the idea cards, the rationale is intuitively appealing to technical people and participants are forced to seek integrative ideas.

However, I find the main danger one of attitude. The mood is less freewheeling than in an interactive method and the ideas can be rather mundane. This makes the technique a good one to follow a more speculative freewheeling session that has had plenty of fantasy but a deficiency of imaginative relevance, which is the hallmark of a successful creativity session.

People	Round 1	Round 2 1st pass to	Round 3 2nd pass to	Round 4 3rd pass to	Round 5 4th pass to	Round 6 5th pass to
P1	P1	P6	P2	P5	P3	P4
P2	P2	P1	P3	P6	P4	P5
P3	P3	P2	P4	P1	P5	P6
P4	P4	P3	P5	P2	P6	P1
P5	P5	P4	P6	P3	P1	P2
P6	P6	P5	P1	P4	P2	P3

Recorded round robin is a scheme for passing on cards among a group of six people, for five rounds, so that no person passes an idea to the same person twice. It is also known as 6:3:5, because in the full version, the six people each convert three ideas and these are modified five times.

Figure 7   Recorded round robin.

WILDEST IDEA

The wildest idea subroutine is a systematic attempt by a brainstorming group to generate ideas from outrageous starting points that have already arisen in the course of the meetings. It is based on the premise that an unusual starting point will help direct thoughts away from the conventional ideas that would in all likelihood have emerged in a more normal idea-generation session.

Although in a brainstorming group there is positive encouragement to develop wishful ideas (hitchhiking), the most speculative ideas are often ignored; that is to say, hitchhiking merely elaborates relatively obvious ideas. If the leader recognizes that this is happening, he or she can select a more unusual ("wild") idea, or ask the group to do so, and request that the group continue brainstorming from that starting point. The customary result is for the wild idea to be developed into a practical one but at the same time retaining an element of unexpectedness by virtue of the unusual starting stimulus. The following examples of wildest idea thinking were obtained in industrial sessions at the Manchester Business School:

> In a session on improving the works canteen, a brainstorming group became rather bogged down with thematic ideas relating to decor. When a wild idea was requested, some said "have topless waitresses." The hitchhiking was instantaneous, and produced an immediate joke—"install a milk machine." This proved more practicable than the idea which triggered it off, and also helped the group to get away from earlier lines of thought.

> In a different session, this time on utilizing waste material, the group had converged on a single line of attack, and had produced a large number of ideas on the principle of recycling the waste back into the system. The most promising idea, however, arose after a suggestion that the waste should be burned to heat the factory. One member of the group (the client) was able to recall that the company, an international one, had recently acquired a subsidiary with an interest in fire-lighters.

In the first example the wildest idea subroutine was introduced deliberately. In the second, the use was spontaneous, clearly a more desirable state of affairs.

There is another situation in which a leader may decide to introduce wildest idea thinking—when the group has failed even to produce the starting stimuli of unusual ideas. Under these circumstances he or she should first ask the group to try to imagine impossible but desirable starting concepts. Once a selection of these is obtained, the procedure becomes similar to the ones described above. The two different approaches are illustrated in Figure 8.

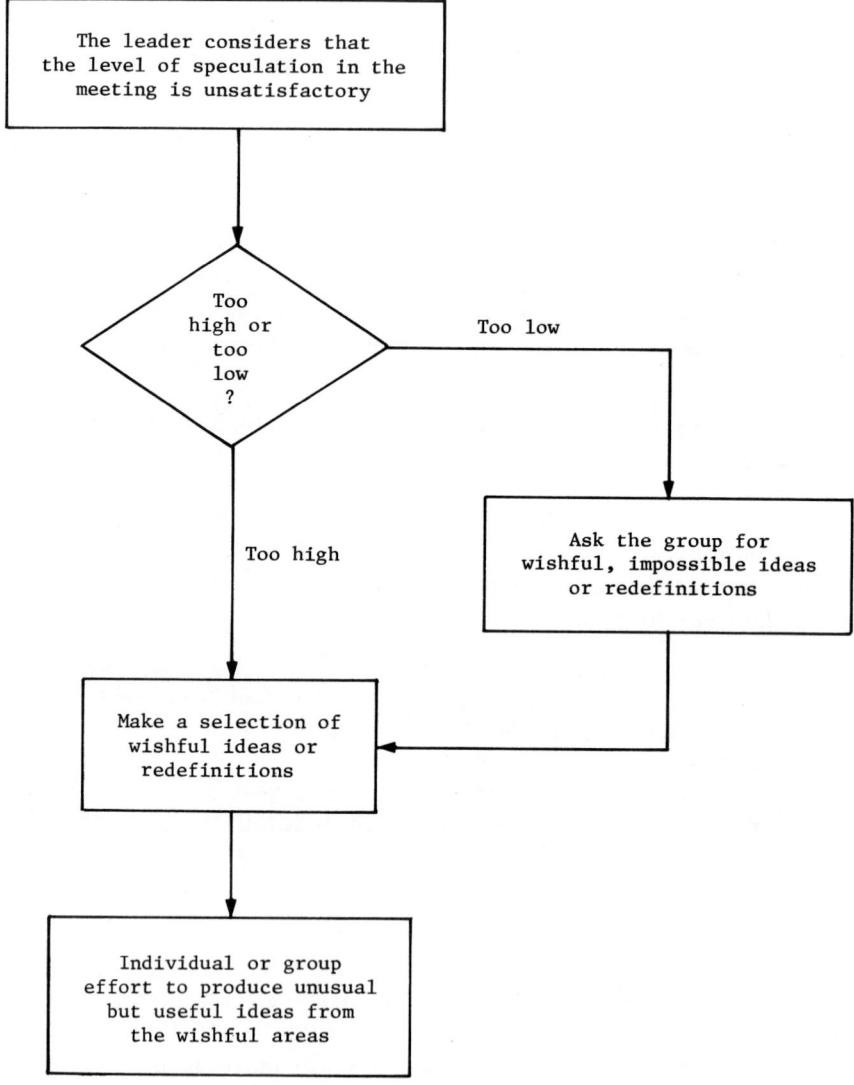

Figure 8   When to use the wildest idea technique.

REVERSE BRAINSTORMING

Reverse brainstorming is the application of Osborn-type procedures to situations in which problems are to be anticipated. At first sight the approach appears very negative, that is, the reverse of "normal" brainstorming. It should be appreciated that postponement of judgment is still the key issue. A typical problem might be, "In what ways might the installation of the new computer cause us problems?" The group should include as many key personnel likely to be involved as possible, but it should also include some disinterested outsiders. Evaluation of the results is relatively easy, as an unanticipated but potentially costly problem is likely to be recognized by "universal acclaim" during the meeting.

COMBINING THE ELEMENTS FOR A CREATIVITY SESSION

The design task for the organizer of a creativity session based on brainstorming is one of selecting and assembling the various subroutines to match the nature of the problem and the group. As each group leader will have an individual influence on the atmosphere of the event, there can be no hard and fast rules about which subroutine follows which. Experience will guide the decision (which may have to be made during the session—as the mood of the group becomes evident).

One powerful piece of advice: It is always interesting to try out new techniques and combinations of subroutines. In the longer term a group meeting should regularly experiment with methodology.

Under more urgent demands—to tackle an industrial problem, for example—with a relatively unskilled group, the key must be to keep the session as simple as possible. Mention during the warm-up that it may be necessary to try one or two methods you cannot explain at the outset, as these will be responses to the unexpected circumstances that emerge during the session.

In general, do not have a group spend longer than half an hour on any one subroutine. If an idea session lasts longer than an hour, take a break when the group feels it needs one—to return later in the day or at the next planned meeting.

The assorted subroutines can be seen as a set of related procedures. Each group tends to reach a certain level of speculation, which will differ according to the procedures and personalities involved. This will be reflected in the types of ideas and their rate of production. The leader can modify the level of speculation as shown in Figure 9. Further understand-

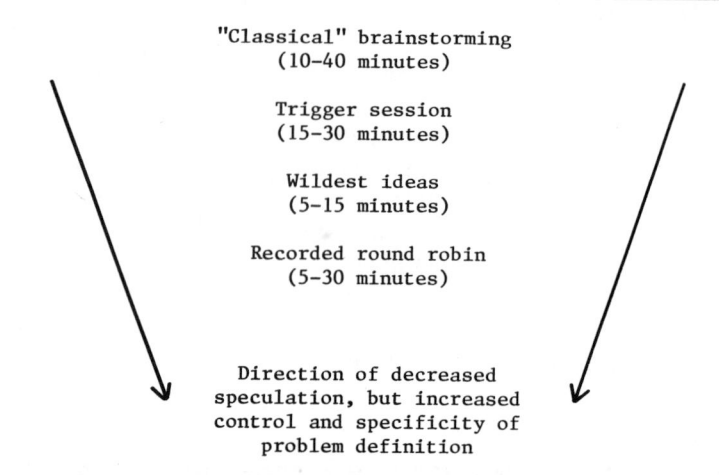

"Classical" brainstorming
(10–40 minutes)

Trigger session
(15–30 minutes)

Wildest ideas
(5–15 minutes)

Recorded round robin
(5–30 minutes)

Direction of decreased
speculation, but increased
control and specificity of
problem definition

Figure 9 Uses of various modifications of brainstorming to vary the level of speculation of the ideas produced.

ing of this principle can be obtained by examining the case study about to follow.

EFFECTIVE EVALUATION OF IDEAS FROM A BRAINSTORMING

At the end of a brainstorming session a large quantity of new data has been accumulated and noted. Before attempting to evaluate the data it is necessary to come to terms with important considerations regarding what a brainstorming can and cannot do. The material produced during an idea-generation session represents starting points or opportunities that the client (problem owner) can accept or reject. The attitude of the client will be a major factor in the subsequent utility of the brainstorming ideas. If a client truly wants to solve a problem, a list of brainstormed ideas will contain some ideas that will be new and trigger an instant response. This is one argument in favor of including the client in the brainstorming session, so that the immediate "gut feeling" reaction can be monitored. The client can indicate to the group those ideas that seem "instant winners" or the best starting points to complete solutions. (One may object that the client's gut feeling could be misleading because of overinvolvement. However, the major block to innovation subsequent to idea generation is nonacceptance rather than technically inadequate ideas. There-

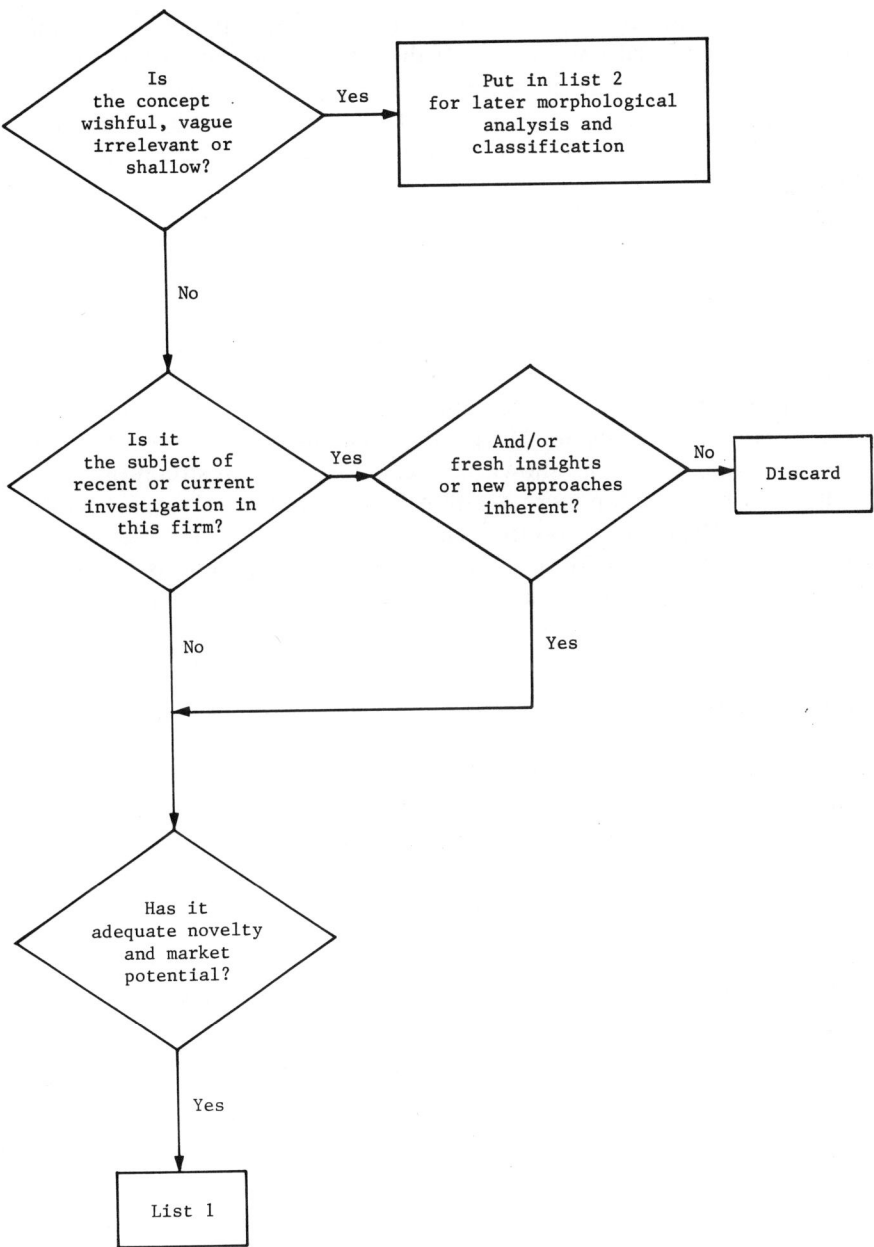

Figure 10 Preliminary screening procedure for product ideas.

fore it is preferable to leave the client as much scope as possible to identify his or her emotional preferences. If the client is present at the meeting (a highly desirable state of affairs, although not advocated by all brainstorming practitioners), he or she is likely to start working on the idea that is most motivating to implement before receiving a formal report of the meeting. The report should then be a summary of the starting points identified by the client. The new material can be circulated to the group so that they can clarify and expand any ideas they wish. The modified version should again go to the client but should not receive wider circulation until it has been rewritten, with the client's short-listed ideas at the start of the document and the speculative ideas confined to an appendix.

Reducing a list of several hundred items to a short list is an example of a restructuring problem. Morphological analysis can always be used, or a screening device as shown in Figure 10. These restructuring techniques are explained in more detail in *Problem-solving through Creative Analysis* by Rickards.[2] It is often felt that such processes risk losing some good ideas. The consolation is that all that is being done is to make a rough separation of stimuli into a concentrated, potentially more valuable subset to be presented to the problem solver.

CASE STUDY: BRAINSTORMING FOR NEW PRODUCT IDEAS

The organization is a small but fast-growing firm, producing and supplying commodities to specialist (trade) customers. As part of a policy of continuous growth the board set up a small group of managers to develop possibilities for new products in the short term (to achieve a significant impact on profit in three years). The new products group, which operated on a part-time basis, met approximately once a week, and comprised technical, financial, and commercial managers of ages ranging from the early 20s to the late 40s. A management consultant was employed to introduce the principles of idea-generation techniques to the group, which at the same time hoped to be able to generate valuable ideas for new products. The first ideas meeting took place in an office of the client firm, equipped with flip charts and easels for recording the ideas, a tape recorder, and writing materials. The consultant acted as leader; the brainstorming marketing manager who had contacted the consultant acted as client.

ORIENTATION SESSION

The subprocedures of the morning's brainstorming activities are shown in Figure 11. The orientation session took the form of a talk about the

SUBPROCEDURE	TIME TAKEN
Orientation session: exercise to build trust and permit flexible speculative thinking	$1\frac{1}{2}$-2 hours
Redefinition of the problem (during coffee break)	5-10 minutes
Idea-generation practice session (using a sugar cube)	5 minutes
Brainstorming - after Osborn	10-12 minutes
Wildest ideas as starting points for original and practical insights	5 minutes
Individual brainstorming ("trigger sessions")	10 minutes
Individual development of ideas from other group members (6-3-5)	5 minutes

Figure 11 The subprocedures in the brainstorming session for new product ideas described in the case study.

principles of creativity and idea generation, illustrated with examples of creativity tests by Torrance[3] and flexibility exercises by de Bono.[4] The group was encouraged to postpone judgment during the production of ideas, again as a preparation for the subsequent brainstorming exercise. Coffee was taken in conference toward the end of the session to avoid a break that might have reduced the value of the preliminary session. At this time a series of redefinitions (about a dozen) was produced, some splitting the problem into subproblems, some redefining it by looking at it in a new way, and some adopting a metaphorical or speculative approach ("How can we print genuine £1 notes for 50p. each?" this being a metaphorical way of thinking about increasing productivity by 100%). Because of the inexperience of the group, discussion began on the merits of the redefinitions produced. The consultant reminded the group of the principle of deferred judgment and also pointed out that redefinitions of a complex problem can be obtained by examining it from different viewpoints and that, far from conflicting with each other, they illustrated different facets of the whole situation. The client was invited to nominate those redefinitions that presented the problem in a new light as far as he was concerned. He selected four redefinitions. These were used as starting points for the idea-generation stages that followed.

The leader explained that the final warm-up exercise before tackling the real problem was a form of practice brainstorming. The group members were reminded of the underlying principle of separating the idea-generation process from any evaluation or criticism stage. They were told that in order to minimize inhibitory criticism of their own ideas as well as those of the other group members, each person should try to be relaxed and prepared to call out whatever ideas come into mind. The leader then asked the members of the group to give as many uses as they could think of for a lump of sugar. (The item was selected arbitrarily as a common object that was at hand.) The first ideas came hesitantly and were rather mundane and closely related to prior thinking about the problem. The leader was able to increase the pace of production of ideas by setting up a rhythm, rather like a conductor of an orchestra, by repeating each idea and adding one or more ideas of his own to keep up the pace. After a minute the group members were producing at a rate of one idea every three or four seconds, which indicated a reduced level of prejudgment by each individual. At this stage some ideas were highly speculative (e.g., "absorbs snake venom"). Afterward the group looked at the flip charts and found instances within the exercise when group members had improved or developed on ideas. The leader congratulated the team and called for a similar level of freewheeling and hitchhiking during the brainstorming of the real problem.

IDEA-GENERATION ACTIVITIES

The leader selected the most general of the redefinitions nominated by the client. This was written down on a sheet of the flip chart together with a précis of the principles on which the meeting was being conducted (see Figure 12).

The message of Figure 12 was on view throughout the session. The leader then repeated the problem redefinition and the group began generating ideas, which were all written on the flip chart and sequentially numbered. (A separate recorder might have been used, but the leader preferred to keep control over the felt tip pen and the group).

He again attempted to increase the tempo of idea production, which was, of necessity, slower than the warm-up exercise because each idea was being recorded manually on the flip chart and because real problems make the postponement of judgment harder. The leader did not introduce his own ideas this time to quicken the pace but relied on asking idea-spurring questions like, "How can we do that? What else might do that? Can we combine the idea with any other?" Although seeking a generally fast tempo, he also was prepared to wait out the occasional silence,

```
Problem redefinition: What extension of our services might
our customers want?

                        RULES

                        Postpone judgments

                        Improve on ideas

                        Let ideas spark off new ideas
```

Figure 12 Starting point for a brainstorming session—problem redefinitions and rules (after Osborn) used in the case study.

believing in the value of the tension that silence produces during a brainstorming session.

About 60 ideas were produced in 10 minutes. Although the group seemed to be postponing judgment, there were not many speculative ideas. The leader therefore let the tempo of idea production drop and then moved into a second variation of brainstorming based on the production of extremely speculative ideas (wildest idea). Each member of the group had to imagine a dream solution to the problem that disregarded the constraints of reality. The fantasy material was written down on sheets of the flip chart, then each idea was briefly treated as a concept to be brainstormed until a realistic idea was developed. The results in one instance had a considerable positive impact on the client and other members of the group (the "gut feeling" response), to such a degree that the postponement of judgment principle was forgotten and the group began to consider how the idea should be progressed. The leader considered this a convenient time to introduce a further modification of brainstorming in case the idea was eventually rejected. (The brainstorming principle is "quantity leads to quality." It was important to avoid the trap of focusing on the first idea that appealed to the group.)

For the next exercise, a trigger session, each member of the group was given a supply of blank filing cards. The redefinitions obtained at the outset of the meeting were shared among the group. Each member was instructed to carry out an individual brainstorming on his problem redefinition, writing down all ideas that occurred to him. A target of 10 ideas per participant with a time limit of five minutes was set. After five minutes each member read out his ideas and attempted to add to his set by hitchhiking on the other ideas being read out. Approximately 40 ideas were produced in the first five minutes, and a further 10 ideas were added in the hitchhiking period. The most comprehensive and potentially valuable list of ideas was provided by the member of the group who had difficulty in cooperating and contributing to the earlier Osborn-type brainstorming.

At this stage a subroutine was introduced—the recorded round robin. The leader gave a brief description of the technique (five minutes).

Then the session again started, with each of the six group members selecting a redefinition of the problem or a subproblem and developing three possible broad solutions of different types, one on each of three cards. The cards were then exchanged among the group to permit modifications and improvements to all the ideas. Each improvement or modification was written underneath the original redefinition. Any completely new ideas sparked off were noted on additional cards to be added to the final list. In the complete version of the procedure the 18 original ideas could have been modified five times, thus producing a minimum of 90 ideas. In this exercise only one exchange of ideas was carried out.

This concluded the first idea-generation session. Before leaving, the group noted those ideas that had appeared to have considerable potential and immediate applicability. Two such ideas were recorded. One (the "instant winner") had been generated in the first brainstorming session; the second came from one of the wildest ideas in the second exercise.

EVALUATING AND PROGRESSING THE IDEAS

The ideas were evaluated initially by the member of the idea-generation team who had commissioned the meeting and who had the most experience with the problem. He circulated a rough draft of all ideas produced among the participants. He also supplied them with a short list of potentially valuable ideas obtained from the rough draft by a screening procedure that eliminated the following: (1) wishful, vague, irrelevant and shallow ideas; (2) concepts that were the subject of current or past investigation; and (3) those of low novelty and market potential (Figure 10). The other members of the original brainstorming group were asked to add further ideas that would pass through the screen after modification to or clarification of the original statement. The liaison man then drew up a list of people who would be primarily concerned with the development and progress of the ideas through the organization. A meeting was arranged with these people, at which time criteria for the acceptance of new ideas were drawn up. The short list had been circulated to these people before the meeting, with a request for any further ideas that they felt should be added to the list. The first short list contained approximately a dozen ideas, and a brief exercise was conducted by the client to evaluate each idea on financial, technical, and organizational grounds. He was also able to build a prototype new product based on one of the new ideas.

SUMMARY

Although brainstorming is presented as a set of almost mechanical devices for groups involved in idea-generation activities, the ideal situation for idea generation would be one in which the principles of brainstorming are being obeyed, without the structure intruding. The different subroutines should be used flexibly with such an understanding.

The actual idea-generation phase is in one sense less important than the preceding preparation phases and the following evaluation phases. Given correct preparation, good ideas will always be produced by a group operating in a "postponed judgment" mode. The large number of ideas generated represent the starting point to successful resolution of the client's problem. A distinction should be made between these ideas—raw materials for further development and a possible solution that must satisfy far more criteria—particularly novelty, relevance (to client), and implementability.

REFERENCES

1. J. Carson and T. Rickards, *Industrial New-Product Development*, Gower Publishing, Farnborough, Hampshire, 1980.
2. T. Rickards, *Problem-solving through Creative Analysis*, Gower Press Limited, Epping, Essex, 1974, pp. 24–27, 43–45.
3. E. P. Torrance, *Guiding Creative Talent*, Prentice-Hall, Englewood Cliffs, N.J., 1962.
4. E. de Bono, *Lateral Thinking for Management*, McGraw-Hill, New York, 1971.

CHAPTER 4

DELPHI METHOD

BERNADINE YOUNG
HAROLD A LINSTONE

Over the past decades sophisticated and attractive processes have been developed for use in long-range planning in many fields. Delphi, one such technique, has been a popular and unique approach. It is designed to utilize group decision-making about what has not yet occurred for situations in which methodologies using the traditional scientific approach are not adequate. Yet controversy in recent years has led to the fear that the technique has lacked acceptable standards.

In light of the increasing needs for good futures research methodology it is important that these discouraging biases about Delphi be reconsidered. The purpose of what follows is to provide an assessment of Delphi, its past development and use, its characteristics and parameters, as well as its applicability and potential for increased use in planning and systems management. This assessment is offered to encourage the reevaluation of Delphi, a proactive approach to group decision-making and futures research.

Although Delphi's roots are in corporate- and government-based futures research, it has been used extensively in other areas and has potential for even broader applicability. Many of these fields have reached a critical point where they must improve planning and systems management. There is simply too great a need for them to discount or underutilize Delphi's unique value. A better understanding of the limitations and parameters of Delphi can improve its value in producing important applied research results.

THE ORIGINAL DELPHI

Delphi was conceived as one of several methods for improving decision-making developed by researchers from the nonprofit Rand Corporation. It was originally designed by Norman Dalkey, Olaf Helmer, and associates at the beginning of the 1950s to collect and utilize expert opinions to determine the effects of nuclear attack by the Soviet Union on national defense.[1] By the end of the 1950s Delphi was beginning to gain attention for uses other than defense problems. The philosophical basis for the expanded use of Delphi was put forth by Helmer and Rescher in their monograph *On the Epistemology of the Inexact Sciences*.[2] They were concerned with justifying the method's limitations and value, as well as encouraging the expansion of the basic knowledge about Delphi.

As a result of such writing, there was stimulation for further studies of Delphi for possible uses in addition to defense projects. Toward this end, Dalkey and Helmer presented the results of new experimentation in a Rand report, *An Experimental Application of the Delphi Method to the Use of Experts*.[3] By this time Delphi was developing into an important technique for governmental and industrial technological forecasting. This is increasingly evident in a later Rand publication, *Report on a Long-Range Forecasting Study* by Gordon and Helmer, which details the benefit of Delphi as a tested tool for long-range forecasting.[4] Such forecasting was needed by government and industry for use in research and development, budget allocations, and trend identification.

As Delphi expanded in its potential application in the late 1960s and early 1970s, researchers from Rand continued to monitor the results of Delphi studies and analyze the process that occurred during a Delphi. The following reports emerged, which presented more detailed and significant data about the methodology:

Analysis of the Future: The Delphi Method, Helmer, 1967[5]

Systematic Use of Expert Opinions, Helmer, 1967[6]

Some Comments on the Problems of Self-Affecting Predictions, Rochberg, 1967[7]

Delphi, Dalkey, 1967[8]

Quality of Life, Dalkey, 1968[9]

Experiments in Group Predictions, Dalkey, 1968[10]

Predicting the Future, Dalkey, 1968[11]

Delphi Process: A Method Used for the Elicitation of Opinions of Experts, Brown, 1968[12]

The Delphi Method: An Experimental Study of Group Opinion, Dalkey, 1969[13]

The Delphi Method, II: Structure of Experiments, Brown, Cochran, and Dalkey, 1969[14]

The Delphi Method, III: Use of Self-Rating to Improve Group Estimates, Dalkey, Brown, and Cochran, 1969[15]

Delphi and Values, Rescher, 1969[16]

The Delphi Method, IV: Effect of Percentile Feedback and Feed-in of Relevant Facts, Dalkey, Brown, and Cochran, 1970[17]

Experimental Assessment of Delphi Procedures with Group Value Judgements, Dalkey and Rourke, 1971[18]

Comparison of Group Judgement Techniques with Short-Range Predictions and Almanac Questions, Dalkey and Brown, 1971[19]

These reports were frequently republished by various scientific and technological journals, giving widespread attention to the Delphi methodology.

By the early 1970s the use of Delphi had been expanded to facilitate communication in a wide variety of areas. Linstone and Turoff list the following types of application[20]:

Gathering current and historical data not accurately known or available

Examining the significance of historical events

Evaluating possible budget allocations

Exploring urban and regional planning options

Planning university campus and curriculum development

Putting together the structure of a model

Delineating the pros and cons associated with potential policy options

Developing causal relationships in complex economic or social phenomena

Distinguishing and clarifying real and preceived human motivations

Exposing priorities of personal values, social goals

Delphi became an important tool for use in applied research in many areas—physical and social sciences, public and business administration, and engineering. It was growing in use from its original purpose because it filled a demand for a technique that could aid in future planning. As such a technique, Delphi proved to be a striking method of utilizing group opinion for decision-making and planning for the future.

The general characteristics of the traditional Delphi were defined by the numerous Rand reports on the subject. It is frequently explained in the literature that the Delphi method, which collected and refined expert group opinions in order to produce group consensus on an issue, is based on the adage, "two heads are better than one."[V] Its main features of anonymity, iteration and controlled feedback, and statistical group responses were intended "to minimize the biasing effects of dominant individuals, or irrelevant communications, and of group pressure toward conformity."[21] The process generally utilized a series of three or four questionnaires. The initial questionnaire simply asked for individual participants' opinions on an issue. The next questionnaire used the opinions listed by participants and asked participants to rank the list. Succeeding questionnaires allowed participants to reconsider their ranking of the list in light of the overall group ranking. Significant and substantial group consensus on the priorities among the items on the list resulted, and divergent opinions could also be pinpointed.[22]

Delphi's growth in popularity for eliciting forecasts or recommendations about the future was encouraged by the continuous reports put out by the Rand Corporation. To those interested in futures research it appeared to be a very simple procedure. Dalkey and Brown's comparison of Delphi to face-to-face conferences using almanac questions strongly indicated that Delphi was a faster and more accurate way of obtaining correct responses.[23,24] Such controlled experimentation on Delphi augmented its increased use in diversified studies.

THE EMERGENCE OF CONTROVERSY OVER DELPHI

As the application of Delphi spread and increased, criticisms emerged. Objective critiques pointed to the following problems: the use of questionnaires included the inherent difficulties of communicating by mail, the selection of participants was inconsistent, a consensus of opinion could not be ascertained in advance, ambiguous results of experiments on Delphi had been reported in some studies, and additional experimentation on the process was needed.[25] The use of Delphi, like any methodological tool, was subject to misuse by researchers. In addition, many questions remained unanswered as to the reasons why Delphi provided consensus, whether the divergent opinions should be given more concern, and how expert opinion was better than other opinions.

In 1974 Harold Sackman presented a harsh criticism of Delphi in a Rand report, *Delphi Assessment: Expert Opinion, Forecasting and Group Process*. He republished his critique in 1975 under the title *Delphi Critique: Expert Opinions, Forecasting, and Group Process*.[26] His skepticism of Delphi was based on the fact that forecasting methodologies had not been refined in terms of a conventional theoretical base. In addition to Delphi lacking a scientific approach, Sackman argued that it was weak in many other ways. According to Sackman's arguments, Delphi lacked value as a methodology because of several inherent problems. These included the subjective definition of experts, the infrequent use of random samples because of the use of experts, the exclusion of benefits of face-to-face confrontations, the inclusion of value judgments, and unmeasured reliability, content, and construct validity. Sackman concluded that Delphi had not met acceptable standards as a research method and should therefore not be used unless drastically improved.

Support for Sackman's observations about the weaknesses of Delphi was offered by Hill and Fowles, who focused on the problems of measuring the reliability and validity of Delphi. They pointed to the lack of standardized procedures for the process, as well as the researchers' pressure for consensus of opinions for use in predicting the future.[27] They agreed with Sackman's basic argument that Delphi lacked a theoretical foundation. Strauss and Zeigler also provided some support for Sackman's critque. They specifically charged that the use of experts was too homogeneous and could stunt innovative thinking. In addition, the emphasis on diminishing divergent opinions could subtract important perspectives.[28] It is significant, however, that while some of the literature agreed with Sackman that Delphi required improvements, it summarily stated that Delphi would be improved and have increasing significance as a tool for soliciting ideas, recommendations, and opinions about the future. This is true of Hill and Fowles and Strauss and Zeigler, all of whom saw the need, in some ways, to improve the traditional Delphi.

Although Sackman's writing is credited with having spurred further study and analysis of Delphi, which improved its applicability and value as a forecasting and decision-making tool, his critique was widely attacked as unsound and unfair. In the autumn of 1975 the journal *Technological Forecasting and Social Change* devoted an issue to defending and reviewing the Delphi technique in light of Sackman's attack. The issue included five papers on various aspects of Delphi that provided a useful and objective critique of Delphi's strengths and weaknesses.

In one of these five papers Scheele asserted that Sackman's critique was "anti-deviationalism when applied to the marketplace of ideas."[29] However, he noted that "there is no validity or correctness [in the use of Delphi], but there is much ambiguity in the statement of Delphi items in many inquiries. . . ." He proposed that quality control of Delphi could be improved by a more careful development of questionnaires.[30] He saw expanded uses for Delphi, and after discussing these possibilities explained that "from a management perspective, [Delphi] has to be better than more meetings."[31]

In another of these papers Coates maintained that, "Sackman ignores the crucial point that Delphi is not a scientific tool, nor is it related to a scientific experiment or a scientifically structured activity."[32] He explained that Delphi was significant as a means of dealing with "judgement and wisdom about the future." He went on to say that, "these are the areas where technology and science are least capable of shedding bright light, and of least utility in either the analysis of an issue or the formulation of an alternative or the proposition of a synthesis. . . . We face a crisis of concepts, ideas, alternatives, diagnoses, foresight, and planning. Delphi is an attempt to deal with all of that."[33] He pointed out that Sackman focused on the traditional Delphi, which is rarely used, and that many modifications have improved the original method.[34]

Goldschmidt also criticized Sackman's arguments against Delphi in one of the issue's articles. He declared that Sackman's methodology for reviewing Delphi was useless because Delphi did not deal with decision-making based on rational knowledge or empirical data. Instead, Delphi is concerned with the type of decision-making that

> requires information that cannot be derived from knowledge, because none exists, nor from empirical study, because this is infeasible or impractical. In these situations, either the decision-maker must rely on his own experience or on the opinions of others. . . . The problem for the decision-maker is how to secure such expert opinion, and more important how to reconcile differences in the opinions he is offered. Delphi is a way of overcoming this problem.[35]

Twiss, reviewing Sackman's book, said that whereas it may have been helpful in stimulating concern for improving Delphi, its rejection of the technique was unwarranted.[36] He specifically criticized Sackman's use of "Standards for Educational and Psychological Tests and Manuals" put out by the American Psychological Society to judge Delphi. Twiss maintained that these standards were inappropriate, making Sackman's arguments appear superficial and sometimes irrelevant.

Jones concluded that Sackman's book presented an unjustified, even dangerous, account, if read alone, of the Delphi technique. Whereas he called 1975 "the year of the great inquest on Delphi," he pointed out that the Rand Corporation's view of Sackman's report had mysteriously not been offered.[37] He concluded that Delphi was basically acceptable but that it might best be limited to use for stating *probable* future directions, instead of predictions.

The most extensive support for the Delphi method in reaction to Sackman was the publication of the book *The Delphi Method: Techniques and Applications,* edited by Linstone and Turoff.[38] While only briefly mentioning Sackman's Rand report in their final chapter and as a bibliographic entry, the book provided a thorough rejection of Sackman's critique. The book also provides abundant data supporting the Delphi method. It points to the ways in which Delphi has been strengthened over the years since the process's development as a defense-related methodology. It provides for the first time a comparative digest of the origins, philosophy, development, modifications, examples of studies. evaluations, and bibliographies on Delphi. Reactions to Linstone and Turoff's book were very favorable. Mahajan praised the book's useful and comprehensive nature.[39] Martino pointed out that its discussion of the philosophical foundations puts criticisms of Delphi in the proper perspective.[40] He said that because the philosophy was not understood by critics, they refused to see that their arguments were unfounded. Martino cautioned that the book is too complicated for use by those not previously acquainted with the Delphi method but that it does provide an excellent summary of how to make questionnaires and avoid the "pitfalls" of Delphi. Twiss pointed to the book's consideration of Delphi as a versatile tool, not restricted to predicting the future where weaknesses are inherent.[41]

Sackman also provided a review of Linstone and Turoff's book.[42] He stated that his basic objections to Delphi are concerned with the infrequent review of the literature on the subject of the Delphi study, the infrequent pilot testing and analysis of questionnaire items, the lack of explicitly stated sampling parameters, and the equivocal results of research on the use of experts and the benefits of anonymity. While he maintained that he was still a "skeptic," he did say that the book offers the best source on the Delphi technique, not mentioning his own controversial book published in the same year.

The period that followed the release of Sackman's 1974 Rand report caused a flurry of efforts to examine the Delphi process through controlled studies and analyses. Experimentation was emphasized in an all-out effort to define the little understood parts of the process. The strengths and weaknesses of Delphi were weighed against each other.

Basically, the inherent disadvantages of Delphi include the possibility of manipulation of the process by researchers, the difficulties of analyzing the interrelatedness of the group's ideas, the tendency toward self-fulfilling prophecies, and the ever-present complications of having to incorporate the effects of the unexpected on the findings.

However, the advantages have continued to evolve as the demand for planning and systems management increases. These advantages include the compatibility and applicability of Delphi to use with computer technology, its effectiveness in allowing individuals to deal with complex problems as a group, its success in providing agreed-upon alternatives for use in long-range planning, its utility in obtaining results when no other methodology is appropriate, and its success in encouraging group and individual consideration of factors that might otherwise be dismissed or neglected in planning futures.

Some other of Delphi's advantages were summarized by Weatherman and Swenson: (1) it has no geographic and scheduling restrictions to get participants together, (2) it is relatively easy to administer, (3) its costs are low compared with conventions or other ways of bringing together group opinion, (4) it facilitates difficult conceptualizations of phenomena, (5) it allows the researcher to focus and simplify topics under discussion, and (6) it provides a useful and interesting means of considering others' ideas.[43]

However, an emphasis on the needs for improvement dominated the literature on Delphi. While the traditional Delphi survived this scrutiny, a variety of modifications of the technique developed, broadening the use of Delphi as a forecasting methodology and for applied research in many fields.

DESIGNING AND MODIFYING DELPHI STUDIES

The recent literature on Delphi agrees that the technique, like all forecasting methods, has limitations that need to be understood in order to produce solid results. It also agrees that Delphi is the best technique, especially in a modified form, for use in formulating concensus on group opinions about the future and for pinpointing areas of strong divergence of opinions. Critiques of Delphi note that misuses have occurred, specifically in the areas of defining and limiting participants to "experts," forcing acceptance of a presupposed viewpoint in the design of the methods, failing to incorporate the disagreements that arise among participants during the process, and drawing conclusions from the data as an accepted prediction for the future. These and similar problems are the fault of the researcher and not of a well-planned Delphi study.

The users of Delphi must take great care in determining its applicability for their respective forecasting needs. In addition to designing the overall procedure, users must be well prepared and knowledgeable about questionnaire design, defining and selecting participants, clarifying hypotheses, applying statistical analyses, and the effects of incorporating modifications in the study's design. The following outline lists the basic steps undertaken in a Delphi study. These overall procedures are more fully developed or altered, depending on the nature of the Delphi being employed and particularly on the availability of computer technology, which transforms the conventional Delphi into a real-time communication conference.

The basic steps of a Delphi study are the following:

1. Survey and analyze the literature of the field or the topic of the study.
2. Identify, select, and elicit the involvement of participants.
3. Design and pretest the pilot questionnaire.
4. Revise and administer the first questionnaire/round one.
5. Provide a statistical analysis of the first questionnaire.
6. Provide feedback to participants' responses and comments.
7. Design, pretest, revise, administer, and then analyze succeeding questionnaires (typically three additional rounds) and provide feedback for each round.
8. Prepare the final analysis report.
9. Share with participants.
10. Apply the results.

Jillson,[44] who provides a progress report that details a Delphi study concerned with drug abuse and the National Institute of Drug Abuse's policy development, provides a flow chart of the study's design, which is generally applicable to Delphi studies and illustrated in Table 1.

Ludlow provides some additional helpful points for those designing Delphi studies[45]:

Respondents will be more receptive if the techniques are tailored to specific groups on the basis of their training and experience.

The administrator should consistently emphasize the distinction between the characteristics of a Delphi interrogation and those of conventional questionnaires and polls.

Table structured by PERSPECTIVE across questionnaire rounds:

PERSPECTIVE	FIRST QUESTIONNAIRE	SECOND QUESTIONNAIRE	THIRD QUESTIONNAIRE	FOURTH QUESTIONNAIRE	FIFTH QUESTIONNAIRE	FINAL SUMMARY
Objectives ("top down")	List of objectives	Rate feasibility and desirability of selected objectives	Re-rate feasibility and desirability of selected objectives			
	List key indicators	Expand key indicators, perform initial rating	Final rating of factors			
Transition Factors ("bottom up")	List factors, indicate direction		Develop policy areas to affect important transition factors which can be influenced by national policy	Identify any contradiction in objectives and policies formulated by the three approaches	Synthesize a consistent and realistic set of national drug abuse policy options	Write a brief summary of the national drug abuse policy options identified by the study, including the normative forecasts for the key indicators
Policy Issue Statements ("political")	Rate selected policy issues; develop for and against arguments	Final rating of selected policy issues; rate for and against arguments				Identify future policy research needs
	Add other important policy issues	Rate and give arguments for and against additional policy issue statements				

Table 1 The Study Design (Reproduced by permission of the publisher from *The Delphi Method: Techniques and Applications*, 1975, edited by Harold A. Linstone and Murray Turoff, Addison-Wesley, Advanced Book Program, Reading, Mass.)

Panelists—particularly those with technical backgrounds—must be *convinced* that judgments often have to be made about issues *before* all facets of the problems have been researched and analyzed to the extent they would like. (For these situations they must be persuaded that their subjective judgments may be a decision maker's most valuable source of information.) . . .

Interpersonal techniques, such as interviews and seminars should be interspersed with the rounds of questionnaires and information feedback.

The source of a suggested item should be identified (for example, panel member number and basic biographical information), taking care not to compromise the anonymity of specific inputs.

Standardized scaled measures should be available to a respondent so that he can qualify his response to specific questions. Such measures are relative competence in a technical area, familiarity with a geographical region, or confidence in an estimate.

If a multidisciplinary approach is desired, respondents should be encouraged to consider all items but to make estimates only on those scaled descriptive phrases with which he feels comfortable. For example, in these exercises it was helpful when respondents indicated their familiarity with a specialized area or the importance of an item even though they did not make probability estimates.

The panelists should decide through their suggestions and evaluations what items should be considered. The criteria for retaining an item for further evaluation should be made clear at the onset of the exercise.

Personal comments and arguments submitted by the respondents should be part of the information feedback.

Linstone presents a "checklist" of areas where designing a Delphi study has been weakened because of lack of insight or clarity about the nature of the process. He elaborates on eight "pitfalls": (1) people are generally concerned with the immediate present and therefore intuitively "discount" the future and have difficulty with decision-making because they do not have a clear grasp of the possibilities of the future; (2) people are generally uncomfortable with uncertainty and may therefore overuse predictions in order to produce certainty; (3) most people prefer simplicity and have trouble dealing with the interrelationship of parts, making it difficult to conceptualize the future in a holistic pattern; (4) expertise does not always produce the best forecasting because of various biases, but some of these biases also apply to laymen; (5) "sloppy execution" by the researcher or the participants, such as the improper selection of participants or lack of participant commitment, may occur; (6) inherent optimism or pessimism of respondents commonly produces "bias toward overpessimism in long-range forecasts and overoptimism in short-range forecasts"; (7) the overuse of Delphi as a favorite tool for forecasting produces possible fallacies about the proper applicability of the method; and (8) the potential for the occurrence of deceptive and manipulative practices by individual researchers is a hazard that Delphi has in common with other methodologies.[46] The manipulation of analyses and communications is an ever-present danger in all societies. It may be intentional or unintentional. The works of Nelson[47] and Rauch[48] give support to the thesis that it is possible to distort data with all types of Delphi—almanac/factual, policy, and decision. These problems must be considered when designing, analyzing, and applying the results of a Delphi study.

A varied literature on the subject suggests that other problems can be controlled by such alterations as using a random sample where appropriate for participant selection, using nonexpert opinion where beneficial, limiting the number of participants to a manageable size or utilizing the aid of a computer for larger studies, pretesting questionnaires to assure clarity of process and the validity of items, using the median or the mode, but rarely the mean, as a measure of central tendency in providing feedback data, providing final results of the questionnaires to the participants; limiting the number of questionnaires to three or possibly four (since more or fewer rounds may have negative effects on results), developing a follow-up study on reasons for failure to respond, planning for panel consistency, and controlling panel fatigue. These ideas are repeatedly discussed by many authors, including the following: McGaw, Browne, and Rees[49]; Huckfeldt and Judd[50]; Weatherman and Swenson[51]; Cyphert and Gant[52]; Pallente[53]; Rasp[54]; Welty[55]; Judd[56]; and Brockhoff.[57,58]

MODIFICATIONS AND APPLICATIONS OF DELPHI

Awareness and effort to control the above problems are necessary and provide the base for modifying the traditional Delphi. In addition, the literature suggests that the Delphi methodology should be combined with other research techniques and technology to design modified Delphi studies. This is specifically discussed by Middendon,[59] Stover,[60] Hill and Fowles,[61] Brockhaus and Mickelson,[62] Linstone and Turoff,[63] and Helmer.[64] Cross-impact analysis and variations of it can be used in studying large systems or interdependent events. Such studies may be complex and cumbersome but can produce a better understanding of the possible future by considering the interdependency of factors. Combining Delphi with morphological or extrapolative analysis helps planners to better conceptualize systemic alternatives as well as to overcome obstacles to the acceptance of these alternative futures. Scenarios, simulations, and model building can be used to aid a Delphi panel's visualization of the alternatives suggested and, then, in deducing from each alternative the probable course of development.

Utilizing trend analysis could be particularly adaptable to educational Delphi studies that desire to plan and control changes which will affect learning. Finally, surveying and analyzing the literature on a subject is strongly recommended for improving the foundation of Delphi designs. A review of the literature on the subject of the Delphi study can encourage innovation on the part of the researcher in designing a particular Delphi, as well as provide useful knowledge about the subject of the Delphi study.

A summary of the recent literature, which gives examples of a variety of applications of Delphi, provides insight into the diversity of the process's utility. Turoff outlines the design of what he called a "policy Delphi," which was used as a tool that utilizes strongly divergent opinions in analyzing and forming policy.[65] Pyke provides information about a large Delphi study that utilized a computer data bank to anticipate 400 technical events.[66] Middendon details the use of a modified Delphi that combines morphological analysis for a study in which decision-making exerted unusual pressure on group harmony because of the effects of the decisions on certain individuals in the group.[67] Derian and Morize present the results of a project that utilizes Delphi to encourage doctors to consider artificial heart surgery.[68] Thompson discusses the use of Delphi in formulating policy for application to the drug field.[69] Witson outlines a study to aid budget planning for a railway anniversary celebration.[70]

The use of Delphi for educational studies about innovation and planning has been encouraged since the late 1960s, especially in higher education (Helmer[71]; Adelson, Alkin, Carey, and Helmer[72]). While increased use is expected, only 19% of 598 Delphi studies reported in a large survey by Brockhaus and Mickelson were in the category of education and public administration.[73] Examples of Delphi's diverse application in educational institutions is provided in the following paragraph.

One study utilizing graduate engineering students first applies almanac questions and then value judgments to illustrate to these students the process of group consensus in opinion formation.[74] Martin and Maynard apply the Delphi technique to goals of private institutions of higher education, using only two questionnaires.[75] They conclude that the use of expert opinion through a Delphi study had provided the same results as other means of obtaining information on the potential for private institutions to change students. Other uses of Delphi in higher education are summarized by Judd,[76] including some dissertation projects concerned with the future of higher education. A Delphi study to examine the obstacles to adopting computers for instructional purposes has resulted in recommendations for planning and implementation.[77] Cyphert and Gant have used Delphi for a large-scale survey of opinions about the University of Virginia's School of Education.[78] One of the most recent areas within education to utilize the Delphi technique in planning has been in student personnel services (Newton and Hellenger,[79] Newton and Richardson,[80] Jonassen and Stripling[81]). A review of a variety of Delphi studies in education is presented by McGaw, Browne, and Rees, who assess the technique as having much value and diverse potential.[82]

In Linstone and Turoff various contributors provide details of applications of Delphi to a variety of corporate and institutional, such as the following studies: policy formation; economic, social, and political trends; concerns for pollution; recommendations on wastewater treatment and disposal systems; regional planning; national drug abuse; corporate environments; plastics and competing materials in the future; steel and ferroalloy industries; aircraft competition; and model building of a transit system. Current Delphi studies, such as in the following cases, provide further illustration of Delphi's applicability, particularly in combination with scientific and computer technology.

An interesting Delphi experience forms the subject of a recent German book by Brockhoff.[83] Using CRT terminals, a real-time Delphi was carried out on the subject of short-term interest forecasts. The participants had access in a terminal to 45 time series of data relevant to the subject. This set of experiments led to the following conclusions:

1. It is feasible to handle a Delphi interaction through computer dialogue, even with untrained participants.
2. The computer dialogue saves time and produces interesting information.
3. Use of the data bank is high if the participant is used to reading professional journals, is unfamiliar with a specific question, and is surprised by it. These factors comprise a yardstick for the estimation of participant expertise.
4. For *short-term* forecasts Delphi is not significantly better or worse than other techniques.

In the People's Republic of China a Delphi is underway on "Technologically Feasible Futures for Earthquake Prediction." A panel of 80

Chinese seismologists is examining the progress of seismology, the possibilities of different prediction methods and their estimated dates of realization, as well as the nature of the barriers to be encountered (Hsu Lee-da, personal communication with Linstone, August 20, 1980).

Forecasts prove not infrequently to be self-fulfilling (or self-defeating). This possibility exists also in the case of Delphi. In a recent Delphi application conducted at the Austrian Academy of Sciences in the area of scientific information and documentation, an attempt was made to apply this principle intentionally, that is, to help in shaping the decision.[84]

We are dealing in this decision Delphi with information systems concentrated in a few large libraries and documentation centers in institutions administered by one central department in the Federal Ministry of Science and Research. The future development in this area will therefore be strongly influenced by the decisions of a few individuals in the federal bureaucracy and main libraries who organize and manage information centers in Austria. The field of interest is quite old, but now confronted with many new technological developments. Unlike mass communication, the field of scientific and technical information is not yet a subject of political involvement.

A group of actual decision-makers in combination with research workers and librarians formed the Delphi panel. Strict anonymity was replaced by quasianonymity: All participants are known, but all statements are anonymous. Four rounds were answered by 91 participants, with the aid of a "reminding procedure" to assure continued participation. Regarding the analysis between rounds, Rauch observes[85]:

> With respect to the formulation and feedback of the intermediate results, the decision Delphi can be a little more complicated, since it becomes necessary to force the participants not only to express their existing opinions, but to generate new ones. On the other hand, it is to some degree easier because for the decision-makers, the problems in question are not theoretical ones. They are confronted with such questions and with all pro and con arguments every day. Consequently, the responses of the decision-makers contained fewer "ifs" and more emotional argumentation than the answers of the scientists. The feedback had much more influence on the decision-makers than on the "experts" and [the former] reached a clear position of consensus or confrontation much more quickly than their colleagues from the universities.

The variety of modifications of the Delphi method mentioned above have evolved as effective and important processes for long-range planning in diverse corporations and institutions. Such modifications are appropriate means of enhancing Delphi and overcoming some of its inherent weaknesses. Linstone points out that while problems exist with any methodology concerned with improving communication or delineating the future, "an honestly executed Delphi" will designate its limitations and therefore be of great importance[86]:

> While the Delphi designer in the context of his application may not be able to deal with, or eliminate, all these problems, it is his responsibility to recognize the degree of impact which each has on his application and to minimize any that might invalidate his exercise. The strength of Delphi is, therefore, the ability to make explicit the limitations on the particular design and its application. The Delphi designer who understands the philosophy of his approach and the resulting boundaries of validity is engaged in the practice of a potent communication process.

The frustration with the failures of systems analysis in dealing with ill-structured problems has led to an interesting new direction of effort—the concept of multiple perspectives.[87] We shall briefly discuss this approach, since it offers an opportunity to see Delphi in a new light.

Science and technology represent the most successful religion of modern times. From Galileo to the Apollo lunar landing, from Darwin to recombinant DNA, the paradigms of science and technology have yielded dazzling triumphs.

These paradigms include the following:

1. The definition of "problems" abstracted from the world around us and the implicit assumptions that problems can be "solved."
2. Optimization or the search for a "best" solution.
3. Reductionism, that is, the study of a system in terms of a limited number of elements (or variables) and interactions among them.
4. A reliance on data and models, and combinations thereof, as modes of inquiry.
5. The quantification of information.
6. Objectively, that is, the assumption that the scientist is an unbiased observer outside of the system he or she is studying.
7. Time movement seen as linear, that is, at a universally accepted pace reckoned by precise physical measurement.
8. Ignoring the individual—a consequence of reductionism and quantification (e.g., use of averages) as well as nonhuman objectivity.

A technology and its environment are typically viewed as a system. Systems analysis tools are considered appropriate and the traditional guidelines for analysis apply. Technical impacts are carefully described and, where possible, quantified. Benefits and costs are calculated. Frequently cause and effect modeling is carried out to study the static and dynamic behavior of the variables that describe the system and its environment. Structural models are illustrative of such tools.[88] System dynamics modeling and decision tree analysis provide other examples.[89] At times the models may drive the analysis; that is, the analyst's modeling background and experience may be instrumental in determining what is analyzed and how. Strong reliance is placed on technical experts as well as on technical reports containing empirical data or theoretical models and data. Rationality is assumed to determine decisions; for example, the alternative with the most favorable benefit–cost relationship will be selected. Figure 1a schematically summarizes the general approach.

The success of this mode of thought and its paradigms—the technical perspectives—has led very naturally to increasing pressure to extend its use beyond science and technology, that is, to society and all its systems. Organizations become cybernetic systems, utility theory determines preferences, decision analysis provides the key to decision-making, policy analysis selects strategies. There is a mathematical theory of war and, of course, "management science." Figure 1b portrays the situation.

Without question, the technical perspective is ideal for well-structured problems in science and technology. However, for ill-structured problems all eight paradigms are unsatisfactory. Referring back to the listing of the paradigms, we note the following:

1. In complex systems that involve human beings we shift problems rather than solve them.
2. Optimization is discarded for a safe-fail strategy.
3. Reliance on reductionism is dangerous where the whole is not the sum of the parts.
4. Modes of inquiry other than those based on data and models prove to be fruitful.
5. Quantification distorts some qualitative inputs and generates a delusion of meaningfulness.
6. Objectivity is a myth, even for the scientist.
7. Time must be differentiated: technological time assumes a zero discount rate, social time is multigenerational and assumes a moderate discount rate, biological time is dictated by an individual's expected life span and position in Maslow's hierarchy of needs.

THE TECHNICAL PERSPECTIVE

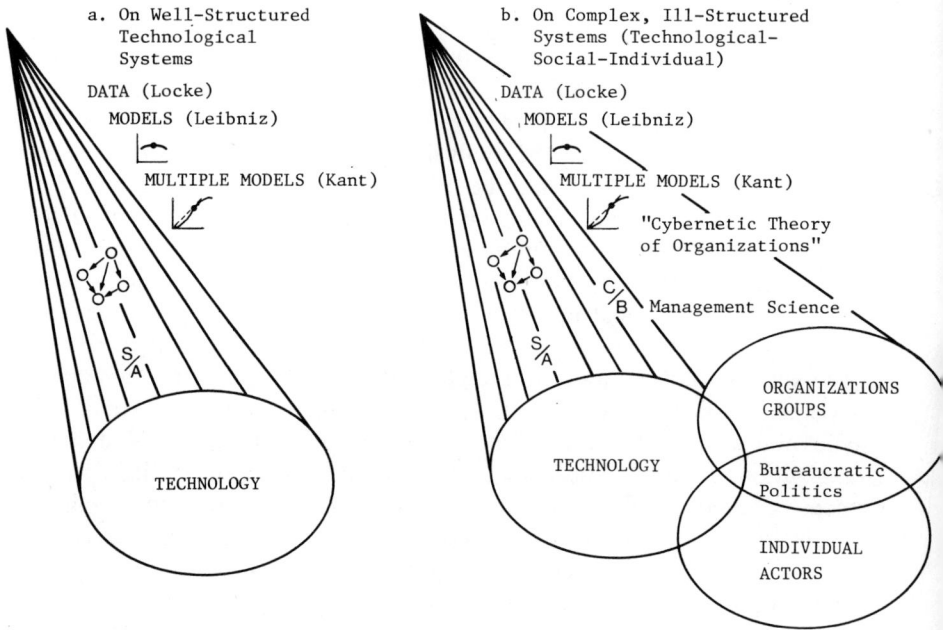

Figure 1   The technical perspective.

8. As to the tendency of ignoring the individual and replacing him or her with types and averages, consider Churchman's comment[90]:

> Economic models have to aggregate a number of things, and one of the things they aggregate is you! In great globs you are aggregated into statistical classes. . . .
>
> Jung says that, until you have gone through the process of individuation . . . you will not be able to face the social problems. You will not be able to build your models and tell the world what to do. . . .
>
> From the perspective of the unique individual, it is not counting up how many people on this side and how many on that side. All the global systems things go out: there are no trade-offs in this world, in this immense world of the inner self. . . . All our concepts that work

so well in the global world do not work in the inner world. . . . We have great trouble describing it very well in scientific language, but it is there, and is important. . . .

To be able to see the world globally, which you are going to have to be able to do, and to see it as a world of unique individual . . . [that is, the ability to] hold conflicting world views together at the same time [and be] enriched by that capability—not weakened by it—that is really complexity.

Using Allison's *Essence of Decision* as a basis, Linstone et al. have extended the concept of the use of multiple models.[91] Table 2 describes the perspectives used and Figure 2 relates them to the preceding discus-

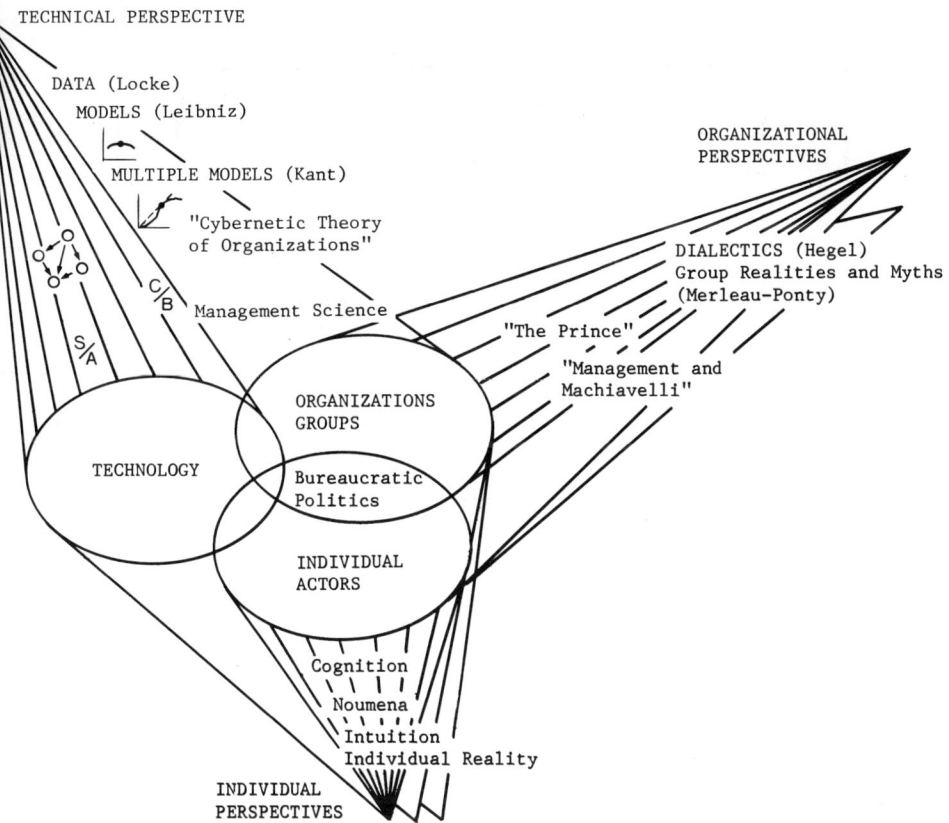

Figure 2 Multiple perspectives, a Singerian inquiring system.

	TECHNICAL (T)	ORGANIZATIONAL (O)	INDIVIDUAL (P)
WELT-ANSCHAUUNG	Science-technology	Organization	Psychology-behavior
CHARACTER-ISTICS	Cause-effect Objective Problem-solving Analysis Prediction Optimization Complete rationality Use of averages, probabilities Trade-offs	Cause-effect & challenge-response Objective & subjective Problem avoidance/delegation Analysis & synthesis Recognition of partial unpre- dictability; action/imple- mentation Satisficing Parochial priorities, incre- mental change Standard operating procedures Factoring/fractionating problems	Challenge-response Subjective Leaders and followers; game-in- process for most Intuition Fear of change and unknown Creativity and vision by few Partial rationality; inner world/ self; Maslow hierarchy of needs Learning Power/influence/dominance
PREFERRED INQUIRING SYSTEM	Lockean-data Leibnizian-model Kantian-multimodel	Hegelian-dialectic Merleau-Ponty-negotiated reality	Intuition-noumena Individual reality
TIME CONCEPT	Technological time Zero discounting	Social time Moderate discounting	Biological time High discounting

Table 2 Linstone Perspectives

sion of the traditional technical perspective. The perspective concept stresses that we must not only consider *what* we are looking at, but *how* we are looking at it. For example, we may study an organization from a technical perspective and see it in terms of system dynamics (i.e., Forrester's industrial dynamic), with information, materials, or dollar flows. but we see it very differently from an organizational perspective. We focus on the power relationships, the competition between subunits, the struggle for budgets, and the unstated goals of the unit. Machiavelli's astute guidelines become more relevant than those of Cyert and March. Most difficult for those trained in the hard or soft sciences and engineering is the individual perspective. Often the only recourse is to draw in persons who come from a very different background, for example, law or journalism. A lawyer is not inhibited by data- and model-based inquiring systems; he or she is comfortable with the dialectics of a court room and in dealing with unique clients. An effective defense strategy is suited to the individual on trial, the particular judge and jury, as much as it is on the "facts" of the case.

It is not surprising that virtually all the tools in the analyst's tool kit are data and model based. *The Delphi method is one of the very few procedures that can bridge the differences among perspectives.* Responses need not be data or model based. They mix the subjective with the objective and the holistic with the reductionist. An individual's intuition or an organization's parochial aims can be introduced with the aid of anonymity. Thus Delphi has an advantage over the interview technique, the most successful procedure to date in obtaining input for the organizational and individual perspectives. A Delphi response does not force an identification of either respondent or perspective. Typically it mixes perspectives, reflecting the different inputs that are mentally integrated by any one person. The unique views of individuals can be probed in Delphi with the aid of feedback, and individuality need not be lost by averaging and in other ways statistically manipulating all responses.

On the other hand, we must recognize certain advantages of the interview over Delphi. The interviewer can alter the line of questions on the spot, based on responses received to the initial questions. He can pick up nuances and body language and read between the lines. In sum, he has greater flexibility. Furthermore, a shrewd respondent may be more inhibited in written responses than in informal conversation even if assured of anonymity; he may simply not trust such assurances. (Of course, such a careful person may also be guarded in oral interview responses.)

Delphi, in sum, does not offer a guarantee that the three perspectives are adequately included. But, unlike data- and model-based techniques, it does permit the introduction of the second and third perspectives.

CASE HISTORIES

Many examples of excellent Delphi designs can be found in the research literature of education, social science, management, engineering, and a number of other fields. Two such cases are considered below as more detailed illustrations of the methodology. These cases provide a fuller overview of how the planning design, operational procedures, and conduct of the process actually look in applied research.

A DELPHI ON THE FUTURE OF THE STEEL AND FERROALLOY INDUSTRY

A three-round Delphi, conducted for the U.S. ferroalloy industry by the National Materials Advisory Board of the Academy of Science and Engineering, was used to systematically examine the effects of technological change on usage trends of ferroalloy.[92] Significant results were applicable for identifying trends and long-range planning in relation to policy issues for alloy industries management. It should be noted that this study applies the Delphi technique to trend extrapolation rather than to events forecasting, which is a departure from the characteristic applications of the earlier Delphi Studies.

Rounds. Each of the three rounds included three sections that present (1) steel usage trends in graphs and related assumptions for extension by respondents, (2) alloys usage trends and related assumptions for extensions by respondents, and (3) future key development trends and assumptions over the next 20 years elicited from respondents.

Participants. Purposeful selection of participants from industry, government, institutes, universities and trade publications was planned rather than random selection. The study involved 34 respondents. Two full-time professionals, with temporary clerical support during peak periods, monitored the study's three rounds. A table of the efforts involved in designing, monitoring, and analyzing the study over 46 weeks is reproduced in Table 3.

Round One. Three questions were posed along with the usage trend graphs on the steel and alloys industries during the 1960s, to be extended to 1985 by respondents: (1) How reliable did the respondents consider this graph extension to be? (2) What key developments did the respondent assume in making his extension? (3) What other developments might result in major revisions in the extension? Figure 3 outlines the formats for key developments that were considered by respondents.

UNITS: WEEKS OR MAN WEEKS*	ELAPSED TIME	SENIOR PROFESSIONAL	PROFESSIONAL	CLERICAL TABU-LATION AND CURVE EXTRAPOLATION	SECRETARIAL (TYPING)
Pre-Round One	4 weeks	4 man weeks	4 man weeks	--	1 man week
During Round One	6 weeks	1 man week	1 man week	--	--
Pre-Round Two	6½ weeks	6 man weeks	--	8 man weeks	2 man weeks
During Round Two	7 weeks	1 man week	3 man weeks	--	--
Pre-Round Three	4½ weeks	1 man week	2 man weeks	5½ man weeks	1 man week
During Round Three	8 weeks	1 man week	1 man week	1 man week	1 man week
Pre-Final Report	10 weeks	4 man weeks	2 man weeks	½ man week	2 man weeks
Totals	46 weeks	18 man weeks	13 man weeks	15 man weeks	7 man weeks

* Man week refers to time spent by male or female

Table 3 Table of Analysis Effort (Reproduced by permission of the publisher from *The Delphi Method: Techniques and Applications*, 1975, edited by Harold A. Linstone and Murray Turoff, Addison-Wesley, Advanced Book Program, Reading, Mass.)

1. Potential Development	2. Likelihood of Occurrence by 1980						3. Impact on U.S. Steel Industry if Development Were to Occur				4. Nature of Impacts (Add if you wish.)
	Very Probable (1-.8)	Probable (.8-.6)	Either Way (.6-.4)	Improbable (.4-.2)	Very Improbable (.2-0)	No Judgment	Strong	Moderate	Slight or None	No Judgment	
2.17 Development of an economical process for the recovery and utilization of Titanium scrap	32	37 A	11	16		5	16	16	53 A	16	Enhance Ti competitive position; steel companies in Ti field will alter product balance; Ti scrap already being used in steel production; reduce Ti price; nonferrous metal prices have tremendous impact on steel demand (i.e. substitution prone) particularly in construction
2.18 Development of an economical process allowing a major improvement in Titanium workability	6	44 A	28 A	11	6	6	6	24 A	53 A	18	Same as 2.17
2.19 More than 20% of U.S. Manganese requirements met by ocean floor mining		22	17	39 A	13	9	11	5	63 A	21	Increased availability of Co and Mn; implies higher cost for Mn and reduced consumption as a result
2.20 U.S. low-grade Manganese ores become economical for meeting 20% or more of U.S. requirements		20	25 A	25	15	15	6	6	59 A	29	Four times cost of imported material; implies disruption of ocean transport or unavailability of foreign sources

A denotes average.

Figure 3 'Format for key developments (Reproduced by permission of the publisher from *The Delphi Method: Techniques and Applications*, 1975, edited by Harold A. Linstone and Murray Turoff, Addison-Wesley, Advanced Book Program, Reading, Mass.)

126

Round Two.   The results from round one were used to update the design of the second questionnaire. For this questionnaire respondents provided scores with each of the assumptions and considerations that were presented. Their scores were numeric, on a scale from 1 through 6, based on the validity or confidence codes presented in Table 4.

Feedback from the first round was provided along with 50% validity or confidence limits attached to the graph curve. Respondents received means and standard deviations for each response to round one, reconsidered individual and overall estimates, and reviewed all the associated reasons given by participants with their previous graph estimates. Respondents then rated each reason between 1 and 6, using the validity scale. They also had the opportunity to describe the nature of those they felt would have the most impact and to provide revised estimates of usage trend graphs. A typical question for round one and two is provided in Figure 4.

New 50% confidence limits were provided; verbal comments were collected; and numerical results were tabulated. Next, statistical calculations of the data were obtained, as shown in Table 5.

Round Three.   This time, sections 1 and 2 were summary information that required no response for participants. Means, standard deviations, and new 50% confidence limits for the graphs were provided for each round. The third section contained information on the percentage distribution of respondents' scores from round two, estimated average scores for each possible development, a summary of all verbal comments, and updated curves for usage trend graphs. Respondents reviewed this and then responded once again as to their preferences for each key development. They also repeated the process of estimating the curves and rating their individual and collective responses. Finally, respondents reconsidered ratings on those assumptions that reflected disagreement among respondents' comments. Reactions to the Delphi process were also requested.

Analysis.   Statistical calculations of the data were provided for all key developments, assumptions that might be reevaluated, and new reasons generated by respondents. These were then associated with the graph curves, which were then reorganized according to the mean validity or confidence scores, in decreasing order. This data was then used to formulate a report to the National Materials Advisory Board for use in long-range national policy planning.

Goldstein also compares the Delphi study with a conventional panel study, carried out simultaneously, and finds substantial reinforcement

Numeric Scale	
	CERTAIN (Average of 1 to 1.5)
1	• Low risk of being wrong
	• Decision based upon this will be wrong because of this "fact"
	• Most inferences drawn from this will be true
	RELIABLE (Average of 1.6 to 2.5)
2	• Some risk of being wrong
	• Willingness to make a decision based upon this
	• Assuming this to be true but recognizing some chance of error
	• Some incorrect inferences can be drawn
	NOT DETERMINABLE (at this time) (Average of 2.6 to 3.5)
3	• The information or knowledge to evaluate the validity of this assertion is not available to anyone--expert or decision maker
	RISKY (Average of 3.6 to 4.5)
4	• Substantial risk of being wrong
	• Not willing to make a decision based upon this alone
	• Many incorrect inferences can be drawn
	• The converse, if it exists, is possibly RELIABLE
	UNRELIABLE (Average of 4.6 to 5)
5	• Great risk of being wrong
	• Worthless as a decision basis
	• The converse, if it exists, is possibly CERTAIN
	NOT PERTINENT (Used to eliminate some assumptions from exercise)
6	• Even if the assertion is CERTAIN or UNRELIABLE it has no significance for the basic issue
	• It cannot affect the variable under question an observable amount
	NO JUDGMENT
blank	• No knowledge to judge this item, but the appropriate individual (expert, decision-maker) should be able to provide an evaluation I would respect

Table 4 Validity or Confidence Scale (Reproduced by permission of the publisher from *The Delphi Method: Techniques and Applications,* 1975, edited by Harold A. Linstone and Murray Turoff, Addison-Wesley, Advanced Book Program, Reading, Mass.)

1. Shown are the boundaries within which 50% of the forecasts fell. After reading column 2, please reestimate. Reliable ____, As good as anyone's ____, Risky ____.

2. Reasons for increasing or decreasing the forecast are presented below. Add others if you wish.

Validity Choice

- Imports in semi-finished form are bound to increase sharply because of lower cost production outside U.S. |
- Current export increase due to Canadian semi-finished imports for re-export after finishing. |
- Re-exports from Canada should decrease as Canada becomes self-sufficient. |
- Semi-finished shipments to Europe will fall as melting begins there. |
- Foreign production facilities for flat stainless are increasing and will be competitive with U.S. in 2 to 3 years. |
- True exports will be confined to special grades where some proprietary position held. |
- U.S. expected to retain competitive position in low nickel stainless. |
- Chromium shortage will boost imports. |
- Improved quality by use of higher purity alloys (e.g., vacuum process) will decrease imports. |
- Since we do not have Ni or Cr, rest of world will supply, ultimately, their own SS. |
- Imports of 3oo-series (Cr-Ni) steel expected to grow. |

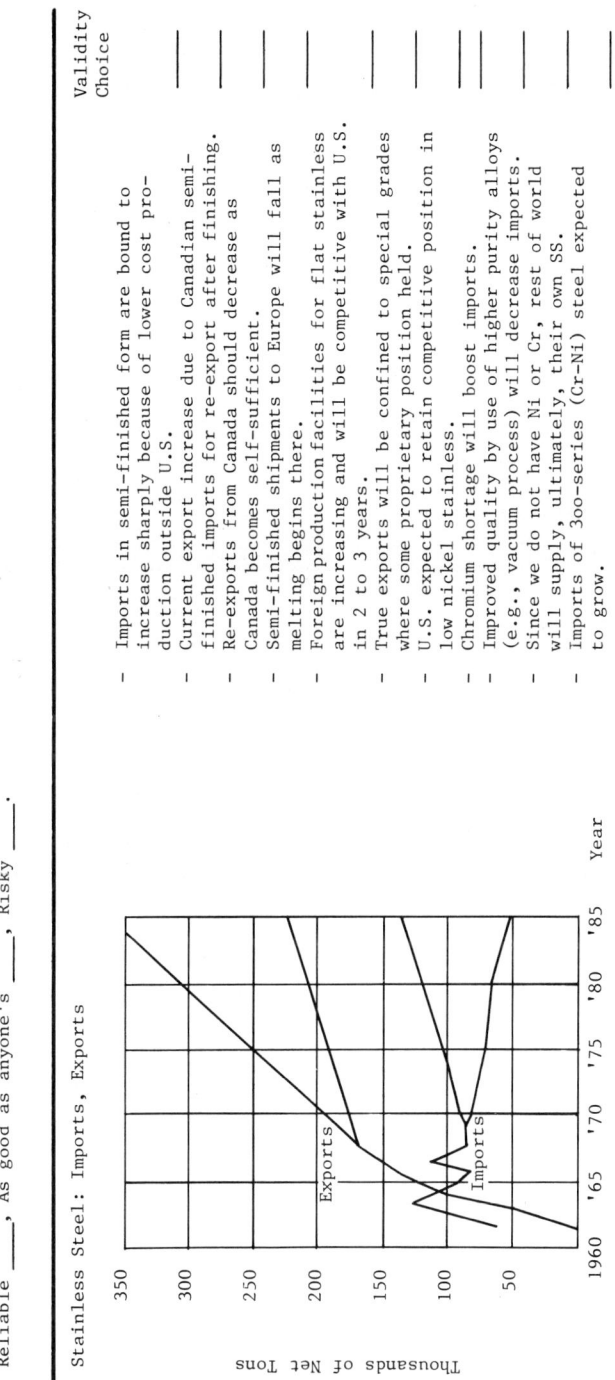

Stainless Steel: Imports, Exports

Thousands of Net Tons

(y-axis: 50, 100, 150, 200, 250, 300, 350)

Exports

Imports

Year: 1960, '65, '70, '75, '80, '85

Figure 4 Form for trend extrapolations. (Reproduced by permission of the publisher from *The Delphi Method: Techniques and Applications*, 1975, edited by Harold A. Linstone and Murray Turoff, Addison-Wesley, Advanced Book Program, Reading, Mass.)

129

QUESTION NUMBER	MEAN	STANDARD DEVIATION
1	3.4	1.0
2	2.6	0.9
3	3.0	1.2

QUESTION NUMBER	VALIDITY CHOICE						NUMBER RESPONDING
	1	2	3	4	5	6	
1	.25		.25	.25		.25	25
2		.33	.33		.33		16
3	.20	.20		.20	.20	.20	19

QUESTION 1

OCCUPATION CATEGORY	VALIDITY CHOICE					
	1	2	3	4	5	6
1. Steel	3	2	1	2	0	1
2. Ferroalloy	1	4	0	0	0	0
3. R & D	3	1	0	1	0	0
4. Government	0	0	1	3	2	0

Table 5 Examples of Statistical Presentations (Reproduced by permission of the publisher from *The Delphi Method: Techniques and Applications,* 1975, edited by Harold A. Linstone and Murray Turoff, Addison-Wesley, Advanced Book Program, Reading, Mass.)

and agreement between the two in their forecasts, as illustrated in Table 6. The following quote summarizes her comparison[93]:

> Based upon the expertise of the [panel], the report is a reliable and comprehensive account of known information and of projections based on this information and on current research and development. In contrast, the Delphi was designed to complement the panel report. The planned approach was to provide an opportunity to indicate uncertainties or disagreements about the subject and to evaluate quantitatively the degree of uncertainty which exists within a large group of experts. The Delphi product attempts to present an awareness of the areas which are subject to differences of view and to highlight the topics which appear to concern the respondent group. . . . Probably the most significant difference between the Delphi and the Committee approaches is the temptation of what the Delphi group could not agree on or what they rejected. Usually the psychological process in a committee of experts tends to eliminate these categories of information in the final report.

In summary, Goldstein, offers the following advice to others running Delphi Studies[94]:

1. When presenting statements for a vote, or synthesizing the respondents' suggestions, be alert for ambivalent wording. Two separate statements may appear as one, leading to confusion as to what should be voted upon. Vague wording or easily misinterpreted wording may also lead to confusion.
2. When editing respondents' comments for clarity, try to preserve the intent of the originator. When editing from round to round, avoid changing a statement so that it has one meaning in round one and another in round two.
3. Lay out the expected processing of the data throughout all the rounds of the Delphi before you finalize the design. You may, by circumstance, be forced later to modify the procedure, but the process of planning ahead will usually turn up any gross problems in your initial questionnaire design and its impact on following rounds.
4. Design the handling of your data so that each response can be processed (or punched for processing) as it comes in. Thus you will not have a frantic rush to analyze all the responses at once when the last tardy return comes in.
5. Keep track of how different subgroups in your respondent group

QUANTITATIVE COMPARISON

PREDICTED FERROALLOY CONSUMPTION IN STEELS AND SUPERALLOYS IN 1980
(IN SHORT TONS OF CONTAINER ELEMENT)

	PANEL REPORT NMAB-276	DELPHI
Chromium	319,260	250,000– 303,000
Cobalt	2,732	3,000– 4,000
Columbium	1,977	1,300– 1,850
Manganese	1,011,235	1,100,000–1,250,000
Molybdenum	21,540	17,400– 21,000
Nickel	124,200	90,000– 115,000
Tungsten	1,850	1,550– 2,600
Vanadium	5,796	5,000– 6,200

QUALITATIVE COMPARISON

TYPE OF ACTIVITY	DELPHI	COMMITTEE
FORM OF INFORMATION	Specific Comments	General Discussion
WEIGHTING OF INFORMATION PROVIDED	Most Rated on Reliability Scale	None
DISAGREEMENT AMONG COMMITTEE MEMBERS OR RESPONDENTS	Indicated in Reliability Score	Eliminated
PRESENTATION OF BACKGROUND INFORMATION	Only as Randomly Generated by Respondents	Thorough and Systematic
RECOMMENDATIONS	Not Specifically Stated	Consensus Recommendations Indicated
RANGE OF INFORMATION PROVIDED	Broad, Reflecting Wide Interest of Respondents	Limited to Specific Committee Subject Area

Table 6 Comparison of the Panel and Delphi Approaches (Reproduced by permission of the publisher from *The Delphi Method: Techniques and Applications,* 1975, edited by Harold A. Linstone and Murray Turoff, Addison-Wesley, Advanced Book Program, Reading, Mass.)

vote on specific items. This can be very useful in analyzing the results and will occasionally produce situations where you wish to let the respondent group know that polarizations or differences based upon background exist.

6. If you are covering a number of fields of expertise, make sure that each field is adequately represented in your group.

7. It should be mandatory that at least two professionals work on monitoring any one Delphi exercise, particularly when the abstracting of comments is a notable portion of the exercise. With two individuals one can always review what the other has done.

8. Pretest your questionnaire on any willing guinea pigs you can find outside your respondent or monitor group. If you have a sponsor, it is useful to go over the design of each round with some of his people before finalizing it.

A DELPHI ON THE FUTURE OF STUDENT AFFAIRS IN HIGHER EDUCATION

A Delphi study involving 10 New England colleges and universities was completed in 1978 at the University of Massachusetts.[95] It looked at future trends in a segment of higher education that is controversial with regard to its effectiveness and contribution to higher education, the lack of congruency between its philosophy and practice, and the need to change student affairs in light of the general climate of change and complexities in higher education. This study systematically encouraged several constituents in higher education to consider ideas about the future of student affairs. It also gathered constituent and overall opinions of the probable directions for student affairs in the next decade for use in policy planning in the field. The results of the study focused on answers to the following questions:

1. What consensus of opinions exists about ideas on student affairs in the next decade, and what are the priorities among the ideas?

2. What divergence of opinions exists about ideas on the future of the field?

3. Are the similarities and differences between subgroups significant?

4. What are the limitations of the results obtained?

5. Are the results of this Delphi applicable and adaptable to planning and decision-making about the future of student affairs?

6. Is the Delphi technique a useful methodology for providing data to use in planning and decision-making in the field?

Rounds. This Delphi employed three questionnaires and a final report of the results to participants for their evaluative remarks. The questionnaires were used to gather and refine ideas about student affairs in the next decade, to provide feedback to individual participants about group responses, to elicit group priorities among ideas about the future, and to develop data on consensus and divergent opinions about the future of the field. Table 7 briefly lists the content, purpose, and use of each questionnaire and the final report of the results to participants.

Participants. A random sample of 10 institutions of higher education was taken from a list of the 111 public and private four-year institutions of New England. Four individuals from each of the 10 institutions were designated (through cooperation with the Presidents' Offices of the schools) for participation according to four subgroups: (1) Dean of Academic Affairs (or equivalent title), (2) President of Student Government (or equivalent student), (3) Dean of Student Affairs (or equivalent title), and (4) Chair of Department of English (or equivalent title). These four subgroups were designated because of their varied relation to issues in higher education and their potential for constructive input on the particular issue of student affairs. A response rate of 70% among these 40 participants was achieved.

Round One. The design of the first questionnaire was based on a literature search for ideas on the trends in student affairs and on the results of the pilot test of the questionnaire. The 42 items for the questionnaire were divided according to the duties and roles of six groups of student affairs personnel: student affairs personnel in general, student affairs administrators, students, student affairs instructors, student affairs specialists, and student affairs consultants. A sample of the questionnaire and its instructions to participants is shown in Figure 5.

Round Two. The results of round one were used to design Questionnaire II, which was composed of two parts. Part 1 contained 12 items from the previous questionnaire that received a group median rating of greatest agreement. In addition, a bogus item was added to test the consistency and thoughtfulness of participants' responses and to consider the effects of manipulation on the study. Participants were also given their individual responses to compare with the group responses and were asked to consider reaffirming their previous rating or changing it and also to rank all 12 items in order of agreement. Part 2 of Questionnaire II was a random list of nine additional ideas suggested by participants through Questionnaire I. Participants were asked to review the random list and to rate and rank each item as they had done in Part 1.

Round Three. Questionnaire III was also composed of two parts, based on the results of the previous questionnaire. Part 1 listed the 10

	QUESTIONNAIRE 1	QUESTIONNAIRE 2	QUESTIONNAIRE 3	QUESTIONNAIRE 4
CONTENT:	-42 items developed from the literature on change in student affairs -a request for additional ideas and comments	-a list of 11 ideas agreed upon by the group + bogus item -feedback appears within the 12 items which provides participants with overall group response to each item for comparison with their own response -a list of 9 items which present new ideas about the future suggested by participants through questionnaire 1	-a repeated list of 10 items from the first part of questionnaire 2 and those items from the second part of questionnaire 2 agreed upon by the group -feedback appears within the items which provides participants with overall group response to each item for comparison with their own response	-an overview of the study which provides group and subgroup responses -a final list of priority items on the future of student affairs -a list of items where significant disagreement occurred -a brief form for final comments and evaluation
PURPOSE:	-to generate a list of items agreed upon by participants -to generate group priorities among the items -to elicit individual ideas not covered in questionnaire 1 -to generate a list of those ideas from the literature with which participants agree	-to allow comparison of individual and group responses -to allow reaffirmation or changes in responses to questionnaire 1 -to generate group priorities among the items -to allow consideration of additional ideas submitted by participants through questionnaire 1 -to generate a list of items with which participants disagree	-to allow comparison of individual and group responses -to allow reaffirmation or changes in responses to questionnaire 2 -to generate group priorities among the items -to generate a list of items with which participants disagree	-to provide participants with a summary of the process -to offer information for possible use by participants -to elicit evaluation of the process
USE:	-to design questionnaire 2 -to formulate a final report -to offer feedback to participants on group responses -to develop data about individual, subgroup and group responses	-to design questionnaire 3 -to formulate a final report -to offer feedback to participants on group responses -to develop data about individual, subgroup and group responses	-to formulate a final report -to offer feedback to participants on group responses -to develop data about individual, subgroup and group responses	-to obtain participants' final evaluation -to offer feedback to participants on group responses -to develop data about individual, subgroup and group responses

Table 7 The Content, Purpose, and Use of Questionnaires and Final Report

QUESTIONNAIRE I

INSTRUCTIONS

The items in the questionnaire refer to ideas suggested by the
literature on student affairs about the future of the field in colleges
and universities. The items are divided into groups which represent
general categories of the future roles which have been suggested. Each
item in a group provides an example of the duties for the roles. Each
item defines the suggested role within the wording of the item.

After reading an item, consider what is stated in relation to what
you think student affairs will be like in ten years. Place a corresponding number of 1 through 5 in the blank space at the left of each
item, using the following scale:

1 - I strongly agree
2 - I agree
3 - I am uncertain whether I agree or disagree
4 - I disagree
5 - I strongly disagree

A. Student Affairs Personnel, in general:

____ 1. Student affairs personnel will work directly with students,
          faculty and administrators to implement student development
          philosophy in curricular and non-curricular programming.

____ 2. Student affairs personnel will frequently undergo evaluations
          of their student development programs.

____ 3. Student affairs personnel will form a single professional
          organization.

____ 4. Student affairs personnel will be included in faculty
          collective bargaining.

____ 5. Student affairs personnel will increasingly use affirmative
          action and equal opportunity type policies in their educational
          and personnel operations.

____ 6. Student affairs personnel will be compensated at a salary
          equivalent to faculty pay.

____ 7. Student affairs personnel will continue to be viewed by the
          rest of the academic community as being of secondary importance
          in the educational process.

Figure 5   Sample questionnaire.

1–strongly agree/2–agree/3–uncertain/4–disagree/5–strongly disagree

A. Student Affairs Personnel, in general: (continued)

____ 8. Student affairs personnel will be almost totally concerned with refining and expanding past services, rather than developing new ones.

____ 9. Student affairs personnel will have primary roles in the process of institutional inclusion of more humanistic educational goals.

B. Student Affairs Administrators:

____ 10. Student affairs administrators will coordinate campus efforts in expanding student development programming.

____ 11. Student affairs administrators will emphasize cost-effectiveness program development because of continued budget constraints.

____ 12. Student affairs administrators will coordinate their internal operations and program accountability, using business management skills such as "management by objectives."

____ 13. More minorities and women will become student affairs administrators.

____ 14. Student affairs administrators will coordinate hiring procedures in student affairs which aim to prevent race and sex discrimination.

____ 15. Student affairs administrators will develop policy to replace the past use of precedent or pressure.

____ 16. The chief student affairs administrator will most frequently be a vice-president of the institution who has direct input into institutional policy making.

C. Students:

____ 17. Students will be decision-makers with respect to the operations of student union facilities, including budgets.

Figure 5   (*Continued*)

1-strongly agree/2-agree/3-uncertain/4-disagree/5-strongly disagree

C. Students: (continued)

____ 18. Student "associates" will participate with faculty and student affairs personnel in student development program planning.

____ 19. Students will be decision-makers with respect to student activities, such as newspaper publishing, fine arts and concert series planning and student organization management, including budgets.

____ 20. Students, along with former students, will totally manage alumni affairs' public relations and fund raising efforts.

____ 21. Students will enhance self-regulation and peer representation through more meaningful rules for student government and judiciary boards.

____ 22. Student involvement as decision-makers, peer advisors, managers, and representatives will result in increased input into institutional policy formation.

____ 23. Large numbers of students will become para-professional workers for student affairs in such roles as student activities managers, curriculum planners, and peer counselors.

____ 24. Large numbers of students will complete internships and practicums in student affairs as part of their educational experience.

D. Student Affairs Instructors:

____ 25. Student affairs instructors will teach such courses as human development, human sexuality, drug and alcohol abuse prevention, career education, family relations, and peer relations.

____ 26. Student affairs instructors will teach human relations skills to administrators, faculty and students.

____ 27. Student affairs instructors will facilitate students' development of leadership and management skills.

____ 28. Student affairs instructors will facilitate students' development of skills such as goal specification, values clarification, and self-development.

Figure 5 (*Continued*)

1-strongly agree/2-agree/3-uncertain/4-disagree/5-strongly disagree

D. Student Affairs Instructors: (continued)

____ 29. Student affairs instructors will facilitate the development of counseling skills for students and faculty who will be doing the majority of student counseling, excluding mental health therapy.

____ 30. Student affairs instructors will teach such physical techniques as relaxation training, yoga, psychoanalytic bodily exercise, and meditation as part of student development programming.

E. Student Affairs Specialists:

____ 31. Student affairs specialists in mental health will increase in numbers, while other types of counseling will be done by faculty and students.

____ 32. Student affairs specialists in career development will coordinate use of community, business and government career placement services, and encourage service expansion.

____ 33. Student affairs specialists in residential living will work for the restructuring of dormitory living, making it a more realistic living experience.

____ 34. Student affairs specialists in computer systems will coordinate the use of more sophisticated equipment for maintaining useful and detailed student records, which will be accessible to students and faculty.

____ 35. Student affairs specialists in psychology will apply knowledge of educational psychology to program planning.

____ 36. Student affairs specialists will combine behavioral theory and humanistic philosophy in program planning.

F. Student Affairs Consultants:

____ 37. Student affairs consultants will assist faculty and students in curricular planning.

Figure 5 (*Continued*)

1-strongly agree/2-agree/3-uncertain/4-disagree/5-strongly disagree

F. Student Affairs Consultants: (continued)

____ 38. Student affairs consultants will arbitrate conflicts and serve as educational-consumer advocates between students and institutions.

____ 39. Student affairs consultants will initiate compliance with new legal rulings on student rights. (For example, due process in academic and personal affairs; search warrants for entering dormitory rooms; and adult stature and privileges.)

____ 40. Student affairs consultants will encourage professional development among faculty, staff, and administrators.

____ 41. Student affairs consultants will facilitate the use of the community (for housing, career development, child care, etc.) as a resource for students, and the use of the college (for cultural events, education, conferences, catering, etc.) as a resource for the community.

____ 42. Student affairs consultants will provide students with continuous assessment of their growth for use in personal and educational planning.

Figure 5 (*Continued*)

Your Ideas

A questionnaire of this type can be confining. Because of this, please add any ideas you hold about the characteristics of student affairs in ten years.

_____

_____

_____

_____

_____

_____

_____

_____

(Please continue on back.)

Comments

Please comment on this questionnaire's format, instructions, background information and the items.

_____

_____

_____

_____

_____

_____

_____

_____

(Please continue on back.)

Figure 5 (*Continued*)

items with the highest rating and ranking among participants, which excluded the bogus item. (The inclusion of the bogus item confirmed that the manipulation of individuals' responses can occur, yet in this case the change in scores was dramatic, but not great enough to continue inclusion, since the bogus item still received the lowest rating and ranking of the final top 12 items.) Participants were also provided with their individual responses and asked to do the following: first, consider reaffirming their rating or changing it; second, briefly explain any response that was different from the group response; and third, rank all 10 items in order of agreement. Part 2 continued the five items of greatest agreement among participants from Part 2 of Questionnnaire II. Instructions were the same as for Part 1, and space for additional comments was again provided.

Final Report. Final feedback was provided on all three questionnaires in a final report of the results. It offered a summary of the study, showing those items for which there was group consensus and those for which there was strongly divergent opinion. In addition, data analysis provided participants with an overview of the significant findings and their implications. While the anonymity of individuals was maintained, participants were informed of the resulting differences and similarities among the four subgroups. Participants were asked to complete an evaluation form as shown in Figure 6 adapted from Scheibe, Skutsch, and Schofer in Linstone and Turoff.[96]

Analysis. The median, first quartile, third quartile, and quartile deviation were calculated for each item to obtain the measure of central tendency and of the variability of ratings appropriate in considering the convergence of opinions that becomes statistically significant in Questionnaire III (see Tables 8 and 9). Next, a more detailed analysis of subgroups' responses was determined through nonparametric analysis of variance using the Kruskal–Wallis H test to analyze the research hypothesis, related to significant differences of opinion between subgroups, through null and alternative hypotheses. The H test was chosen because of its applicability to studies with several small related groups and with an ordinal scale level of measurement and because it allows the generalization of significant results, using chi square distribution tables, as representative of larger groups.

In summary, Young provides the following evaluative remarks that can be considered in designing Delphi studies:

1. In this study an upswing in enthusiasm occurred with the second questionnaire, and participants took time to write lengthy comments, which they did less of for the first and third rounds. This curve might

RESULTS OF GROUP EVALUATION

Group preferences among statements are CAPITALIZED and the group medians are given for each statement.

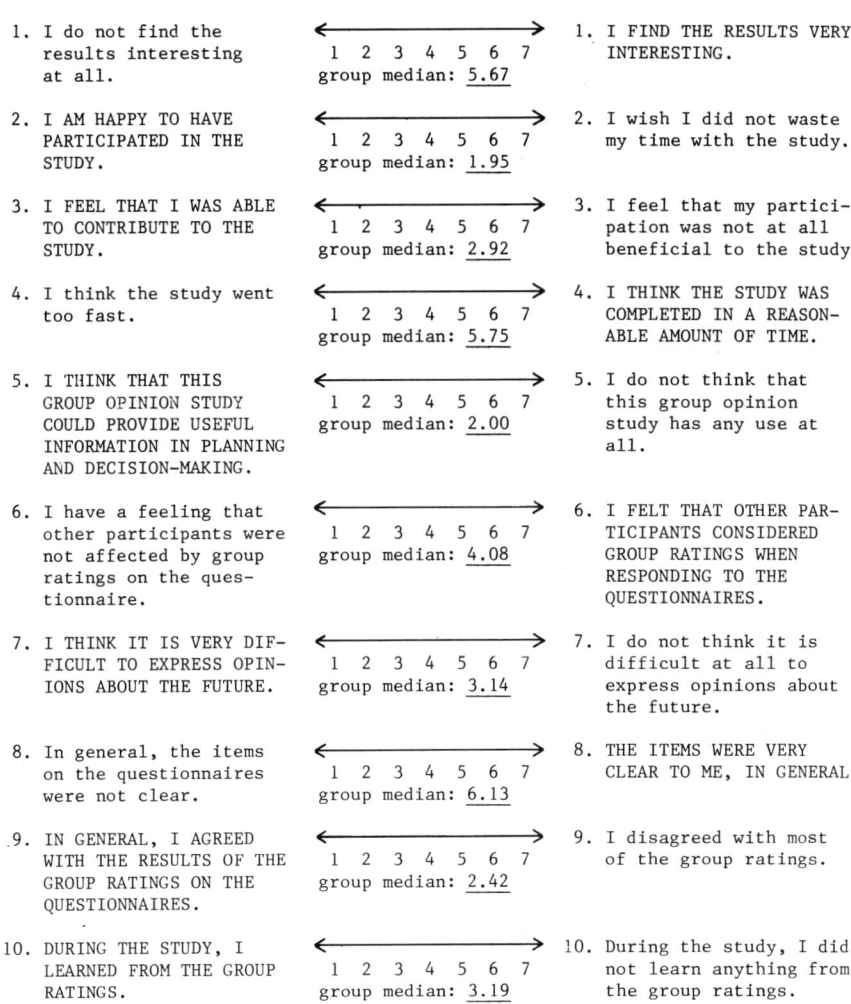

1. I do not find the results interesting at all.

group median: 5.67

1. I FIND THE RESULTS VERY INTERESTING.

2. I AM HAPPY TO HAVE PARTICIPATED IN THE STUDY.

group median: 1.95

2. I wish I did not waste my time with the study.

3. I FEEL THAT I WAS ABLE TO CONTRIBUTE TO THE STUDY.

group median: 2.92

3. I feel that my participation was not at all beneficial to the study.

4. I think the study went too fast.

group median: 5.75

4. I THINK THE STUDY WAS COMPLETED IN A REASONABLE AMOUNT OF TIME.

5. I THINK THAT THIS GROUP OPINION STUDY COULD PROVIDE USEFUL INFORMATION IN PLANNING AND DECISION-MAKING.

group median: 2.00

5. I do not think that this group opinion study has any use at all.

6. I have a feeling that other participants were not affected by group ratings on the questionnaire.

group median: 4.08

6. I FELT THAT OTHER PARTICIPANTS CONSIDERED GROUP RATINGS WHEN RESPONDING TO THE QUESTIONNAIRES.

7. I THINK IT IS VERY DIFFICULT TO EXPRESS OPINIONS ABOUT THE FUTURE.

group median: 3.14

7. I do not think it is difficult at all to express opinions about the future.

8. In general, the items on the questionnaires were not clear.

group median: 6.13

8. THE ITEMS WERE VERY CLEAR TO ME, IN GENERAL.

9. IN GENERAL, I AGREED WITH THE RESULTS OF THE GROUP RATINGS ON THE QUESTIONNAIRES.

group median: 2.42

9. I disagreed with most of the group ratings.

10. DURING THE STUDY, I LEARNED FROM THE GROUP RATINGS.

group median: 3.19

10. During the study, I did not learn anything from the group ratings.

11. DURING THE STUDY, I DID NOT FEEL I NEEDED TO TALK WITH THE OTHER PARTICIPANTS.

group median: 3.00

11. I would have liked to talk to the other participants during the study.

Figure 6 Results of group evaluation

	ITEM # ↓	ORDERED ITEM # ↓	ORDERED ITEM #	
	QUESTIONNAIRE 1 ↓	QUESTIONNAIRE 2 ↓	QUESTIONNAIRE 3	FINAL RANK
First Quartile Quartile Deviation Third Quartile	(14) 1.10 .53 2.15	(1) 1.28 .46 2.21	(1) 1.20 .48 2.15	 1
First Quartile Quartile Deviation Third Quartile	(32) 1.41 .48 2.36	(2) 1.38 .47 2.31	(2) 1.20 .52 2.23	 2
First Quartile Quartile Deviation Third Quartile	(13) 1.63 .40 2.43	(4) 1.50 .47 2.43	(3) 1.20 .52 2.23	 3
First Quartile Quartile Deviation Third Quartile	(18) 1.67 .38 2.43	(5) 1.70 .35 2.40	(4) 1.28 .49 2.25	 4
First Quartile Quartile Deviation Third Quartile	(19) 1.10 .69 2.48	(6) 1.20 .61 2.42	(5) 1.50 .39 2.28	 5
First Quartile Quartile Deviation Third Quartile	(1) 1.58 .45 2.48	(9) 1.56 .44 2.44	(8) 1.56 .44 2.44	 6
First Quartile Quartile Deviation Third Quartile	(39) 1.54 .47 2.48	(7) 1.56 .39 2.33	(6) 1.69 .33 2.36	 7
First Quartile Quartile Deviation Third Quartile	(10) 1.58 .45 2.48	(8) 1.56 .41 2.38	(7) 1.50 .37 2.24	 8
First Quartile Quartile Deviation Third Quartile	(2) 1.31 .72 2.75	(11) 1.50 .57 2.64	(10) 1.56 .41 2.38	 9
First Quartile Quartile Deviation Third Quartile	(11) 1.58 .51 2.61	(10) 1.20 .61 2.42	(9) 1.61 .37 2.34	 10

Table 8 Comparison of Quartiles for Convergence, Part 1, Top 10 Items

	ITEM # ↓	ORDERED ITEM # ↓	
	QUESTIONNAIRE 2 ↓	QUESTIONNAIRE 3 ↓	FINAL RANK
First Quartile	(2) 1.63	(1) 1.63	
Quartile Deviation	.52	.44	
Third Quartile	2.67	2.50	1
First Quartile	(4) 1.93	(2) 1.85	
Quartile Deviation	.62	.58	
Third Quartile	3.17	3.00	2
First Quartile	(6) 1.93	(4) 1.83	
Quartile Deviation	.79	.54	
Third Quartile	3.50	2.90	3
First Quartile	(1) 1.83	(5) 1.77	
Quartile Deviation	.90	.87	
Third Quartile	3.64	3.50	4
First Quartile	(7) 1.83	(3) 1.83	
Quartile Deviation	.76	.71	
Third Quartile	3.36	3.25	5

Table 9 Comparison of Quartiles for Convergence, Part 2, Top Five Items

be considered in designing the rounds to enhance commitment to the task at the appropriate time.

2. Table 10 illustrates changes in responses.[97] As is expected for the Delphi technique, divergent opinion was much more evident in the first questionnaire than in the final one. It is important to focus on the existence of individual and subgroup divergent opinions when considering the findings of a study.[98]

3. In designing introductory remarks or letters, as well as instructions to questionnaires, sensitivity to subgroups' and individuals' attitudes toward change, in addition to attention to convergence of group opinions, can provide invaluable information which, if overlooked, could undermine the accuracy of findings.[99]

PART 1 (TOP TEN ITEMS)	RESULTS OF QUESTIONNAIRE 1	RESULTS OF QUESTIONNAIRE 2	RESULTS OF QUESTIONNAIRE 3
PERCENTAGES*			
1's (strongly agree)	26.07	25.71	26.07
2's (agree)	51.43	55.36	61.79
Total 1's + 2's	77.50	81.07	87.86
3's (uncertain)	17.50	17.14	10.35
4's (disagree)	4.64	1.07	1.43
5's (strongly disagree)	.36	.72	.36
Total 4's + 5's	5.00	1.79	1.79

* refers to percentage of 280 ratings from 28 participants to 10 items

PART 2 (TOP FIVE ITEMS)	RESULTS OF QUESTIONNAIRE 2	RESULTS OF QUESTIONNAIRE 3
PERCENTAGES*		
1's (strongly agree)	8.57	9.29
2's (agree)	50.00	55.71
Total 1's + 2's	58.57	65.00
3's (uncertain)	22.14	20.71
4's (disagree)	17.86	13.57
5's (strongly disagree)	1.43	.72
Total 4's + 5's	19.29	14.29

* refers to percentage of 140 ratings from 28 participants to 5 items

Table 10   Percentages of Responses With the Likert-Type Scale of 1 to 5

4. As has already been mentioned, manipulation of individual response is possible and therefore a threat to the integrity of the study. Every effort should be made to assure that manipulation by monitors cannot occur. For example, questionnaire results could be calculated by a small group of impartial outsiders.[100]

5. Because the Delphi process results in changes in individual ratings, a problem may arise in cases where no opinion really exists but participants concur with the intent of the process and record an opinion that is inconsistent over time and does not take into account new information. The addition of space for comments on questionnaires can lessen this factor.[101]

6. Low return rates of questionnaires and fewer comments on Questionnaire III may indicate "fatigue," overused or poorly designed questionnaires, or a lack of commitment to the issue of change. Other mediums, such as audiovisual materials, should be considered.[102]

7. The number of additional ideas on change offered by respondents relates to commitment, attitudes toward change, creative thinking, and so on. Such input or lack thereof is a major concern for Delphi studies.[103]

8. The most important measure of results in a Delphi study relates to the generation and consensus of ideas. Therefore a thorough literature search and careful attention to the design of the questionnaire and process are paramount in achieving significant results.[104]

9. Delphi is also useful in generating new study topics, which can be more specific in purpose. Related to this study on student affairs, the following examples were indicated as topics for further research:

   (a) Planning and adopting a "student development" model.

   (b) Ascertaining effects on students of increased responsibility and other student development changes.

   (c) Overcoming problems of student involvement (or lack of it) in curricular and noncurricular committee work.

   (d) Improving student–faculty–administrator communication during crisis and struggle situations.

   (e) Considering student affairs staff development and training.

   (f) Identifying aspects of institutionalized racism and sexism in student affairs functions, personnel, and development.

   (g) Implementing the role of faculty as student development educators.

   (h) Examining problems of resource changes and utilization for the future of student affairs.

(i) Considering current business and regulatory aspects of student affairs as separate entities from student affairs operations.

FINAL WORDS

Delphi successfully weathered the storm over its usefulness as a forecasting methodology. This was partly due to the fact that Delphi is a useful tool for helping researchers provide valuable results. In addition, Delphi has undergone beneficial modifications since its invention for defense planning. Many of these changes were the results of efforts by pioneers in the development of Delphi, but many more have resulted from suggestions by planners in a wide variety of fields interested in improving the Delphi technique.

Delphi has been used extensively in operational and applied research, particularly in the areas of physical and engineering sciences. In recent years it has been used increasingly in social sciences, public administration, and higher education.[105] While the drawbacks or pitfalls of Delphi need to be understood by researchers, the advantages of Delphi are also significant to understanding the value of the process.

A review of the literature reveals that there are a variety of conditions that make Delphi an appropriate method for use by researchers, and Linstone and Turoff list them as follows[106]:

The problem does not lend itself to precise analytical techniques but can benefit from subjective judgements on a collective basis.

The individuals needed to contribute to the examination of a broad or complex problem have no history of adequate communication and may represent diverse backgrounds with respect to experience or expertise.

More individuals are needed than can effectively interact in a face-to-face exchange.

Time and cost make frequent group meetings infeasible.

The efficiency of face-to-face meetings can be increased by a supplemental group communication process.

Disagreement among individuals is so severe or politically unpalatable that the communication process must be referred and/or anonymity assured.

The heterogeneity of the participants must be preserved to assure validity of the results; i.e., avoidance of domination by quantity or by strength of personality (bandwagon effect).

Since the use of Delphi has spread in recent years, a summary of characteristics for typical Delphi studies includes the following: 60% of the studies used between 5 and 40 participants and the remainder were much larger studies; 75% of the studies were completed in 8 months or less; the great majority of the studies used three or fewer questionnaires; 43% of the studies cost under $5000; and the majority of users of Delphi stated that it was chosen over other methods because of its ability to gather expert opinion and bring convergence to diverse group opinions.[107] These characteristics are not necessarily totally positive, but the literature concludes that scrutiny of Delphi has encouraged and maintained quality standards among most researchers. Delphi has not only become a favorite tool for futures research but its limitations are better understood and its advantages are fully respected.

The use of Delphi has increased, particularly in fields such as education, public administration, and social sciences, which anticipate substantial changes in the future. Delphi has applicability to these and other fields, not only as a forecasting tool but even more appropriately as a means of bringing constituents or subgroups together to consider group opinions, obtaining group consensus for use in thinking about and planning for alternative futures, and developing methods to enable subgroups to consider the complexities of group opinions and future decision-making.[108] While planning for the future is complicated by the fact that it cannot be solidly based on empirical techniques, future analysis is necessary and very much in demand. Delphi, with its advantages and disadvantages, has filled the gap that exact sciences research cannot fill in using extrapolations about what can be done to encourage, improve, and plan for what has not yet occurred.

Delphi will continue to evolve in response to the needs for better planning and systems management. Five years ago Linstone and Turoff defined Delphi as "a method for structuring a group communication process."[109] The link between Delphi and computerized conferencing remains crucial today. Most Delphi activity is applied rather than methodological. Clients want results, not academic papers. Its use has spread as far as the People's Republic of China.

There is a real need for more experimentation on Delphi: our understanding of its internal dynamics as a structured human communication process is very limited. With or without this knowledge, we are rapidly becoming a network nation. Electronic mail, computer conferencing networks, electronic movement of money, and computerized medical diagnosis are just a few of the innovations that are about to change our lifestyle. By 2000–2010 about 50% of all phones will have terminals.[110]

We can thus expect a plethora of structured communication concepts as a wide public domesticates computers and terminals—and today's

Delphi may well be on the way to joining its Greek ancestor as a historic relic.

REFERENCES

1. Norman Dalkey and Olaf Helmer, *The Use of Experts for the Estimation of Bombing Requirements—A Project Delphi Experiment*, RM–727–PR, Rand Corporation, Santa Monica, Calif., 1951.

2. Olaf Helmer and Nicholas Rescher, *On the Epistemology of the Inexact Sciences*, R–353, Rand Corporation, Santa Monica, Calif., Feb. 1960.

3. Norman Dalkey and Olaf Helmer, *An Experimental Application of the Delphi Method to the Use of Experts*, RM–727–PR, Rand Corporation, Santa Monica, Calif., 1962.

4. T. J. Gordon and Olaf Helmer, *Report on a Long-Range Forecasting Study*, 2982, Rand Corporation, Santa Monica, Calif., 1964.

5. Olaf Helmer, *Analysis of the Future: The Delphi Method*, P–3555, Rand Corporation, Santa Monica, Calif., 1967.

6. Olaf Helmer, *Systematic Use of Expert Opinions*, P–3721, Rand Corporation, Santa Monica, Calif., 1967.

7. R. Rochberg, *Some Comments on the Problems of Self-Affecting Predictions*, P–3735, Rand Corporation, Santa Monica, Calif., 1967.

8. Norman Dalkey, *Delphi*, Rand Corporation, P–3704, Santa Monica, Calif., 1967.

9. Norman Dalkey, *Quality of Life*, P–3805, Rand Corporation, Santa Monica, Calif., 1968.

10. Norman Dalkey, *Experiments in Group Predictions*, P–3820, Rand Corporation, Santa Monica, Calif., 1968.

11. Norman Dalkey, *Predicting the Future*, P–3948, Rand Corporation, Santa Monica, Calif., 1968.

12. Bernice Brown, "Delphi Process: A Method Used for Elicitation of Opinions of Experts," *ASTME Vectors*, Vol. 3, No. 1, 1968, pp. 4–8.

13. Norman Dalkey, *The Delphi Method: An Experimental Study of Group Opinion*, RM–5888–PR, Rand Corporation, Santa Monica, Calif., 1969.

14. Bernice Brown, S. Cochran, and N. Dalkey, *The Delphi Method II: Structure of Experiments*, RM–5957, Rand Corporation, Santa Monica, Calif., 1969.

15. Norman Dalkey, B. Brown, and S. Cochran, *The Delphi Method III: Use of Self-Rating to Improve Group Estimates*, RM–6115–PR, Rand Corporation, Santa Monica, Calif., 1969.

16. N. Rescher, *Delphi and Values*, P–4182, Rand Corporation, Santa Monica, Calif.. 1969.

17. Norman Dalkey, B. Brown, and S. Cochran, *The Delphi Method IV: Effect of Percentile Feedback and Feed-in of Relevant Facts*, RM–6118–PR, Rand Corporation, Santa Monica, Calif., 1970.

18. Norman Dalkey and Daniel Rourke, *Experimental Assessment of Delphi Procedures with Group Value Judgements*, R–612–ARPA, Rand Corporation, Santa Monica, Calif., 1971.

19. Norman Dalkey and Bernice Brown, *Comparison of Group Judgement Techniques*

with Short-Range Predictions and Almanac Questions, R–678–ARPA, Rand Corporation, Santa Monica, Calif., 1971.

20. Harold Linstone and Murray Turoff, Eds., *The Delphi Method: Techniques and Applications*, Addison-Wesley, Reading, Mass., 1975.

21. Norman Dalkey, *The Delphi Method: An Experimental Study of Group Opinion*, RM–5888–PR, Rand Corporation, Santa Monica, Calif., 1969, p. v.

22. Brown et al., op. cit.

23. Norman Dalkey, *The Delphi Method: An Experimental Study of Group Opinion*, RM–5888–PR, Rand Corporation, Santa Monica, Calif., 1969.

24. Dalkey and Brown, op. cit.

25. Richard Weatherman and Karen Swenson, "Delphi Technique," in Stephen Hencley and James Yates, Eds., *Futurism in Education: Methodologies*, McCutchan Publishing, Berkeley, Calif., 1974, pp. 97–114.

26. Harold Sackman, *Delphi Critique: Expert Opinions, Forecasting, and Group Process*, D. C. Heath, Lexington, Mass., 1975 (published in 1974 by the Rand Corporation under the title *Delphi Assessment: Expert Opinions, Forecasting and Group Process*).

27. Kim Q. Hill and Jib Fowles, "The Methodological Worth of the Delphi Forecasting Technique," *Technological Forecasting and Social Change*, Vol. 7, No. 2, 1975, pp. 179–192.

28. H. J. Strauss and L. H. Zeigler, "Delphi Technique and Its Uses in Social Science Research," *Journal of Creative Behavior*, Vol. 9, No. 4, 1975, pp. 253–259.

29. D. Sam Scheele, "Consumerism Comes to Delphi," *Technological Forecasting and Social Change*, Vol. 7, No. 2, 1976, p. 216.

30. Ibid., p. 218.

31. Ibid., p. 219.

32. Joseph F. Coates, "In Defense of Delphi," *Technological Forecasting and Social Change*, Vol. 7, No. 2, 1975, pp. 193–194.

33. Ibid., p. 193.

34. Ibid., p. 194.

35. Peter Goldschmidt, "Scientific Inquiry or Political Critique? Remarks on Delphi Assessment," *Technological Forecasting and Social Change*, Vol. 7, No. 2, 1975, pp. 195–213.

36. Brian Twiss, "The Delphi Debate," *Futures*, Vol. 8, No. 4, 1976, pp. 357–358.

37. H. Jones, Book review of Harold Sackman's *Delphi Critique*, *Long Range Planning*, Vol. 9, No. 4, 1976, p. 95.

38. Linstone and Turoff, op. cit.

39. V. Mahajan, Book review of Harold Linstone and Murray Turoff's *Delphi Method*, *Job Marketing Research*, Vol. 13, No. 3, 1976, pp. 317–318.

40. Joseph Martino, Book review of Linstone and Turoff's *The Delphi Method*, *Technological Forecasting and Social Change*, Vol. 8, No. 4, 1976, pp. 441–442.

41. Brain Twiss, Book review of Linstone and Turoff's *The Delphi Method*, *Long Range Planning*, Vol. 9, No. 4, 1976, pp. 94–95.

42. Harold Sackman, "A Skeptic at the Oracle," *Futures*, Vol. 8, No. 5, 1976, pp. 444–446.

43. Weatherman and Swenson, op. cit., p. 112.

44. Irene Jillson, "The National Drug-Abuse Policy Delphi: Progress Report and Findings

to Date," in Harold Linstone and Murray Turoff, Eds., *The Delphi Method*, Addison-Wesley, Reading, Mass., 1975.

45. John Ludlow, "General Applications: Delphi Inquiries and Knowledge Utilization," in Harold Linstone and Murray Turoff, Eds., *The Delphi Method*, Addison-Wesley, Reading, Mass., 1975, pp. 120–121.

46. Harold Linstone, "Eight Basic Pitfalls: A Checklist," in Harold Linstone and Murray Turoff, Eds., *The Delphi Method*, Addison-Wesley, Reading, Mass., 1975, pp. 574–586.

47. Bradley W., Nelson, "Statistical Manipulation of Delphi Statements: Its Sources and Effects on Convergence and Stability," *Technological Forecasting and Social Change*, Vol. 12, No. 1, 1978, pp. 41–60.

48. W. Rauch, "The Decision Delphi," *Technological Forecasting and Social Change*, Vol. 15, 1979, pp. 159–169.

49. B. McGaw, R. K. Browne, and P. Rees, "Delphi in Education—Review Assessment," *Australian Journal of Education*, Vol. 20, No. 1, pp. 56–76.

50. Vaughn E. Huckfeldt and Robert C. Judd, "Issues in Large Scale Delphi Studies," *Technological Forecasting and Social Change*, Vol. 6, No. 1, 1974, pp. 75–88.

51. Weatherman and Swenson, op. cit.

52. Frederick Cyphert and Walter Gant, "The Delphi Technique: A Case Study," *Phi Delta Kappan*, Vol. 36, No. 5, 1971, pp. 272–273.

53. J. J. Pallente, "Delphi Technique for Forecasting and Goal Setting," *NASSP Bulletin*, Vol. 60, 1976, pp. 86–89.

54. Alfred Rasp, "Delphi: A Decision-Maker's Dream," *Nation's Schools*, Vol. 93, 1973, pp. 29–32.

55. Gordon Welty, "Some Problems of Selecting Delphi Experts for Educational Planning and Forecasting Exercises," *California Journal of Educational Research*, Vol. 24, 1973, pp. 129–134.

56. R. C. Judd, "The Use of Delphi in Higher Education," *Technological Forecasting and Social Change*, Vol. 4, 1973, p. 173.

57. Klaus Brockhoff, "The Performance of Forecasting Groups in Computer Dialogue and Face-to-Face Discussion," in Harold Linstone and Murray Turoff, Eds., *The Delphi Method*, Addison-Wesley, Reading, Mass., 1975.

58. Klaus Brockhoff, "Delphi Progrosen in Computer-Dialog," J. C. B. Mohr, Tubingen, Federal Republic of Germany, 1979.

59. W. H. Middendon, "A Modified Delphi Method of Solving Business Problems," *IEEE Transactions on Engineering Management*, EM-20, 1973, pp. 130–133.

60. John Stover, "Suggested Improvements to the Delphi/Cross-Impact Technique," *Futures*, Vol. 5, No. 3, 1973, pp. 308–313.

61. Hill and Fowles, op. cit.

62. William Brockhaus and John F. Mickelsen, "An Analysis of Prior Delphi Applications and Some Observations on Its Future Applicability," *Technological Forecasting and Social Change*, Vol. 10, No. 1, 1977, pp. 103–110.

63. Linstone and Turoff, op. cit.

64. Olaf Helmer, "Problems in Futures Research: Delphi and Casual Cross-Impact Analysis," *Futures*, Vol. 9, No. 1, 1977, pp. 17–31.

65. Murray Turoff, "The Design of a Policy Delphi," *Technological Forecasting and Social Change*, Vol. 2, 1970, pp. 149–170.

66. D. L. Pyke, "A Practical Guide to Delphi," *Futures*, Vol. 2, No. 2, 1970, pp. 143–152.
67. Middendon, op. cit.
68. Jean-Claude Derian and Françoise Morize, "Delphi in the Assessment of Research and Development Projects," *Futures*, Vol. 5, No. 5, 1973, pp. 469–483.
69. Loran T. Thompson, *A Policy Application of Delphi Technique to the Drug Field*, Rand Corporation, Santa Monica, Calif., 1973.
70. L. S. Witson, "A Unique Event and Delphi Long Range Planning," *Technological Forecasting and Social Change*, Vol. 10, No. 1, 1977, pp. 79–83.
71. Olaf Helmer, *The Use of the Delphi Technique in Problems of Educational Innovations*, P–3499, Rand Corporation, Santa Monica, Calif., 1966.
72. M. Adelson, M. Alkin, C. Carey, and O. Helmer, "The Education of Innovation Study," *American Behavioral Scientist*, Vol. 10, No. 7, 1967, pp. 8–27.
73. Brockhaus and Mickelsen, op. cit.
74. L. R. Doyon and T. V. Sheehan, "Classroom Exercise in Applying the Delphi Method for Decision Making," *Socio-Economic Planning Science*, Vol. 5, No. 4, 1971, p. 363.
75. Leslie Martin and Kian Maynard, "Private Institution of Higher Education: An Application of the Delphi Technique," *Intellect*, Vol. 102, 1973, pp. 129–131.
76. Judd, op cit.
77. Harold D. Stolovitch, "Focus on the Future: From Delphi to Delphi," *Viewpoints*, Vol. 52, No. 2, 1976, pp. 9–20.
78. Cyphert and Gant, op. cit.
79. F. B. Newton and G. Hellenger, "Assessment of Learning and Process Objectives in a Student Personnel Training Program," *Journal of College Student Personnel*, Vol. 15, No. 6, 1974, pp. 492–497.
80. F. B. Newton and Robert L. Richardson, "Expected Entry-Level Competencies of Student Personnel Workers," *Journal of College Student Personnel*, Vol. 17, No. 5, 1976, pp. 426–430.
81. E. O. Jonassen and R. O. Stripling, "Priorities for Community College Student Personnel Services During the Next Decade," *Journal of College Student Personnel*, Vol. 18, No. 2, 1977, pp. 83–86.
82. McGaw, Browne, and Rees, op. cit.
83. Klaus Brockhoff, "Delphi Prognosen in Computer-Dialog," J. C. B. Mohr, Tubingen, Federal Republic of Germany, 1979.
84. Rauch, op. cit.
85. Ibid.
86. Harold Linstone, "Eight Basic Pitfalls: A Checklist," in Harold Linstone and Murray Turoff, Eds., *The Delphi Method*, Addison-Wesley, Reading, Mass., 1975, pp. 574–586.
87. Harold Linstone et al., "The Use of Alternative Decision Models for Technology Assessment," NSF Project, Futures Research Institute, Portland State University, Portland, Ore., 1980.
88. Harold Linstone et al., "The Use of Structural Modeling for Technology Assessment," *Technological Forecasting and Social Change*, Vol. 14, No. 4, 1979.
89. A. L. Porter et al., *A Guidebook for Technology Assessment and Impact Analysis*, North Holland, New York, 1980, p. 370.

90. C. W. Churchman, "A Philosophy for Planning," in H. A. Linstone and W. H. C. Simmonds, Eds., *Futures Research: New Directions*, Addison-Wesley, Reading, Mass., 1977.

91. Harold Linstone et al., "The Use of Alternative Decision Models for Technology Assessment," NSF Project, Futures Research Institute, Portland State University, Portland, Ore., 1980.

92. Nancy Goldstein, "A Delphi on the Future of Steel and Ferroally Industry," in Harold Linstone and Murray Turoff, Eds., *The Delphi Method*, Addison-Wesley, Reading, Mass., 1975, pp. 210–226.

93. Ibid., pp. 223–226.

94. Ibid., p. 226.

95. Bernadine Young, *Perspectives on the Future of Student Affairs: A Modified Delphi Study in Higher Education*, University Microfilm International, Ann Arbor, Mich., 1978.

96. M. Scheibe, M. Skutsch, and J. Schofer, "Experiments in Delphi Methodology," in Harold Linstone and Murray Turoff, Eds., *The Delphi Method*, Addison-Wesley, Reading, Mass., 1975, pp. 262–287.

97. Young, op. cit., p. 181.

98. Ibid., p. 191.

99. Ibid., p. 193.

100. Ibid., pp. 195, 196.

101. Ibid., p. 196.

102. Ibid., pp. 196–217.

103. Ibid., p. 196.

104. Ibid., pp. 198–201.

105. Brockhaus and Mickelsen, op. cit., p. 103.

106. Linstone and Turoff, op. cit., p. 4.

107. Brockhaus and Mickelsen, op. cit., p. 109.

108. Timothy Weaver, "The Delphi Method," *Phi Delta Kappan*, Vol. 3, No. 5, 1971, pp. 267–272.

109. Linstone and Turoff, op. cit., p. 3.

110. S. R. Hiltz and M. Turoff, *The Network Nation: Human Communication via Computer*, Addison-Wesley, Reading, Mass., 1978.

CHAPTER 5

INTERPRETIVE STRUCTURAL MODELING (ISM)

JOHN N. WARFIELD

Interpretive structural modeling (ISM) is a computer-assisted, interactive learning process whereby structural models are produced and studied.[1,2] Learning takes place through group interaction, with the help of the computer, under the leadership of a skilled facilitator. The vehicle for the learning that takes place during the process is the accumulation of information needed to construct an interpretive structural model or "map(s)" through group discussion that occurs in response to a series of questions posed by the computer through suitable display means, such as a large screen or several video terminals.

The computer functions are governed by ISM software (which is *not* to be confused with the process itself). The software has been developed in several versions, but the two most commonly used are the first generation programs[3] developed at the Battelle Memorial Institute, Columbus, Ohio, and the second generation programs[4] developed by the University of Dayton.

In some applications the primary benefit from the process accrues to the participants in the form of the learning that takes place as a consequence of its use. In this instance the participants are presumably enabled to explain and carry out various actions or take various decisions as a consequence of having taken part in the process. In other applications the primary benefit from the process may stem from the product, the map(s) that demonstrate structural relationships among a set of elements representing some theme or issue.

The ISM process can also be perceived as a building block for larger processes. For example, it has been described as one component of a systematic approach to "idea management,"[2] where its function is to structure information that has been developed through other components of an idea management process.

The ISM process is beginning to be used as one component of a larger process of "conceptual design."[5] In this context it may be used in several different design steps, such as structuring the mutual compatibility of a set of options, structuring priorities on design decisions, and structuring preference among a set of options.

The ISM process is an outgrowth of a historical sequence of developments[6] that might be thought to have begun with the idea of the syllogism, developed as a vehicle for deductive logic by Aristotle (ca. 384–322 B.C.). Leibniz (1646–1716) articulated the need for "a universal scientific language and a calculus of reasoning for the manipulation of it," and urged that graphic representations of thoughts be used in place of the characters of ordinary prose. Leibniz used circles as an aid to logical reasoning, and Euler (1707–1783) used them to study the syllogism, though this approach is usually attributed to Venn (1834–1923).

DeMorgan (1806–1871) was apparently the first to symbolize the important concept of relation in any general way. As he once noted in regard to one of his works, "the general idea of relation emerges and for the first time in the history of knowledge the notion of relation and relation of relation are symbolized." The relevance of relations to learning and modeling can hardly be overemphasized, and the ISM process is founded on the concept of emphasizing the relation(s) being used to study a theme or issue and model it.

C. S. Peirce (1839–1914) was apparently the first to recognize the importance of transitive relations. He also developed a graphical language along the lines advocated by Leibniz, but it did not receive much attention.

The modern concept of relation was enunicated by the 20-year-old Norbert Weiner in 1914, after he concluded that its treatment in the *Principia Mathematica* was unsatisfactory.

In 1921 Kuratowski provided a definition of *relation* based on Weiner's work that was instrumental in connecting the theory of relations to the theory of sets more directly.

A milestone in the use of graphics as a language for working with relations occurred with the publication by Harary, Norman, and Cartwright[7] in 1965 of a book on "structural modeling," in which basic ideas were brought together in one place and the use of digraphs as a way to model various phenomena was illustrated.

More recently, Atkin[8,9] has developed a theory that relates to the interpretation of two-level hierarchies in terms of the structural relationships involved and has applied this to illuminate the game of chess, to study social conditions in cities, and to analyze numerous facets of the University of Essex. The method has also been applied to help solidify the diagnosis of the rare Behcet's disease.[10]

Recently work has been reported[11] on how to program a computer to generate well-organized, readable digraphs[12] including more than 500 elements.

Along with these technical developments, the prolonged study of human development by Piaget[13] has served to illuminate the learning aspects. In his book *To Understand is to Invent* (1973) Piaget mentions these matters:

If logic itself is created rather than being inborn, it follows that the first task of education is to form reasoning (p. 49).

. . . a very clear-cut law of evolution: that all mathematical ideas begin by a qualitative construction before acquiring a metrical character . . . (p. 102).

The comprehension of elementary mathematics depends on the formulation of qualitative structures . . . and the more the preliminary formation of the logical functions is facilitated, the greater the receptivity to mathematical instruction at every level (p. 8).

This brief historical review is intended to point out that the use of graphical methods based on relations, and especially the use of digraphs based on transitive relations, is in a direct line of scientific thought that has been evolving for over 2000 years. The availability of the computer is not, however, the only reason why it is now possible to take advantage of this approach. Advances in the understanding of group performance and group facilitation are also necessary and have been forthcoming, and these advances together with the availability of the computer make possible the use of methods such as ISM.

The basic premise underlying ISM can be described in the following manner: Because people are fundamentally limited (ref. 2, Chap. 3) in terms of how much information they can conveniently and effectively work with, and because issues and systems are becoming very complex and interlocked, recourse to the assistance of a computer and a group facilitator will help ease the difficulties that arise from the combined impact of complexity and human-bounded rationality.

The purpose of the ISM process can then be described as "to extend capacity to define complex systems and enhance interdisciplinary efforts to communicate about system improvement."

However, this basic purpose does not capture the entire scope of possible applications of ISM. This process has been developed from a very fundamental perspective to make it applicable to a wide variety of contexts. Hence, whereas it is primarily intended as a group learning tool whereby groups can construct models, it can also be used by an individual in such mundane chores as maintaining a regularly updated personal task schedule.

In this chapter we first present an overview of ISM, stressing the people, equipment, and substantive content that is dealt with in its use. We then discuss the process output, the maps that represent the results of the structuring effort carried out in the group setting. With an understanding of maps, some of their general features, and how they are read, one can more readily appreciate the next discussion, which deals with the ISM process theory. In this section we have attempted to simplify the presentation of the theory by narrowing the mathematics to that most closely allied with the theory of sets. The matrices used in previous discussions[2] are omitted entirely in deference to explanations that rely on set theory.

Next there is a discussion of types of past activities that have been supported by the use of ISM, with special emphasis given to the types of relations. General types of relations are mentioned, and the specific types used in the applications are given. The benefits and limitations of the process are then discussed and a list of references is given.

AN OVERVIEW OF ISM

An overview of ISM is presented with the aid of Figure 1. The process can be described briefly in terms of three dimensions:

A. People.
B. Equipment.
C. Substantive content.

PEOPLE

The people involved with the use of ISM can generally be described under one or more of the following categories:

Broker.

Facilitator.

Technician.

Participants.

Observers.

The broker is a person who understands that there is a context requiring study, learning, and organization. For example, the context might be that the Federal Government has passed a law requiring that the U.S. Forest

---

INTERPRETIVE STRUCTURAL MODELING PROCESS

A. PEOPLE

    Broker
    Facilitator
    Technician
    Participants
    Observers

B. EQUIPMENT

    Computer (containing ISM software)
    Dedicated telephone line
    Coupler
    Display Unit(s)
    Computer Terminal with hard copy capability
    Standard classroom or meeting facilities

C. SUBSTANTIVE CONTENT

    I.   Context (a Theme, Issue, or Problem)
    II. Elements and Contextual Relations(s)
    III. Votes
    IV. Relation Maps
    V.   Interpretation

Figure 1 Process overview: ingredients.

Service do participative planning for the future use of forest and grassland resources. It might be required that the plan establish how national, regional, state, and private goals and objectives would be related to the future allocation of resources, with specific numerical targets to be ultimately established for five-year intervals over several decades. Moreover, the broker is the person who identifies the participants and encourages them to take part in the ISM process as a means of developing substantive content related to the context. The broker will also see to the financing of the work to be done and will make arrangements to acquire the services of a facilitator and a technician.

In some past applications one person has been broker, facilitator, and technician. This is unusual and requires a very versatile person. One person who has filled all three roles is Dr. Brian Carss of the University of Queensland, who undertook to develop policy for in-service teacher training.[14]

The facilitator is a person who is very skilled in helping groups work together. But this is only one requirement that the facilitator must satisfy. In addition, the facilitator must be familiar with the ISM process, so that certain decisions can be made as the process advances. Above all, the facilitator needs to be motivated primarily by the desire to see the group succeed in its work, and to behave accordingly. Without a capable facilitator the ISM process is not likely to be very useful.

The technician is a person who understands what equipment is needed for the ISM process, knows how to operate a computer terminal, knows how to use the ISM software, understands the need for reliable operation of equipment when a group is present, keeps a low profile, and does not interfere in the work of the group.

The technician also needs to understand how to work with the process output, so as to make it available to the group in a useful form. This requires that the technician have an understanding of the nature of the maps produced (but not of their substantive content).

The participants are chosen because they are believed to have some combination of the following attributes:

Knowledge of the context to be studied and structured.

Capacity to contribute to the application of results after the ISM process has been completed.

Awareness of sources of relevant information concerning the context to be studied.

Political sensitivity.

Capability of representing the views of some constituency in an articulate, nondemagogic manner.

Capacity to read and engage in focused dialogue.

Of these, the last is essential, while the others are quite desirable.

For best results it will usually be desirable that there be not more than eight participants. This raises the question of what to do in a situation where the number of interested parties who desire to take part exceeds this number. There are three possible approaches for dealing with this condition:

Additional persons may be designated as observers and may watch and hear all of the activities.

Persons may trade off, with one person serving as participant while another serves as observer, then exchanging roles at one or more points in the process.

In the ultimate, a city or region can be "wired" (as in the experiments in Columbus, Ohio), where persons can see, hear, and vote by means of two-way cable. While this would *not* provide a good means for full participation in a discussion, it *would* provide for full participation in the voting that goes on in the process. By this means thousands of people could participate.

Lendaris[15] has extensively discussed role characteristics.

EQUIPMENT

As mentioned, it is sometimes possible to conduct an ISM process application without a computer. However, this is the exception rather than the rule.

Normally a computer is required. The computer will contain ISM software.[3,4] The computer is time shared in most applications and is used only a very small percentage of the time. At present ISM software is installed on a variety of computers. In most instances a large, high-speed computer is preferred for reasons to be explained later.

Usually the computer will be far from the locale where the process is being conducted; then a telephone line or some other communications link is needed that must be available constantly throughout the process. This telephone line, for instance, will be connected to a local computer terminal with hard-copy printout capability by means of a coupler. Normally acoustic couplers are used.

In the past display units of two types have been used. One type is the video terminal (like a television receiver in appearance), which is capable of displaying information to the participants. Usually several of these are used to assure that all participants can readily read the display. Another type of display is the Advent screen. This screen is large enough so that normally only one will be required. The display is projected by means of a lens system that is part of the display system.

In addition to the equipment mentioned, it will be desirable to have a comfortable situation for the participants, with access to the usual meeting facilities. A blackboard is usually helpful, and a large empty wall space can be very useful for displaying the maps that are produced.

SUBSTANTIVE CONTENT

The term *substantive content* refers to the information that is being dealt with in the process and the way in which it can be treated. In Figure 1 we suggest, by Roman numerals, that there be a phasing associated with the substantive content.

In the first phase a context is identified. The context might be interpreted through a theme[16] consisting of a few descriptive sentences, as an issue that is at least somewhat focused by a brief description, or as a problem to be investigated. The product of this first phase is always a short written description occupying less than one page. The description does not deal with resolutions or solutions but is designed as a focusing statement for the work to be carried out.

In the second phase a set of elements is identified, along with one or more contextual relations that appear to be of primary interest in exploring the element set. Experience has shown that the elements may be of many types and that numerous contextual relations may be drawn upon. However, in a typical exploration the number of elements will usually be much larger than the number of types of contextual relations.

Let us mention here some specific examples of the kinds of elements and contextual relations that have been used.

Example 1. The elements are statements of objectives. The contextual relation is, "should help achieve." A typical statement that would be explored would be of the form, "The attainment of objective A (where objective A would be spelled out) should help achieve objective B (where objective B would be spelled out)."

A statement of the type just given would appear on the display unit.

The participants should discuss this statement until no further discussion appears warranted, then a vote would be taken by the facilitator to obtain the sense of the participants. The majority voting rule would be used. The result of the vote would be entered into the computer.

Example 2.   The elements would be proposed municipal projects. The contextual relation could be, "should have a higher priority than." A typical statement for exploration might read, "The proposed municipal swimming pool should have a higher priority than the renovation of the River Road Sewage Plant." Again discussion and voting would take place.

Example 3.   The elements might be economic parameters and the contextual relation might be, "is a mathematical function of." A typical statement might read, "The rate of savings from personal income is a function of the prevailing interest rate on certificates of deposit." As before, discussion and voting would follow the appearance of such a statement on the display unit.

Example 4.   The elements might be particular milestones in the development of a system. The contextual relation might be, "should occur before." Then a typical statement might read, "Development of a digital receiver should occur before selection of a transmission channel."

As has been mentioned, the selection of an element set and one or more contextual relations completes the second phase of the ISM process. This is followed by the third phase, in which the participants view the statements posed by the computer, discuss them, and vote upon them. The voting represents the third phase of the ISM process.

The fourth phase of the process occurs when the entire element set has been explored through the selected contextual relation. At this point the computer makes available the organizational data needed to construct a map of the relationships that have been uncovered during the third phase.

In the fifth phase of the process what is done will be contingent upon an assessment and interpretation of the map(s) produced so far. This phase is described as an interpretation phase, but it may well occasionally involve one or more iterations of the preceding phases.

Before discussing the fifth phase we digress momentarily to address a philosophical issue. The ISM process is a learning process first and a production process second. One can conceive of very simple artificial examples where there would never be a need for iteration. If the element set consisted of the integers from 1 to 10 inclusive and the contextual relation was, "is less than," we could imagine that a structuring effort would produce a map that simply shows the natural progression of the integers. One could simply look at the map and verify that the integers are in the correct order. But it would normally be true that participants would have known before the process began how the results would turn out, and no learning about the relative magnitudes of the numbers would have taken place in the process. Hence there would be no need to iterate.

In general, however, substantial learning takes place in the process. It is possible, for example, that the context statement may even be determined to have been ill stated. In one application of ISM it was concluded by the group that the context that they initially believed to be aptly stated had no substance to it whatever. In less traumatic experiences it is more common that the participants determine that certain originally perceived relationships require modification. This means that the map(s) produced may require modification.

Other reasons may also suggest map amendment. Sometimes the information needed to respond to a statement is factual in nature but is unavailable to the participants. Rather than disrupt the process, an initial response may be made that later should be changed, after the factual information has been collected.

The process itself may be responsible for defects in the map(s) produced. The process is designed to try to minimize the time required from the participants. The primary process characteristic involved in time minimization is the use of the computer to draw logical inferences from participant responses. To take an absurdly simple example, if the participants agree that Tom is older than Jerry and that Jerry is older than Sally, the computer would infer that Tom is older than Sally and would save participant time by not asking whether Tom is older than Sally. It is important to understand the cumulative effect of computer inference on group time. If, as in some applications, the computer is able to infer 80% of the responses, the participants can complete in one working day what would require a week to carry out if all responses were sought. As in all designs, trade-offs are made. The use of computer inference in the way described places on the participants the burden of checking the maps that result from the use of the process. Checking and amending maps normally takes *much* less time than would be required if all possible questions were asked.

Amendment software is available[4] to facilitate changes in maps that are produced. This software would be used in the interpretation phase.

Sometimes maps are too complex to interpret readily. In this event several additional steps are possible. One such step involves gathering weights from participants on selected relationships.[2] Now, instead of asking whether two elements are related, the question posed deals with a scale of intensity of relationship. Weighting is used only when it is absolutely necessary to facilitate interpretation, because it is much more time consuming and difficult than making a yes or no assessment. Also, weighting is usually restricted to a small subset of elements.

Still more interpretive aids are described in the detailed theory of ISM. These will not be described here, as they require an intimate knowledge of process theory.

Recent publications[11,12] indicate strongly that computer construction of the maps produced in the ISM process will eventually be possible. This will avoid the manual labor presently required to construct the maps for participant viewing and to construct amended maps that may be required as a result of interpretation.

PROCESS OUTPUT: MAPS

The output of the ISM process consists of one or more maps. The long name for these maps is "interpretive structural models." However, the short term has obvious advantages in discourse. Where necessary to distinguish such maps from geographic maps the term *relation maps* is recommended.

A detailed understanding of such maps is valuable, not only because they are the products of ISM process application but also because such an understanding is needed to allow a broader discussion of the ISM process to be adequately developed and to help the reader develop an appreciation for process features.

A map can be viewed as analogous to a prose discussion in that it can be "read."[6] Since a map is a new form of communication, it is necessary to learn how to read a map in order to take best advantage of this form of expression. For this purpose we introduce an elementary map and describe how it is read.

Figure 2 shows a map using the contextual relation "weighs less than."[17] You will notice that the map contains six elements and that these are numbered for ready reference in the discussion.

One of the statements that can be read from the map is, "A feather weighs less than a ball point pen." It is possible to encode this statement as follows: 1R2. The symbol 1 represents the element "a feather." The symbol R represents the contextual relation "weighs less than." The symbol 2 represents the element "a ball point pen." Evidently other statements could be encoded uniquely in a similar way.

Notice also that the map can be described as a linear hierarchy containing levels. In this respect it would be similar to an organization chart

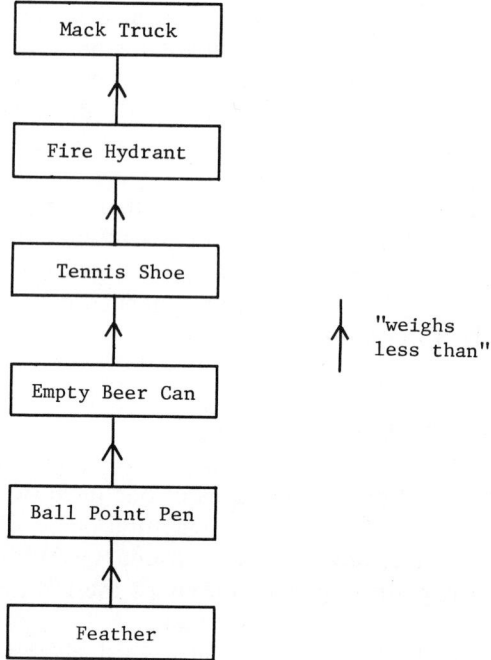

Figure 2 Example of a relation map using the relation "weighs less than."

for an organization having only one person or division in each level of the hierarchy. The description *linear* refers to the fact that all of the elements lie on a single line from one end of the hierarchy to the other.

Since the numbering of the levels normally differs from the numbering of elements, we may as well reflect this in discussing Figure 2. We say that element 6 lies at level 1, element 5 lies at level 2, and so on, with element 1 lying at level 6.

WALKS REPRESENT PROSE STATEMENTS

An intuitive idea of what is meant by a "walk" on a map can be gained by supposing that the map is located on a flat surface and that you can stand on any of the several elements and walk along the direction of the arrows to any other element. Every possible such movement generates a distinct walk. The total number of walks on a map is countable, since we deal only with finite maps. In Figure 2 there are 15 walks. *Every walk corresponds to a prose statement.* Thus the map in Figure 2 is described as having a *content* of 15 statements.

The statements clearly can be written out in detail, or they can be expressed in the symbolic form illustrated earlier. In this latter form the entire set of statements is given by 1R2, 1R3, 1R4, 1R5, 1R6, 2R3, 2R4, 2R5, 2R6, 3R4, 3R5, 3R6, 4R5, 4R6, 5R6. For review, we note that the symbolic statement 3R6 represents the prose statement, "An empty beer can weighs less than a Mack truck."

Notice that the general pattern used in writing the set of statements is to read all of the statements originating at level 6 first, then read all those originating at level 5, and so forth. Clearly, it would also be possible to begin by reading all those originating from level 2 first, then take those originating from level 3 next, and so on. When a group is examining such a map as a collective effort, it is advisable to establish a reading convention before beginning to read the map.

While most maps from applications so far have *not* yielded linear hierarchies, the simplicity of the linear hierarchical structure lends itself to some observations concerning the graphical advantage that can be gained through the use of maps to present relational information.

Suppose that the number of levels in a linear hierarchy is given by n. Then one can compute the number of statements in the content of the map from the formula $n(n - 1)/2$. A 30-element map would then have a content of 435 statements, whereas a 60-element map would have a content of 1770 statements. One can readily be convinced of the graphical communication advantage by first constructing a 60-element map to portray the content and then writing out the 1770 statements that would convey the entire prose content of the map.

All of the relation maps that are produced from the ISM process correspond directly to what are called *digraphs* in the mathematical theory of graphs. In the mathematical language the elements are referred to as *vertexes* of the digraph, and the lines with arrows are referred to variously as *edges* or *arcs*.

It is very important to appreciate that the correspondence between digraphs and relation maps is obtained through these operations:

Association of each element from the element set being considered with one vertex of a digraph.

Association of a contextual relation with every walk on a given digraph (thus if several contextual relations are explored with the same element set, there will be a distinct digraph corresponding to each contextual relation).

Association of a statement with every walk on a digraph, where the statement involves the elements appearing at the origin and termination of the walk (respectively) along with the contextual relation.

These three types of associations are all that are required to transform a relation map to its underlying digraph or to go from a digraph to the corresponding relation map.

If, however, the map or digraph does not yet exist, one other type of association is needed. This association involves the substantive knowledge required to specify whether or not a walk exists on the digraph. If the general form of statement appearing on the display before the participants in an ISM process is uRv, where u and v are elements and R is the contextual relation, one can assign a binary variable x_{uv} to this question and make the following associations:

$x_{uv} = 1$ is a mathematical way of associating the perceived truthfulness of uRv with a mathematical space.

$x_{uv} = 0$ is a mathematical way of associating with a mathematical space the condition that xRv is not known to be true (which can mean either that xRv is known to be false or that its verity remains to be determined).

This fourth type of association allows the results of a vote on a question to be translated into a binary representation. The answer to each question posed by the computer is thereby translated into an instruction to the computer to either assign a walk from u to v on the map (when $x_{uv} = 1$) or *not* to assign such a walk (when $x_{uv} = 0$). This simple condition is the information that the technician supplies to the computer through the terminal following each vote that is taken during the ISM process.

CYCLIC ELEMENTS

Sometimes some of the elements on a map will be cyclic elements. If it is true both that $x_{uv} = 1$ and that $x_{vu} = 1$, so that a pair of elements is related in both directions, we say that the elements are cyclic. When a group of several elements is such that every member of a group is related in both directions to every other member of the group, we suppress the arrows within the group. A simpler portrayal is used that represents the conditions just mentioned. Figure 3 shows a three-element cycle and the normal way of portraying the cycle. The bullets preceding the element names are a reminder that the elements are in a cycle. Also, the bullets are helpful to the eye in enabling one to read a map easier than would otherwise be the case, since it is not necessary for one to sort out the typewritten statements.

Figure 4 shows a map produced by a high school biology class.[17] It can be seen that this map contains one two-element cycle. The translation

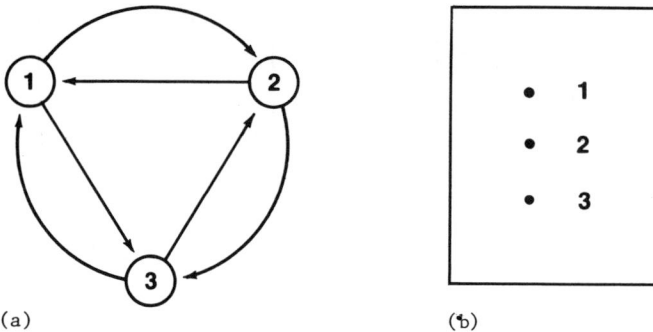

Figure 3 Two ways to portray a three-element cycle: (*a*) showing all edges, (*b*) simplified portrayal.

or content of the cycle is as follows:

Building power plants helps achieve increased industrialization.

Increased industrialization helps achieve building power plants.

TYPES OF MAPS

There are several types of maps, classified according to structural features.[18] A single element unconnected to anything else is called an *isolated element*. If all the elements of a map are isolated, the map is called an *array*. We have already described a map that consists of a *cycle* in the preceding section. A map that contains walks but no cycles is called a *hierarchy*. We have already described a special type of hierarchy, the linear hierarchy. A map such as that illustrated in Figure 4 is described as a *mixed structure*. A mixed structure is the most general type: It contains at least one cycle, and if each of its maximal cycles is replaced with a single "proxy" element, what remains will be a hierarchy. All of the types mentioned can occur in applications.

PROCESS THEORY

While the process theory has been explained with the aid of matrices,[2] it is possible to present it in a somewhat simpler manner. Our basic approach is to present the theory in three parts. The first part gives the fundamentals of *digraph models*. This part is presented in a relatively formal mathematical style, free of reference to applications. In this way the theory remains adaptable to a variety of purposes. This part is by far

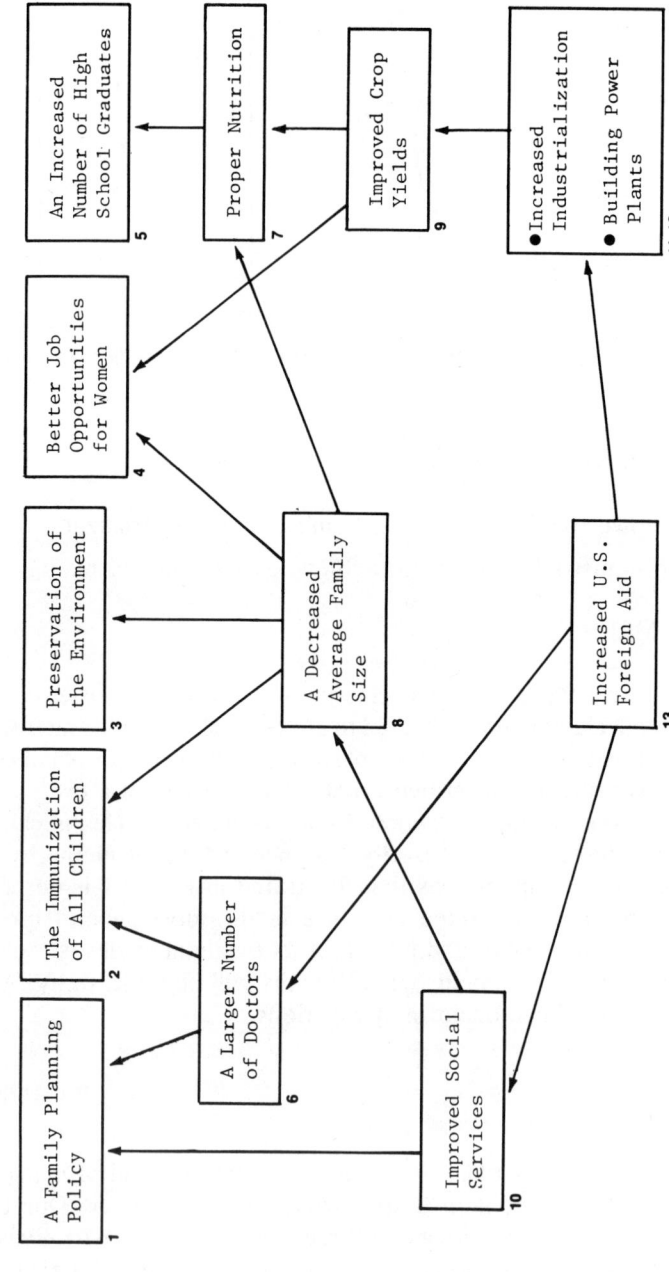

Figure 4 Example of a relation map using the relation "helps achieve" (developed by a high school biology class).

the longest of the three parts because it presents most of the terminology and the fundamental theorems related to digraph models that are relevant to ISM.

The second part, much shorter, gives the *fundamentals of digraph-based models*. Digraph-based models are those models whose structures are representable by digraphs. A real-world modeling situation that is amenable to digraph-based modeling can be essentially mapped into the digraph model theory by a few simple associations.

The third part, also quite short, discusses the *development* of digraph-based models and presents the fundamental ideas that are the basis for ISM.

Following this presentation, we discuss some aspects of interpretation that may be useful in applications.

FUNDAMENTALS OF DIGRAPH MODELS

We begin by assigning to any positive integer the name *element*, anticipating that eventually elements will be associated with these integers. We call the set $I_n = \{1, 2, 3, \ldots, n\}$ the *element set*. We may on occasion also refer to I_n as an index set or vertex set.

We call the set $E = I_n \times I_n = \{e_{ij}\} = \{\langle i, j \rangle\}$, $(i, j \in I_n)$ the *cell set*. The cell set is the Cartesian product of the element set with itself. The notation $\langle i, j \rangle$ is a reminder that the pair of elements i, j is *ordered*. We call e_{ij} a *cell* of E. We say that a cell has an *image* $\mathrm{Im}(e_{ij})$. Evidently $\mathrm{Im}(e_{ii}) = e_{ii}$ for all $i \in I_n$.

Any subset of E is called a binary relation. For brevity we use the shorter term *relation* to represent a subset of E. We say that E is the *universal relation* on E and note that E is the only relation on E that includes all the cells of E. We use the notation $\emptyset$ to represent the *empty relation*. An empty relation contains none of the cells of E. We say that r_1 is the *reflexive relation* on E, where r_1 contains every cell of E that is its own image.

We define two relations r and r_c in E as follows:

1. $r \subseteq E$.
2. $r_c \subseteq E$.
3. $r \cup r_c = E$.
4. $r \cap r_c = \emptyset$.

We say that r and r_c are *complements* of each other and that together they are a pair of *complementary relations* on E. Finally, we note that the pair of relations constitutes a *partition* of E.

Now let r be distinguished from r_c by the requirement that the reflexive relation on E is always included in r, so that $r_1 \cap r = r_1$ and $r_1 \cap r_c = \emptyset$. Then we call r a *primary* relation on E and r_c a *secondary* relation on E. We expect that a primary relation and a secondary relation will have somewhat different properties. Also, we note that given either we can determine the other.

Given any primary relation r, there is induced by r on E a partition Π_e called the *cell-classification partition*. This partition has at most five *blocks* expressed by

$$\Pi_e = r_1; r_2; r_3; r_4; r_5$$

where the following is true:

1. r_1 is the reflexive relation on E, including every cell of E that is its own image.
2. r_2 contains every cell of r whose image is *not* contained in r.
3. r_3 contains every cell of r whose image is also contained in r, except for those cells contained in r_1.
4. r_4 contains every cell of r_c whose image is *not* contained in r_c.
5. r_5 contains every cell of r_c whose image is also contained in r_c.

Then it should be clear that

1. $r = r_1 \cup r_2 \cup r_3$.
2. $r_c = r_4 \cup r_5$.

Also it should be clear that the image of any cell contained in r_2 will be contained in r_4, and conversely.

Let e_{ij} and e_{uv} be any pair of cells in E. We say that this pair of cells has *inner incidence* if and only if $j = u$. We say that the pair has *outer incidence* if and only if $i = v$.

Let e_{uj} and e_{jv} be any pair of cells of E having inner incidence. To such a pair there is a unique cell of E called the *closure cell* of the pair, expressed as $\mathrm{Cl}(e_{uj}, e_{jv}) = e_{uv}$.

Let r^* be a relation in E. We say that r^* is *transitive* if and only if r^* contains all the closure cells of all the pairs of the cells of r^* that have inner incidence; otherwise r^* is nontransitive.

Evidently the following relations are transitive:

1. The universal relation E.
2. The empty relation $\emptyset$.
3. The reflexive relation r_1.

THEOREM 1.

To every nontransitive relation t there corresponds a unique relation δ such that (1) $t \cup \delta = r^*$ is transitive and (2) for all ψ such that $t \cup \psi$ is transitive, $\delta \subseteq \psi$.

Proof. Since t is nontransitive, it must contain one or more pairs of cells having inner incidence whose closure cell(s) are not contained in t. Let δ' be the set of all such closure cells. Let $t' = t \cup \delta'$. If t' is transitive, the conditions of the theorem are met by setting $r^* = t'$ and $\delta = \delta'$. If not, apply the argument a second time, letting δ'' be the unique set of closure cells of t'. Then let $t'' = t' \cup \delta''$. If t'' is transitive, then $\delta = \delta' \cup \delta''$, and the conditions are met with $r^* = t''$. If not, the argument can be repeated until transitivity is achieved, since the set I_n is finite and the limiting relation E is known to be transitive.

That condition (2) is satisfied is assured, because in forming δ we append only cells to t that are essential to its transitivity. Q.E.D.

The *union* of two transitive relations may be transitive or be nontransitive. The *intersection* of two transitive relations is transitive.

A closure cell is described as *reflexive* when and only when the pair e_{uj} and e_{jv} to which the closure cell corresponds has outer incidence.

Let r^* be a relation in E. We say that r^* is *semitransitive* if and only if r^* contains all the closure cells of all the pairs of r^* that have inner incidence, except that r^* does not necessarily contain *any* reflexive closure cells. When a type of relation can be either semitransitive or not, depending on the circumstances, we describe it as *mesotransitive*.

THEOREM 2.

If r^* is transitive, then it is semitransitive, but the converse is not true in general.

THEOREM 3.

If r is transitive, then we have the following:

1. r_2 is transitive.
2. r_3 is semitransitive.
3. r_4 is transitive.
4. r_5 is mesotransitive.
5. r_c is mesotransitive.

Adjacency Digraphs Induced by a Primary Relation r. Let r be a primary relation on E. Then we say that $D_A(r)$ and $D_A(r_c)$ are the *adja-*

cency digraphs induced by r on E, where

$$D_A(r) = \{I_n, r_2 \cup r_3\}$$

$$D_A(r_c) = \{I_n, r_4 \cup r_5\}$$

The members of I_n are called the *vertices* of the digraphs. The members of $r_2 \cup r_3$ are called the *edges* of $D_A(r)$, and the members of $r_4 \cup r_5$ are called the *edges* of $D_A(r_c)$.

Walk on a Digraph.   A *walk* of length $n(n > 0)$ on a digraph is an ordered sequence of n edges of the digraph, such that the following is true:

1. Every edge in the sequence has inner incidence with the edge immediately preceding it (except the first, which has nothing preceding it).
2. No two edges in the sequence are the same.
3. No two edges in the sequence have outer incidence, except, possibly, the first and last edges in the sequence.

If the first edge in the sequence is e_{ij} and the last edge in the sequence is e_{uv}, then the following are defined:

1. i is called the *origin* of the walk.
2. v is called the *termination* of the walk.
3. The ordered pair $\langle i, v \rangle$ is called the *name* of the walk (or, for short, "the walk").

Co-Named Walks.   If two or more walks on a digraph have the same name, they are called *co-named walks*.

Statement and Content of a Relation.   Let r^* be a relation in E and let $e_{ij} \in r^*$. Then, by a *statement of a relation r^** we mean the expression ir^*j, where $i \neq j$. The *content* $C(r^*)$ of a relation r^* is the set of all statements of the relation r^*.

The content $C(r)$ of a *primary* relation r is given by the set of all statements of the relation $r_2 \cup r_3$. The content of a *secondary* relation r_c is given by the set of all statements of the relation $r_4 \cup r_5$.

Statement of a Digraph.   Let $\langle i, v \rangle$ be a walk on a diagraph. Then we mean, by a statement of the digraph, the expression iwv, where $i \neq v$, and where the relation symbol w indicates the existence of the mentioned walk on the digraph.

Content of a Digraph.   The content $C(D)$ of a digraph is given by the set of all the statements of the digraph.

THEOREM 4.

A necessary and sufficient condition that a primary relation r with adjacency digraph $D_A(r)$ be transitive is that the content of r be the same as the content of $D_A(r)$, with $r = w$.

THEOREM 5.

If t is a nontransitive relation with transitive closure r^* and $D_A(t)$ is the adjacency digraph of t, then the content of $D_A(t)$ is identical with $C(r^*)$, with $r^* = w$.

THEOREM 6.

If $\{t_k\}$ is the set of *all* nontransitive relations that have the same transitive closure r^*, then all members of the set of adjacency digraphs $\{D(t_k)\}$ have the same content.

Antecedents and Succedents. Let r^* be a relation on E and let ir^*j be a statement of the relation. Then we say that i is an *antecedent* of j and j is a *succedent* of i. Then for every element $i \in I_n$ a given relation r^* induces an *antecedent set* $A(i) \subseteq I_n$ consisting of all members of I_n that are antecedents of i. Similary, there is induced a *succedent set* $S(i)$ $\subseteq I_n$ consisting of all members of I_n that are succedents of i.

Levels of a Digraph. A transitive primary relation r induces an ordered level partition Π_L on I_n into m *digraph levels*, with

$$\Pi_L = L_1; L_2; \cdots ; L_m$$

where the following is true:

1. Any element $i \in I_n$ is also contained in L_1 if and only if $S(i) \cap A(i)$ $= S(i)$.
2. Any element $i \in I_n$ is also contained in L_j if and only if

   (a) i is not contained in $\displaystyle\bigcup_{k=1}^{k=j-1} L_k$

   (b) $S'(i) \cap A(i) = S'(i)$, where

   (c) $S'(i) = S(i) - S(i) \cap \displaystyle\bigcup_{k=1}^{k=j-1} L_k$

A vertex of a digraph is called a *top-level vertex* if and only if it is a member of L_1. A vertex of a digraph is called a *bottom vertex* if it has no antecedents at any higher-numbered level of the digraph.

q-Adjacent Sets. Two sets are said to be q-adjacent if they have q + 1 members in common. Two sets are described as disjoint if and only if they are (-1)-adjacent. If sets are q-adjacent ($q > 0$), then they are also $(q - 1)$-adjacent, $(q - 2)$-adjacent, and so on down to 0-adjacency.

q-Connected Sets. Two sets α and β are said to be q-connected if there is an ordered sequence of sets beginning with α and ending with β such that every side-by-side pair of sets in the ordered sequence is at least q-adjacent.

THEOREM 7.

A necessary and sufficient condition that a digraph shall consist of a single part is that the antecedent sets of the top-level vertexes of the digraph are 0-connected.

Theorem 7 can be used to determine from a given relation r^* whether the adjacency digraph $D_A(r^*)$ consists of a single part or not and, if it does not, to determine the number of parts of the digraph.

THEOREM 8.

If i and j are any two vertexes in the same level of a digraph $D_A(r)$, where r is a transitive primary relation, then either (1) e_{ij} and e_{ji} are *both* contained in r_3 or (2) e_{ij} and e_{ji} are *both* contained in r_5.

In a given level of a digraph a vertex is described as *isolated* if there is no walk from that vertex to any other vertex in the same level. If there is a walk from any vertex i in a certain level to any other vertex j in that same level, then, as Theorem 8 shows, there must also be a walk within the level from the vertex j to the vertex i. The vertices i and j satisfying this latter condition are described as *cyclic*.

It is then evident that any transitive primary relation r induces on each of its levels a partition of the level into two blocks, one of which contains all the isolated vertexes of the level and the other of which contains all the cyclic vertexes of the level.

If a vertex i is cyclic within a level L^*, the set comprised of $i \cup A_{L^*}(i)$, where $A_{L^*}(i)$ consists of all of the antecedents of i in level L^*, is called a *maximal cycle* of the transitive relation r and of the digraph $D_A(r)$.

A transitive relation r thus induces on each of its levels a partition of the level into one block consisting of all of the isolated vertexes of the level and as many other blocks as there are maximal cycles within the level.

Array. A digraph having only one level, all of whose members are isolated, is called an array.

Hierarchy. A digraph having at least one walk and no cyclic vertices is called a hierarchy.

Linear Hierarchy. A digraph $D_A(r)$ is described as a linear hierarchy when and only when $r_2 \neq \emptyset$ and $r_3 = r_5 = \emptyset$.

THEOREM 9.

If $D_A(r)$ is a linear hierarchy, so is $D_A(r_c)$, and conversely.

THEOREM 10.

If $D_A(r)$ is an array, $D_A(r_c)$ consists of one maximal cycle.

Digraphs with Mixed Structure. A digraph $D_A(r)$ is said to have mixed structure if r_3 is not empty (i.e., the digraph contains at least one cycle) and, in addition, either r_2 is not empty or r_5 is not empty or both.

Any digraph with mixed structure can be condensed through the use of *proxy vertices.* We note that every vertex in a given maximal cycle has the same antecedent set and the same succedent set. A proxy vertex is a vertex that replaces all of the vertexes of a maximal cycle and which is assigned the same antecedent and succedent sets as the vertexes in the maximal cycle (except that members of the maximal cycle itself are excluded from these sets). If every maximal cycle in a digraph with mixed structure is replaced with a proxy vertex, a new digraph is formed from the original one. The new digraph is called a *condensation digraph.* If the digraph with mixed structure had only one level, the condensation digraph will be an array; otherwise the condensation digraph will be a hierarchy. If the digraph with mixed structure is $D_A(r)$ and r_5 is empty, the condensation of $D_A(r)$ will be a linear hierarchy.

Span of an Edge of a Hierarchy. If the origin of an edge of a hierarchy lies in level j of the level partition and the termination of that edge lies in level $j - k$, then we say that the edge has *span* k.

Essential Edge of a Hierarchy. Noting that every edge of a hierarchy is also a walk of the hierarchy, we say that an edge of a hierarchy is essential if and only if it has no co-named walks. Clearly every edge whose span is 1 has no co-named walks, hence every edge with span 1 is essential.

Partition of Edges by Span. Suppose that r is a transitive primary relation whose adjacency digraph $D_A(r) = D_A(r_2)$ is a hierarchy. Then r induces, via the level partition Π_L, another partition Π_s on the edges of $D_A(r)$ according to the span of edges with

$$\Pi_s(r_2) = E_1; E_2; \cdots ; E_u$$

where u is the longest span of any edge of $D_A(r_2)$ and E_j consists of all edges of D_A having span j.

We define $r' \subseteq r_2$ to be the set of essential edges of r_2, and we call r' the *skeleton relation* of r_2. Then there is a partition of the edges of r' that assigns its edges to blocks according to their spans, that is,

$$\Pi'_s(r') = E_1'; E_2'; \cdots ; E_v'$$

where v is the longest span of any edge of $D_A(r')$ and E_j' consists of all edges of $D_A(r')$ having span j.

THEOREM 11.

The content of $D_A(r')$ is equal to the content of $D_A(r)$. The digraph $D_A(r')$ is called the *skeleton digraph* of r. To find the skeleton digraph, it is sufficient to find the partition Π'_s and take its edges as the edges of the skeleton digraph. The block E_j' is found from the block E_j by removing from E_j any edges that have co-named walks in the digraph $D_A(r)$.

Content of a Digraph With Mixed Structure.   The content of a digraph with mixed structure can be found as follows:

1. Define a proxy element for each maximal cycle.
2. Determine the condensation digraph.
3. Determine the skeleton digraph.
4. Find the content of the skeleton digraph.
5. To every statement in the content of the skeleton digraph that involves one or more proxy elements substitute, in turn, each of the cyclic elements for which the proxies stand.
6. Find the content of each maximal cycle of the mixed structure.
7. The content of the digraph with mixed structure is the union of the results of the preceding three steps, except that all statements containing proxy elements are deleted.

Increment Relation.   Let $D(r') = D_1$ be the skeleton digraph of a transitive primary relation r, and let $D(r_c') = D_2$ be the skeleton digraph of r_c. Since it is possible that the contents of the two digraphs may overlap, we define the increment relation

$$\Delta r = C(D_1) \cap C(D_2)$$

THEOREM 12.

If r_s is empty, then (1) D_1 and D_2 are linear hierarchies and (2) $\Delta r = \emptyset$.

THEOREM 13.

If r_c is semitransitive, the increment relation is empty.

THEOREM 14.

If r_5 is semitransitive, r_c is semitransitive.

THEOREM 15.

If r_5 is not semitransitive, then (1) r_c is not semitransitive and (2) the increment relation is not empty.

FUNDAMENTALS OF DIGRAPH-BASED MODELS

Let S be a set of (usually prose) elements, and let S be *associated* in a one-to-one way with I_n. Let R be a transitive contextual (usually prose) relation and let R_c be its contextual complement. Let R be *associated* with r and let R_c be associated with r_c. Let D_1 and D_2 be the skeleton digraphs of r and r_c, respectively, with increment relation Δr. Let M_1 and M_2 be maps found by replacing the members of I_n with the members of S, where the edges of the maps correspond to edges of D_1 and D_2, respectively, with the edges interpreted in statements as representing R and R_c, respectively.

Then we say that a model M is *digraph based* if M consists of the maps M_1 and M_2 and the increment relation Δr, the latter possibly being empty in special cases. Statements of digraphs D_1 and D_2 correspond to statements of the maps M_1 and M_2.

If S and R are prose expressions with proper syntax, then the set of statements formed from S, R, and R_c corresponds directly to the corresponding statements of the maps M_1 and M_2 when the increment relation is empty. Otherwise the statements formed from S, R, and R_c correspond to the statements of the map M_1 and those statements of M_2 that are not also statements of the increment relation.

Correct and Complete Maps. A map M_1 (and the digraph D_1) is described as *correct* if every statement of M_1 is valid as interpreted through S and R. The map M_1 is described as *complete* if it is correct and, in addition, the map M_2 is correct (except for those statements of M_2 that are included in the increment relation) as interpreted through S and R_c.

Content of a Digraph-Based Model. The content of a digraph-based model is the set of all statements read from the complete map M_1 and the correct map M_2, except for those statements from M_2 that are also included in the increment relation.

DEVELOPMENT OF DIGRAPH-BASED MODELS

The development of digraph-based models requires the prior definition of the element set S and choice of the transitive primary contextual relation R. The definition of the complement contextual relation R_c is based on the definition of R. Model development is then based on a series of *assertions* concerning individual statements.

Assertion.   If $s_i \in S$ and $s_j \in S$, then the following are designated as assertions:

1. $x_{ij} = 1$.
2. $x_{ij} = 0$.

The first of these is equivalent to a statement of the form, "s_iRs_j is true," and the second is equivalent to a statement of the form, "s_iRs_j is not known to be true."

An assertion of the first type indicates that the primary relation holds between two members of the set S. An assertion of the second type may be weaker, since it expresses either that the secondary relation *does* hold between the ordered pair of elements or that it *may* hold between them. This distinction is an arbitrary one, made to reflect realities of human knowledge so that uncertainty is not interpreted as truth through the primary relation.

Transitive Embedding.   Let us suppose that there is available, in a computer, a storage space called r space where cells of E can be stored for the purpose of building up gradually the relation r.

Then, by transitive embedding we mean the gradual accumulation of elements of r in r space through the accumulation of assertions.

Transitive embedding can be carried out by either of two algorithms described in ref. 2: the scanning algorithm and the bordering algorithm. Both algorithms incorporate provisions for forming transitive closure from partial results and embedding the transitive closure automatically in r space. Also, both algorithms incorporate criteria for sequencing the development of assertions that are intended to make the embedding process efficient (i.e., to require fewer assertions in order to develop r).

Map Computation.   When r is contained in r space, the map computation can be carried out using the level partition concept, to identify not only the levels but also the maximal cycles (if any) in each level. The skeleton maps are computed and the necessary information is generated to make map construction possible.

Map Construction.   In most applications so far, maps are constructed manually, based on computer output. However, map construction can be automated to provide constructions that will normally be superior from those that are developed manually in several respects. First, they will typically have fewer edge crossings; second, they can be constructed faster; and third, they can be constructed with a much smaller likelihood of errors in construction.

Weighted Embedding. The presence of large maximal cycles on a map may be a source of difficulty in interpretation. When this is the case, weighted embedding may be useful. This consists of modifying the structure of the map to facilitate interpretation, through the use of weights applied to the edges of maximal cycles. Weighted embedding itself consists simply of supplying numerical weights of intensity to the edges of a maximal cycle. Beyond that point threshold data may be supplied that enables the computer to eliminate certain edges in the maximal cycle, thereby changing the maps that are computed and constructed.

Cycle Study Sequencing. Sometimes a maximal cycle will be sufficiently large that its interpretation via weighted embedding is not clear. In this event it may be worthwhile to study smaller constituent cycles in a systematic sequence. This can be done through the development of the geodetic cycles that are contained in a maximal cycle. The geodetic cycles themselves form an inclusion hierarchy, which offers a possible cycle study sequence.[2]

INTERPRETATION OF DIGRAPH-BASED MODELS

The interpretation of digraph-based models can be managed in several ways. The simplest approach is to inspect the map of the primary relation by reading its content through the reading of statements associated with walks on the map. As the statements are read, the validity of the statements is examined and the meaning of the structural relationships discussed.

If the primary map is judged to be correct, the secondary map should be read to assess whether the primary map is complete. If the primary map is not complete, amendments to the primary map are made (and the computer can determine how to amend the secondary map and the increment relation).[6]

Thus the first step in map interpretation is to assure that the primary map is complete. Following *completeness testing*, it may be possible to understand and interpret the primary map just as it stands, and no further work may be required.

At times, however, a detailed understanding may be required. For this purpose methods developed by Atkin[8,9] may be applied to explore adjacent levels of a map. With the Atkin method two adjacent levels are examined either globally or locally. The global exploration involves the method of q-analysis, which establishes structural features that can be employed in map interpretation. The local exploration involves computer analysis of local irregularities in the structure, which can be interpreted as forces within the system being modeled.

In some instances the establishment of structure will be just the first step in a larger modeling program. Once the structure is understood, quantitative measures and equations may be introduced that relate the elements of S. An example of a field where this approach can be (and has been) applied is econometric modeling where the structural approach provides not only a means of organizing a quantitative model but also lends considerable insight into how to solve the model for numerical results.[19]

APPLICATIONS SUPPORTED BY THE ISM PROCESS

In presenting a discussion of applications that have been supported by the ISM process, the following objectives are sought:

To illustrate the variety of applications to date.

To illustrate the variety of users to date.

To classify the applications according to the kinds of maps that have been produced.

To clarify the kinds of relations that have been used in developing maps and the kinds that have *not* been used (since the latter are also of importance in applications).

It is believed that this approach will best serve a variety of readers whose interests may be as varied as have been the applications of the ISM process.

In addition to the foregoing, a few selected applications will be discussed in greater depth.

KINDS OF MAPS PRODUCED

The maps produced so far in the application of ISM can be classified into five categories:

Intent structures.

Priority structures.

Attribute enhancement structures.

Structures of a process.

Mathematical dependency.

The kinds of maps are listed in the order of the frequency with which they have been used in 50 example applications. In these applications 19 intent structures were developed, 16 priority structures, 8 attribute enhancement structures, 6 structures of process, and 1 structure of mathematical dependence.

We will discuss briefly how these kinds of structures are distinguished.

Intent Structure. An intent structure is a map for which the element set consists of statements of objectives. The relation may be written as, "should help achieve." A typical statement to be discussed in developing assertions for a map would be of the form, "Attainment of Objective A should help achieve attainment of Objective B."

Priority Structure. A priority structure is a map for which the elements are to be ranked according to relative preference, relative intensity, or some other (normally judgmental) attribute. An example statement would be, "The Ninth Street sewer project is at least as high in priority as the Swanson Park Swimming Pool enlargement."

Attribute Enhancement Structure. An attribute enhancement structure has typically been developed around a set of elements that can be described as problems or opportunities. The relations used are usually "would enhance" or "would aggravate." An example statement would be, "Increasing the housing allowance would enhance the retention of personnel."

Structure of a Process. A process structure usually involves elements that can be described as activities, events, decisions, or some combination of these. An example statement would be, "Event A should precede event B," illustrating the use of "should precede" as a structuring relation.

Mathematical dependence. A map of mathematical dependence is one whose elements are quantifiable, and an example relation for such a map can be expressed by "is a function of." A typical statement would have the form, "A is a function of B."

Table 1 matches users to kinds of structure for 50 applications that have been analyzed to gain a statistical feel for uses. Of the 50 applications, 41 were domestic, with the other 9 being distributed as follows: Japan, 3; Australia, 2; Africa, Brazil, Canada, and Mexico, 1 each. The

	I.S.	P.S.	A.E.S.	S. of P.	M.D.	Total
Government, Federal	5	2	3	1	0	11
Government, Regional	3	1	2	0	0	6
Government, State	2	4	0	0	0	6
Government, County	0	1	0	0	0	1
Government, Municipal	1	3	1	0	0	5
University	5	1	0	3	1	10
Research Institute	2	1	0	0	0	3
Citizen Group	0	1	0	0	0	1
Individual	1	1	0	0	0	2
Corporation	0	1	1	1	0	3
High School Class	0	0	1	1	0	2
Total	19	16	8	6	1	50

I.S. = Intent Structure
P.S. = Priority Structure
A.E.S. = Attribute Enhancement Structure
S. of P. = Structure of Process
M.D. = Mathematical Dependence Structure

Table 1   Matching Users to Kind of Structure

process has also been used in India and Saudi Arabia, but only in work-shops as far as is known.

TYPES OF RELATIONS

Before discussing specifics of applications we will mention a typology of relations and discuss how this typology relates to the ISM process. In every use of the process one or more contextual relations must be selected for use in developing the map(s). Experience has shown that this is an area in which users are not prepared. Also, the wide variety of choices available is itself a source of some disorientation. By classifying relations into types, most of this difficulty can be alleviated.

Five types of relations appear to be of greatest significance in inter-relating elements in applications:

Comparative relations, by which one element is compared to another. Examples: A is larger than B, A is heavier than B, A is prettier than B, A is more important than B.

Definitive relations, which express commonly understood facts or logical conditions. Examples: A is an attribute of B, A is included in B, A implies B.

Influence relations, by which the influence of one element on another is expressed. Examples: A causes B, A is a partial cause of B, A enhances B, A aggravates B, A increases B.

Spatial relations, by which the relative orientation of elements is expressed. Example: A is north of B, A is above B, A is to the right of B.

Temporal relations, in which elements are organized according to their respective positions on a time scale. Examples: A precedes B, A should precede B, A follows B, A overlaps B (in time).

Within some of these categories a further distinction can be made according to whether the relation expresses quantitative or qualitative matters. For example, one could use a comparative relation to express a statement like, "A is at least 10 pounds heavier than B." However, this type of distinction is so application specific that there does not appear to be any great merit in pursuing this point further.

In the applications given in Table 1 the comparative relations are typically associated with the development of priority structures. The influence relations are typically associated with intent structures or attribute enhancement structures. Temporal relations are typically associated with structures of a process.

Definitive relations and spatial relations have not been used in the applications mentioned in Table 1, at least not as the contextual relations that make up part of the ISM process. Nevertheless, in preparing the element set for use in the ISM process, definitive relations are inevitably assumed simply because set membership is itself a definitive relationship. Similarly, spatial relations have not been used in the applications mentioned in Table 1, although they are useful in automating the machine construction of maps.

Since the ISM process is based on the use of a transitive primary relation, the choice of any relation should be made with the aim of finding one that is transitive. To illustrate the subtlety that may be involved, suppose that you were interested in structuring the days of the week according to a temporal relation.

If you chose the relation "immediately precedes" to structure the days of the week (or the months of the year), you would find that this relation is not transitive. Though Monday immediately precedes Tuesday and Tuesday immediately precedes Wednesday, it is not true that Monday immediately precedes Wednesday. On the other hand, if you chose the relation "precedes" to do the structuring, it would be seen to be transitive.

SPECIFICS OF APPLICATIONS

Now we will present specifics of applications of ISM for each type of structure. In each instance we identify the user and give a capsule description of the application.

Intent Structures

1. *A Board of the National Research Council.* To organize the objectives of a study of engineering manpower and education policy.
2. *Office of Environmental Education.* To organize the objectives of environmental education into a structure that would reflect the mission of environmental education.
3. *Department of the Interior.* To organize objectives for energy planning.
4. *U.S. Forest Service.* To organize objectives for using forest and grasslands wisely during a period of several decades into the future.
5. *U.S. Fish and Wildlife Service.* To organize ecological objectives for injection into mineral and energy development processes.
6. *U.S. Forest Service.* To develop regional resource intent structures compatible with federal law.
7. *Rapides Area Planning Commission, Louisiana.* To organize transportation goals.
8. *Louisiana Areawide Planning Districts.* To organize development goals as a means of qualifying for federal assistance.
9. *Ohio Environmental Protection Agency.* To organize objectives of the agency as represented in various legislative acts.
10. *Utah Division of State Lands and Forestry.* To organize forest planning objectives for compatibility with regional and federal plans.
11. *Central Ohio Transportation Authority.* To organize transportation management goals.
12. *University of Alabama, Huntsville.* To organize university goals.
13. *Franklin University, Columbus.* To organize university goals, mission, and policies.

14. *Department of Humanities, University of Virginia.* To clarify department mission.
15. *Department of Engineering—Economic Systems, Stanford University.* To clarify program purpose.
16. *Sociology.* To organize the goals of sociology, as a means of sharpening the definition of the discipline.
17. *Urban Round Table.* To organize thoughts about a possible national urban policy.
18. *Research Institute.* To organize objectives of staff development.
19. *Corporation.* To organize customer needs and wishes, as a means of showing the customer how the company's computers could be used to meet customer requirements.

Priority Structures

1. *U.S. Fish and Wildlife Service.* To rank order objectives contained in the intent structure as a means of focusing future effort.
2. *Brazilian PLANALSUCAR Agency.* To rank order agency efforts in advancing Brazil's alcohol fuel program.
3. *Mexican Farmers.* To rank order possible obstacles to transfer of methane generator technology to Mexican farms.
4. *Queensland, Australia.* To rank order activities in in-service teacher training through formal policy development.
5. *State of Louisiana Planning Office.* To rank order developmental constraints relative to coastal zone management.
6. *Louisiana Office of Science, Technology, and Environmental Policy.* To rank order obstacles to investment in a LANDSAT (satellite) Access System.
7. *Louisiana Urban Studies Institute.* To rank order energy resources.
8. *City Council, Cedar Falls, Iowa.* To put priorities on planned future municipal projects.
9. *City Council, Kent, Ohio.* To rank order budget line items, in preparation for a budget reduction.
10. *Planning Commission, Dayton, Ohio.* To rank order factors in the improvement of the city.
11. *University of Virginia Committee.* To rank order moving arrangements in transferring several departments to a vacant building.
12. *Research Institute.* To rank order transportation projects.

13. *Citizen Group, Dayton, Ohio.* To rank order factors that would increase neighborhood safety and improve neighborhood vitality.
14. *Individual.* To profile the individual in terms of priorities on learning disabilities, as a way to help plan an educational program for the learning disabled person.
15. *Corporation.* To rank order factors related to future markets.
16. *Genesee County, Michigan.* To rank order budget line items.

Attribute Enhancement Structures

1. *National Science Foundation.* To assess how components of methodology could reinforce one another in carrying out technology assessments.
2. *Interagency Futures Research Committee.* To explore societal trends affecting the future of American governance.
3. *U.S. Coast Guard.* To explore how various policies could be developed that would improve personnel retention.
4. *Environmental Protection Agency.* To explore regional issues related to the use of coal in the Rocky Mountains area.
5. *University of Dayton.* To explore means of restoring the Sahel region of Africa, overcoming the effects of ill-conceived intervention that did not take ecological factors into account.
6. *Several Municipal Study Groups.* To study the mutual influence of a set of factors relating to inner-city investment.
7. *High School.* To carry out a classroom learning exercise about factors in developing nations.
8. *Corporation.* To explore interactions in product development programs.

Process Structural Models

1. *Office of Environmental Education.* To portray the operation of the educational system as a process.
2. *University of Queensland.* To develop learning sequences for a programmed textbook.
3. *McGill University, Department of Chemical Engineering.* To develop process models for chemical engineering designs.
4. *University of Virginia, Department of Engineering Science and Systems.* To portray sequences in a set of systems engineering methodologies.

5. *Corporation.* To conduct planning for system development.
6. *High School.* To study energy flows in food webs.

Mathematical Dependence Model

1. *University.* To analyze, structure, and see how to solve an econometric model.

The classifications of these applications are somewhat arbitrary, and their primary function is to give a feeling for what was done. The computer was used in about 80% of these applications. In the remainder the ISM process was not used in full. Instead, the process concepts were used, with structuring being done manually.

TWO CASES

Two short descriptions of cases may help illuminate the application of ISM. The first case is an example of its use in Kent, Ohio.[20] The acting city manager's preliminary 1979 budget contained proposed general fund expenditures of $10.2 million but anticipated revenues of only $8.3 million. The City Council wanted an orderly, comprehensive way of making expenditure reduction decisions. They requested assistance from two faculty members of Kent State University, James Coke and Carl Moore.

Coke and Moore applied the Nominal Group Technique (NGT, discussed in Chapter 8 of this book) and the Delphi method (Chapter 4) as a means of developing and screening the list of elements to be prioritized.

In the NGT session the Kent City Council identified 49 items as candidates for budget reduction. These were expressed in sufficient detail that the city manager could attach dollar figures to them.

The NGT results provided the basis for a Delphi questionnaire. In the first round of the Delphi activity 10 new elements were added. Also, ratings were applied to the elements on a scale of 1 to 5. In the second round of the Delphi activity the initial results were conveyed to the council, and ratings were again conducted. These results were used to arrive at the final list of elements for the ISM exercise.

In the ISM exercise the council was exposed to a series of questions of the following general form:

In reducing expenditures, should the council choose to reduce programs in the planning department by 30%, before choosing to eliminate overtime snow removal?

After the structuring was completed the facilitators were able to recommend three groups of items to the council. Priority 1 items were clearly expendable; Priority 2 items were expendable, with reservations; Priority 3 items were those that most of the council did not want to cut, except as a last resort.

In reporting on the work Coke and Moore noted that, "ISM has some useful side benefits. For one, the debate on each question frequently sharpens issues for further intensive discussions. This occurred several times in the Kent City Council's session. ISM also draws out points of difference without leading to protracted bickering. Lastly, a great deal of group learning takes place in an ISM session."

A video tape, approximately half an hour long, documented the highlights of this particular case.

A somewhat similar project was subsequently carried out by Coke and Moore in Genesee County, Michigan, and this too has been documented on videotape through the National Association of Counties, Washington, D.C.

The second case is a project carried out by faculty of the University of Dayton, originally under the leadership of Brother Ray Fitz, who is now president of that institution.

In this second case the Strategies for Responsible Development (SRD) group at the University of Dayton demonstrated how ISM can be used as part of an approach to planning large-scale rural development projects.†

Since 1974 SRD has been involved in research and educational programs that focus on distributive justice and Third World development. One of its major efforts is a seven-year project in the Republic of Niger, one of the countries hardest hit by the Sahelian drought of 1968–1974.

The idea for this integrated agricultural development project originated at a conference held in 1975 in Toronto. At that time SRD formed a consortium with the Institute for the Study and Application of Integrated Development (ISAID) of Toronto. The consortium made a feasibility study in Niger, selected a project site at Chikal, and started planning for a rural development project. The project was named "Projet Tapis Vert" (PTV), a reference to the green carpet over the desert mentioned in the Koran.

Projet Tapis Vert has been planned in three phases. During the first phase, in 1977, surveys were made of the geology, topography, soils, vegetation, hydrology, and social organization of Chikal. Several needs became apparent—all were interrelated: erosion control, soil regenera-

---

† The following description was contributed by Ms. Karen O. Crim and Ms. Joanne Troha, members of the SRD group at the University of Dayton.

tion, a community health program, the identification of drought-resistant crops, new sources of protein, cash crops, sources of energy, and the integration of women in the village development.

Phase II is the present three-year (1978–1981) test and demonstration phase. During this time activities proposed in the study phase are being evaluated in terms of technical feasibility and social acceptability.

During Phase III, beginning in 1981, an attempt will be made to encourage the villagers to adopt practices found successful in the PTV tests, and the North American team will complete the transfer of control of the project activities to the villagers and the Nigerian government.

Planning for an integrated agricultural development project posed a sizable challenge for the interdisciplinary consortium. The group needed to find an effective way to handle the complexities of a human ecosystem and to encourage the participation of group members in developing a project plan for the Sahel. At SRD's suggestion the group decided to apply a systems approach to project planning that used a number of tools, including ISM. The major tasks in this planning approach were problem definition, project design, and documentation of the project plan. Interpretive structural models helped the group accomplish the first two tasks.

ISM in Problem Definition

The first step in any planning effort is to assess the present situation and to describe the major cause and effect relationships of the problem. Low agricultural productivity in Niger has four interrelated causes, namely, (1) severe winds and water erosion, (2) poor soil quality, (3) poor land management, and (4) primitive agricultural technology and practices. For a better understanding of these causes, and their effects, the consortium began extensive research, made a field trip to Niger, and complemented these studies with a series of modeling exercises.

First, with the help of ISM, the group developed a dynamic model that showed the course of the ecological undermining of the Sahel over the last 15–20 years. It also gave a long-term projection of what would happen should the cycle of aridification continue. The next step was to develop a scenario of a desirable future for the Sahel. A second ISM exercise was undertaken here to construct a "normative" model of the future for the group to discuss. It consisted of desirable policy outcomes that together described what a sustainable human ecosystem in Niger should include. This application of ISM proved to be a valuable learning experience. The project team became more aware of the importance of distinguishing between types of elements; for example, intervention elements cannot be mixed with outcome elements. The group also learned to criticize the logic and consistency they used in thinking about large, complex systems.

A third ISM exercise moved the group from a broad outline of the desired future to a more specific description. This time the project team extracted submodels from the total system and examined them in greater detail. Determining what issues the project should address completed the problem definition. The group compared the logical or extended future of the system against the desired future, producing a list of needs. ISM helped show how the needs were related.

ISM in Project Design

A clear problem definition had been prerequisite to the actual designing of Projet Tapis Vert. Now, with the major issues before them, the project team could begin to design a general strategy for regenerating the Sahelian ecosystem.

The consortium views a project as a system of activities intended to bring about a change in the community's quality of life, that is, to improve an existing system. In turn, each activity is drawn up in a way that specifies how it transforms resources into a product or service that contributes to the change. The new system of agricultural production was to be integrated and sustained at the village level. By creating a structural model of project outcomes, the group was able to clarify the assumptions, logic, sequencing, and input requirements that were to make up the project design.

First the project team generated some 140 elements, divided into five categories, by completing the statement, "An important seven-year outcome at Chikal (Niger) is . . . ." The exercise clarified values held by the group, revealed assumptions, and helped identify weaknesses in the background, experience, and technical knowledge of the project team. SRD was interested in sequencing their input resources and resulting outcomes, so a means–end relationship was used to structure the model:

> Does the project team believe that
> OUTCOME A
>   is a means to
> OUTCOME B
>   in most cases?

The result was five subsystem models that reflected the major purposes of PTV. The interconnection algorithm in ISM facilitated joining the subsystems. The group then amended the total system structure, which included looking for ways to simplify the model. Simplifying the structure made it easier to explore and interpret the planning implications of the models. It also made it possible to use the models in the funding proposal as explanations of the underlying logic of the project.

Documentation

The final task was to document the results of the problem definition and project design into a comprehensive project plan. The notes, assumptions, definitions, and relationships that emerged during the ISM exercises helped the project team select many of the design elements. The team added a narrative summary, schedule, personnel plan, and budget to complete their plan for Projet Tapis Vert.

BENEFITS AND LIMITATIONS OF ISM

The expression "benefits and limitations of ISM" used to title this section is potentially very misleading. This is because the ISM process itself is always a part of a more embracing process, and any benefits or limitations are tempered by and strongly dependent upon the more embracing process within which the ISM process is contained. If the process is deemed to be successful in an application (as it usually is), we can be sure that the success was to a considerable measure a consequence of the skills of the facilitator of the process and the knowledge and attitudes of the participants. If the process is deemed a failure (as has been true in a very limited number of situations), observation suggests that the primary reason for its failure lies in the insensitivity of the person who was attempting to facilitate the process.

If the process is deemed successful, this will usually be because either or both of two things have occurred:

A substantial amount of learning took place as a consequence of use of the process.

The map(s) produced have an evident utility in the context within which it or they were produced.

These are the most important considerations in assessing the benefits of the ISM process. They can be elaborated in more detail, particularly in terms of the substantial time saving that goes on in the process when compared with what would have to be done without its use. Also, the quality of the thought and communication that is usually stimulated by the process can be considered to be one of its benefits, though this benefit clearly depends upon the participants for its full realization.

In assessing limitations, once again limitations are well envisaged in terms of the larger context within which the process is applied. So far as we are aware, the ISM process has been criticized (at least verbally, if not in print) in terms of the following aspects:

The map(s) that is produced may be sensitive to the sequence in which the elements are examined (one person has spoken of the "first-element syndrome").

There exist certain paradoxes related to the sequence of voting that make the results of voting suspect.

The process requires that the primary contextual relation be transitive and thus does not provide a means for working with intransitive relationships.

The process statements require binary assertions, but many questions are more complicated than that, hence unrealistic responses are sometimes requested.

The process is very tiring and demands great concentration from the participants.

The process is time consuming (one person has complained that the process may require as much as two hours!).

Some of the statements that the computer requests the group to address are boring.

The scanning method of transitive embedding is boring because it repeatedly involves the same elements in a series of questions, and this may induce the participants to be careless in their responses.

The voting procedure assigns equal weights to the answers of each participant, though some may be better informed than others.

The process is qualitative and does not allow for the incorporation of factual knowledge from research while allowing off-the-cuff responses.

Let us discuss briefly each of these potential limitations. First of all, we note that the structural modeling by any means involves the sequencing of choices, in regard both to the order in which elements are examined and the order in which assertions are made. This will be true whether the ISM process is used or not. The origin of these limitations does not lie in the ISM process itself, nor in any other modeling process of which we are aware. Rather, it is a limitation imposed by the nature of the human being and by the way in which information flows. It does not flow by the gallon like a liquid but, rather, it flows discretely like letters coming from a typewriter. Thus the proper way to deal with these limitations is to build into any modeling process a means of compensating for their impact. In other words, the limitations are not denied but are accepted and designed for. The way in which these two matters, the "first-element syndrome" and the voting paradoxes, are dealt with is by iteration in the process. But iteration is not included solely to take account of these features.

Since ISM is *designed* to be a learning process, it is part of the design presumption that learning will occur, and, as a consequence, some of the assertions that are made in developing a map might later be judged to be unsatisfactory. This is the principal reason for expressing as part of the process design the need to examine the first map that is produced systematically to determine whether there is a need for any change. By the time the entire map has been produced, it can be presumed that a more holistic image of whatever is being mapped has been attained, and this allows the participants to review the map and revise it as necessary. This review and revision process is greatly facilitated by making it possible to read the entire content of the cell set by means of walks on maps. Also the revision is facilitated by amendment software, which is part of the ISM software.

The limitation to transitive relations should be considered in the light of what purpose structural models are intended to serve. If there is to be any significant benefit from such models, it must stem from the fact that some conditions from the object world of study are very similar to conditions that are inherent in the choice of the structural devices that are used to represent that object world. The digraph-based models are of interest because of their directionality and connectivity, and the fact that they model what may be rather long chains of relationships.

But rather long chains of directional relationships are themselves fundamentally transitive from a structural point of view. This can be seen by recognizing that if there is a digraph walk from A to B and another from B to C, there is inherently a walk from A to C. *Structurally, the digraph is a transitive form.* Thus it does not seem to make any sense to base structural models on digraphs unless the relations to be represented are themselves transitive.

Secondly, many of the relations of importance in the world are transitive. All of our measurements of time and space involve the fundamental assumption of transitivity in the relationships, and we cannot imagine a world in which this condition is not satisfied in time and space relations.

Thirdly, transitive relations, *because of their transitivity*, are among the most difficult ones to work with intuitively because they may generate long strings of logical relationships.

Finally, transitivity is at the heart of those forms of human reasoning that have been studied for over 2000 years, as we have indicated earlier, although this was not recognized, so far as we can tell, until about a century ago.

If an unrealistic response is asked for in the ISM process, the participants or facilitator should recognize that this reflects a deficiency in the element set (usually) or in the contextual relation chosen (rarely). Such a deficiency clearly should be remedied if the map that is produced is to be satisfactory. This is readily done by rephrasing the elements to make them sensible in terms of the structuring or, if necessary, by choosing a different contextual relation. If there *is* a limitation in this area, it can only be because the participants and the facilitator have been conditioned to blindly follow the methodological process, or because the software design has not made it easy to amend elements or relations.

The process is definitely tiring and does demand that the participants concentrate and offer their opinions and knowledge in an environment of sharing, exploration, and refinement. This limitation can presumably be best dealt with by limiting the length of any one sitting to between two and four hours. The software does make it possible to stop the process at any point and resume it from that point, so that if fatigue becomes a problem the facilitator will detect it and interrupt the process, to continue it at a later time. On the other hand, the absolute amount of time required is not particularly related to the process in a negative sense but, rather, is a measure of the difficulty of organizing the substantive content of the theme or issue.

It is not possible to judge in advance how long a time will be required to complete the use of the ISM process. It is not even possible to accurately judge the amount of time required to produce an initial primary map. This is because the amount of time required depends partly on the amount of discussion that goes on (which is encouraged as part of the learning experience), partly on the amount of inference that the computer is able to produce as a consequence of the assertions made by the group, partly on the embedding method that is used, and partly on the constant availability of the computer and other hardware.

A rule of thumb for generating an estimate of the time needed to attain the first map has been offered[2]:

$$T \text{ (hours)} = \tfrac{1}{600}e^2 p^{0.5}$$

where e is the number of elements in the element set and p is the number of participants in the modeling process. Because of the importance of other factors mentioned, the use of this equation to estimate the time is subject to substantial error.

The question as to how much time it is worthwhile to spend in the use of the process clearly must be related to the theme that is being explored, the significance of the knowledge gained, the value of the map(s) produced, and the time required by alternative approaches to gaining the same information. The benefits of team building that may accrue from the process may also be a factor.

If boredom is a problem for participants, it may occur because they are not highly motivated to achieve the results that the process is intended to produce. Alternatively, it may be a consequence of fatigue, which the facilitator should try to detect, as mentioned previously. In any case, boredom seldom seems to be a problem.

The scanning method places every element before the participants early in the process to give an opportunity to assimilate the element set and correct any elements that are clearly poorly stated. However, some may prefer to use the bordering method, which offers greater variety in element presentation.

The voting procedure is easy to change. If desired, individuals could be assigned weights to their votes, either for all questions or for selected questions, or they could even weight their own votes according to the confidence they place in their conclusions. However, it should be remembered that the ISM process is intended to be a learning process, and it is the participants that make it so by the discussions they present on matters that they know about or on which they have strong beliefs. Thus it appears that a weighted voting system might work against dialogue and debate and reduce the process to less-well-informed button pushing.

The knowledge that is applied in generating assertions for use in developing structural models is supplied by the participants. Nothing prevents them from using all forms of available information, and the amendment software makes it possible to amend a structure at any time as long as sufficient incentive exists and as new knowledge becomes available.

There is one truly significant limitation to the ISM process as it can presently be carried out. This is the limitation posed by the combinatorics of themes or issues being explored. Let us briefly explore this limitation, for although it will be shared in nature with any modeling method, there are some features of the ISM process to which this limitation can be explicitly related.

Suppose the element set S contains n members. Then, if a complete primary structure is to be attained, it is necessary to have $n(n - 1)$ assertions. For sufficiently large n this number can be approximated as n^2, and we will make this approximation in what follows.

These n^2 assertions suffice to distinguish that one structure out of a set containing $2^{n(n-1)}$ members. For $n = 4$ the number of possible structures is 4096, and the one that is produced is thereby distinguished from the other 4095. For $n = 8$ the number of possible structures has grown to approximately 140,000 trillion. It seems intuitively evident that the likelihood of producing, without a systematic process, the one structure that is closest to or coincides with complete and correct information is extremely small. Thus one aspect of structuring can be called the capacity to *discriminate* in picking out that one from such a large number. This is a human limitation, but its impact begins to be felt eventually, even with machine assistance.

If, as is often true, a group of participants requires about two minutes on average to produce one assertion, it can be seen that for n^2 assertions one would require $2n^2$ minutes to develop the initial map with the process. Then, to structure an element set with 100 elements, one would require over eight 40-hour weeks of effort if the group had to make *all* of the assertions without the benefit of computer inference. As the number of elements grows larger, eventually the time requirements tend to become prohibitive when the group is making all the assertions. The amount of time required to perform these operations is considerably diminished through the use of ISM. ISM theory and computer software provide for:

Computer calculation of *transitive closure*.

Repetition of this process to calculate *inferences*.

Computer selection of statement sequencing using a criterion to select each statement to try to assure that each inference is possible.

Thus, with an element set having 100 members, the actual time required might be more like one week rather than eight. But eventually, even with computer assistance, the time will become prohibitive for large enough element sets.

In the foregoing discussion it was presumed that the limiting factor in arriving at an assertion was the average time consumed in discussion by participants. But this may not be the limiting factor. As the number of elements grows large enough, the time required to compute the most desirable next statement will also rise, to the point where it becomes the limiting factor. It may be possible at that point to dispense with the strategy of having the computer calculate what the next statement should be and simply take statements in an arbitrary order, within the general context of the embedding theory. Then there will still be a limit, this having to do with how long it takes the computer to calculate the transitive closure and carry out the necessary bookkeeping associated with this.

None of the foregoing limitations appear to have been encountered in applications so far, but these applications have typically involved less than 50 elements. In only one application known to the writer did the number of elements exceed this number. In this application the number was approximately 160; a partial structuring was carried out with the ISM process, and smaller subsystems were then individually structured. This avoided the more extensive effort that would be needed with such a very large set.

SUMMARY

In this chapter we have sought to present a description of a process called Interpretive Structural Modeling (ISM). We began with a brief description of the process and showed that it is a natural outgrowth of a long series of steps in the evolution of human thinking about relationships among elements. We gave the basic premises of the process and its principal purposes.

We then gave an overview of the process with respect to three primary dimensions: the participants, the equipment and facilities, and the substantive content (as it evolves in a series of steps in the process).

We discussed in considerable detail the nature of the output or product of the process. The product is one or more relation maps showing how a set of elements is related in terms of a specific contextual relation. We indicated that walks on the maps represent prose statements, and we illustrated the graphical advantage of maps over the corresponding information presented in prose form. We mentioned that some elements on maps are cyclic and indicated how these are represented. Also, we mentioned the major types of maps.

Then we gave the bulk of the process theory, presenting this first in terms of the fundamentals of digraph models. Next we discussed the fundamentals of digraph-based models and followed this by discussing how such models are developed. Finally we discussed the interpretation of digraph-based models. All of the ISM maps are digraph-based models.

We then discussed the kinds of activities that have been supported by the ISM process, the kinds of maps that have been produced, and the kinds of contextual relations that have been used. We then tabulated 50 applications of ISM according to the different kinds of structures that were produced and indicated what types of organizations were involved in these applications and the central purpose of the application.

Two cases of application of ISM were discussed at greater length. These cases are particularly well documented. The Kent, Ohio, application to municipal budget reduction is described in the literature and on videotape. Projet Tapis Vert has been described in a series of reports and technical papers.

Finally, we discussed the potential benefits and limitations of ISM and emphasized that the success or failure of the process often depends on factors external to the procedures recommended. These include the complexity of the issue being explored, the skill of the facilitator, and the knowledge and attitudes of the participants in the process.

REFERENCES

1. J. N. Warfield, *Structuring Complex Systems*, Battelle Monograph Number 4, Battelle Memorial Institute, Columbus, OH, 1974.
2. J. N. Warfield, *Societal Systems: Planning, Policy, and Complexity*. Wiley, New York, 1976.
3. E. Brill, G. Clendening, and D. Malone, *The Interpretive Structural Modeling System* (*User's Guide and Programmer's Manual*), Battelle Columbus Laboratories, Columbus, OH, February 1975.
4. R. L. Fitz, D. R. Yingling, J. B. Troha, and K. O. Crim, *Computer Implementation of Interpretive Structural Modeling*, University of Dayton Research Institute Report UDR-TR-79-79, Dayton, OH, September 1979.
5. J. N. Warfield, "Systems Planning for Environmental Education," *IEEE Trans. Syst. Man Cybern.*, Vol. 9, No. 12, Dec. 1979, pp. 816–823.
6. J. N. Warfield, "Complementary Relations and Map Reading," *IEEE Trans. Syst. Man Cybern.*, Vol. 10, No. 6, June 1980.
7. F. Harary, R. F. Norman, and D. Cartwright, *Structural Models: An Introduction to the Theory of Directed Graphs*, Wiley, New York, 1965.
8. R. H. Atkin, *Mathematical Structure in Human Affairs*, Crane, Russak, New York, 1974.
9. R. H. Atkin, *Combinatorial Connectivity in Social Systems*, Birkhaüser, Basel and Stuttgart, 1977.
10. M. A. Chamberlain, "A Study of Behcet's Disease by q-Analysis," *Int. J. Man Mach. Syst.*, Vol. 8, No. 5, Sept. 1976, pp. 549–566.
11. K. Sugiyama, S. Tagawa, and M. Toda, "Effective Representations of Hierarchical Structures," International Institute for Advanced Study of Social Information Science, Research Report No. 8, Tokyo, Sept. 1979.
12. J. N. Warfield, "Crossing Theory and Hierarchy Mapping," *IEEE Trans. Syst. Man Cybern.*, Vol. 7, No. 7, July 1977, pp. 505–523.
13. J. Piaget, *To Understand is to Invent*, Grossman, New York, 1973.
14. B. W. Carss and L. D. Logan, "An Example of Participatory Decision-Making Applied to Primary In-Service Education in Queensland," *Admin. Bull.*, Vol. 7, No. 5, 1976.
15. G. G. Lendaris, "On the Human Aspects in Structural Modeling," *Tech. Forecast. Soc. Change*, Vol. 14, 1979, pp. 329–351.

16. D. W. Malone, "An Introduction to the Application of Interpretive Structural Modeling," *Proc. IEEE*, Vol. 63, No. 3, March 1975, pp. 397–404.

17. K. O. Crim, "Use of Interpretive Structural Modeling in Environmental Studies at the Senior High Level," University of Dayton, Report on Grant G007700611, Dayton, OH, April 1979.

18. J. N. Warfield, "Some Principles of Knowledge Organization," *IEEE Trans. Syst. Man Cybern.*, Vol. 9, No. 6, June 1979, pp. 317–325.

19. A. El-Mokadem, J. N. Warfield, D. Pollick, and K. Kawamura, "Modularization of Large Econometric Models: An Application of Structural Modeling," *Proceedings of the 1974 IEEE Conference on Decision and Control*, IEEE, New York, 1974, pp. 683–692.

20. J. G. Coke and C. M. Moore, "Group Processes for Making Public Expenditure Reduction Decisions," Chapter 6 in *Managing Fiscal Retrenchment in Cities* (H. J. Bryce, Ed.), Academy for Contemporary Problems, Columbus, OH, March 1980.

CHAPTER 6

ISSUE-BASED INFORMATION SYSTEM (IBIS)

DONALD P. GRANT

IBIS is an acronym for *issue-based information system*. The IBIS was developed by Werner Kunz and Horst Rittel at the Studiengruppe für Systemforschung in Heidelberg, West Germany, to serve as an argumentative procedure for decision-making groups to use in the coordination and support of debate over political decisions, including those occurring in planning and design.

The IBIS can be used as a procedure for arguing design, planning, and policy making decisions on any topic. The IBIS procedure rests on the assumption that all these activities are political in nature and serves to develop a discourse over an initially unstructured topic or problem area. Swanson (1977) lists transportation in Los Angeles, economic "stagflation," energy issues, and fair employment practices in a single firm as topics that the IBIS procedure might address. He states that the function of the IBIS is to raise and dispute issues from multiple points of view or positions and to offer arguments and counterarguments in order to develop an appreciation of problem situations by the decision-makers involved.[1]

The IBIS procedure structures, monitors, and records debate and develops a data base as the debate proceeds.

The characteristics of the IBIS encourage thoroughness in problem coverage and assist in making the course of argument explicit and retraceable. The IBIS can be used as a procedure for deliberation and decision either by groups or by individual decision-makers wishing to look at problems from multiple points of view. The IBIS can be used by an individual decision-maker to simulate conflicts of interest among multiple clients to a problem, either when actual clients cannot communicate with each other or as a means of anticipating future conflicts before they occur to the actual clients.

The components of an IBIS are the following: issues, positions, arguments, and references; questions of consensus, answers, and supporting evidence; topics that group several issues and questions; and forms, tables, matrices, and trees that relate various components with each other in several ways.

Some concepts that relate to the IBIS are the Social Technology[2]–Delphi approach,[3] the concepts of Kantian, Singerian, and especially Hegelian inquiring systems described by Churchman,[4] and the concept of user participation in planning and design, widely discussed during the past decade. A fundamental concept in the IBIS approach is Rittel's "second generation" in design and planning methods, in which the user/client is assumed to be more expert about his own values than any planner or technical expert representing the user/client could ever be.[5–7]

Criticisms of the IBIS approach are that it is easy to smother arguments under mountains of bookkeeping, that the prospect of being frank (and accountable) about one's decision bases can be uncomfortable or even dangerous and may prevent some people from participating in any argumentative planning procedure, and that some people simply find formally structured procedures uncomfortable and/or unacceptable.

Arguments for the IBIS procedure are that it encourages thoroughness, exposes underlying values, and makes it possible to communicate, record, and argue decision bases. A related argument is that public exposure and the recording of people's decision bases tend to inhibit unethical or excessively selfish arguments from being set forth.

A methodology for IBIS development is described by Swanson.[8] Dehlinger and Protzen[9] discuss the application of IBIS to planning issues, as does Grant.[10] Mann[11] suggests the combination of the IBIS with a gaming approach. Swanson[12] proposes the use of the IBIS for business management information systems. An annotated bibliography on the IBIS is found in Grant.[13]

The basic assumption underlying the development and use of the IBIS is that planning, design, and policy making are best modeled as examples of argumentation. The use of the IBIS in environmental design has been referred to as an argumentative model of design, and involves Rittel's concept of a "symmetry of ignorance" between the designer and the client,[14] in which decisions and choices are organized and structured. Kunz and Rittel state the purposes of the IBIS as being to support coordination and planning processes in political decision-making by identifying, structuring, and settling issues and by providing relevant information at the appropriate times.[15]

The design and planning activities supported by the IBIS include analysis, problem definition, ideation, and selection. The large-scale IBIS developed to serve the German Federal Republic as an Environmental Planning Information System (UMPLIS) is also intended to serve as a problem anticipator and monitor in the area of problem recognition and as an evaluator of planning efforts.

LIMITATIONS OF THE APPROACH

Grant provides a summary of criticisms of the IBIS that describes some of the limitations of the approach.[16] The IBIS has been subjected to criticism as a purely logical decision-making method, as well as on a variety of other grounds.

Some people view decision-making as being made up of illogical or alogical as well as logical components and criticize purely logical approaches to decision-making as excluding significant human values or characteristics. The IBIS is a basically logical approach to decision-making, being based on the weighing of arguments for one position against the arguments for another position in order to reach a decision. A negative criticism of the IBIS form is that, being a basically logical procedure, it does not readily accommodate illogical or alogical components. An argument counter to this criticism is that illogical or alogical components can be included as well as logical components, provided that they are stated in a communicable form as arguments for or against positions on issues. A different counterargument is that illogical or alogical components have no place in planning or design in any case and that if the procedure being followed makes it difficult to include them, then that is all for the good. It is not the purpose of this critique to take a position on the relative merits of these arguments. The purpose is, rather, to set forth the various critical arguments for the reader's consideration.

This criticism relates closely to the prior criticism of the IBIS as a purely logical solution format. The formal structure of the IBIS requires the statement of and recording of issues, positions, and arguments.

A negative criticism in this regard is that planning and design deliberations are likely to involve many themes simultaneously, not only the explicitly stated issues, and that the requirement for having everything "on the table" is likely to take away from the process of decision rather than add to it. A positive criticism in this regard is that since planning and design affect many people's interests, all factors should be explicit and "on the table" and should not be considered at all if they cannot be made explicit and stated openly. The latter positive criticism assumes that the factors to be considered in a decision should be substantive and essential factors, and not such underlying factors as ego maintenance, "king of the hill," the displacement of tensions, or political trade-offs. The basic nature of this criticism in both its positive and negative forms is that the IBIS tends to require explicitness and encourages a monothematic discussion centering on the theme as stated, without its implicit multithematic undertones. The positive critique is that this is how things ought to be. One form of a negative critique is that this is not how things ought to be. Another form of a negative critique is that this is not how things are or ever will be, regardless of how they ought to be.

In a conflict of interest there is every incentive for the outsider to engage in a public and explicit debate, and every incentive for the insider not to engage in such a debate. A negative critique of the IBIS approach is that only those with no power are likely to be willing to participate. A positive critique is that the formal and recordable structure makes it obvious who will not participate in a planning or design argument and that this may aid in forcing participation through the authority of the law or public opinion. A related point is that participation is often an attraction in itself and that this may overcome the reluctance to participate in some cases. A neutral comment is that reluctance to participate in deliberation and decision is a permanent dilemma of democratic procedures and that the IBIS format might either help or hinder efforts to overcome this delemma. A factor discouraging participation in argumentative processes for planning and design is the "information overload" factor: Many people regard their lives as already being sufficiently complicated without adding to the complication by participating in anything that can be avoided.

Those with access to means of specious persuasion are likely to want to exercise the advantage that such means provide them with. A negative criticism of such approaches as the IBIS and Delphi is that those with a vested interest in specious persuasion, in such forms as speaking ability, charismatic personality, authority, or personal prestige, are not likely to be willing to participate in any process that prevents them from exercising their advantage. A counterargument is that, first, if such persons do participate, the IBIS format decreases the potential for specious persuasion, and, second, if such persons refuse to participate, the IBIS provides a powerful means for publicly simulating their points of view stripped of the advantage of the means of specious persuasion. A related argument is that possessors of the means of specious persuasion who do not wish to participate may be forced to do so by law or public opinion, and that in such cases it is helpful to have a structured format which tends to neutralize specious means of persuasion and to keep the discussion on substantive issues.

The comment is sometimes made that the IBIS is more like politics or policy making than it is like planning in the technical or administrative sense of the word. This comment is sometimes made with the intent of demonstrating beyond all doubt that the IBIS is a desirable approach to planning. This comment is also sometimes made with the intent of decisively proving that the IBIS is not planning at all. This raises the issue of what planning is, and when it begins. A similar issue can be raised with respect to design at other scales.

A negative criticism of the IBIS is that it seems to rest upon the naive assumption that decision-making should be out in the open and based on motives that can be declared in public. The negative critique is that this is not how things are or ever will be, and that therefore the IBIS is an inadequate planning tool because it does not conform to reality. A positive critique is that the use of a structured format like the IBIS may help to make things open and publicly debatable, and that that is how things ought to be.

One of the major positive points in a critique of the IBIS is that it encourages thoroughness in the consideration of a problem and its potential solutions. A negative version of this same point is that the format may encourage thoroughness for its own sake, to the detriment of important points or at the price of having the discussion bog down. The extreme statement of this criticism is that the format might open the way for pathological, self-destructive thoroughness.

A positive critique is that the IBIS format forces the "objectification" of the reasons for taking a position. *Objectification* is defined as the act of making understandable and communicable. A negative response to this point of view is that such forced objectification might lead to the rejection of the whole procedure by powerful interested parties. This point relates to matters already discussed under the critiques with regard to specious persuasion and incentives to participation.

Planners at the technical or administrative levels may regard value-based discussions in the IBIS or other similar formats as a threat to their own prerogatives and refuse to cooperate in necessary ways. Such a refusal to cooperate might be either open and obvious or covert and subtle. One important form of noncooperation is to refuse to provide the necessary data and documents or to provide the wrong material or material in the wrong form. Such refusals or reluctance to cooperate on the part of people at the technical or administrative levels may stem from a simple resistance to new procedures as well as from feelings of being threatened.

One of the appeals of the IBIS to many people is the promise of freshness. Each new issue potentially restructures the nature of the problem at hand and that of the information system that is being generated ad hoc in the course of the discussion. Several negative criticisms can be constructed on the subject of the institutionalization or bureaucratization of the IBIS as a standing information system. The overall positive critique is that the IBIS does tend to generate an information system ad hoc; the overall negative critique is that the IBIS can easily degenerate into an institutionalized standing information system with its own experts and misuses. The IBIS can become a complicated ritual that guarantees permanent employment for its priests.

Since the IBIS has specific procedures and recording processes to be administered, there is the possibility that the planner can become adept at manipulating the discussion in conformity with his own values and interests.

It is easy to think of reasons for maintaining things as they are. There might be a tendency for accumulated arguments to scare off potential innovators and discourage adventurousness.

The simple fact of detailed, step-by-step organization might well scare off half of humanity. This point also relates to a possible psychological incompatibility with many people's problem-solving outlooks.

A positive point is that the IBIS facilitates recording the course of a discussion and of the reasons for taking certain positions. This is useful in situations of delegated authority, of legal review or lawsuit, and of simple forgetfulness over time. The recording function also facilitates the resumption of a discussion after an interruption and further facilitates the entry of a new participant into a discussion by making the past record of the discussion available to him. A negative aspect of this feature of the IBIS is that the knowledge that a discussion is being recorded for future reference may inhibit some participants from stating their real reasons for decisions or from contributing to the discussion.

The process of stating many different arguments for and against positions on an issue almost inevitably leads to the simulation of many different points of view, even when an IBIS is being constructed by only one person. In group applications the consideration of points of view of people not participating is facilitated and encouraged. The IBIS can thus be a useful tool for problem-solving, considering the points of view of multiple clients, including those who have not yet been identified, do not yet exist, or cannot speak for themselves. A negative aspect of this point is that it is easy to bring so many points of view to bear that action becomes difficult or impossible.

The IBIS format encourages the anticipation of the consequences of taking each position set forth. The proponents and opponents of a given position construct a set of predicted consequences in the course of offering arguments for and against a position. A negative side of this aspect of the IBIS is that decision and action may be rendered difficult.

A positive aspect of the IBIS and other systematic methods of decision-making is that their use carries a set of indirect benefits parallel to the direct benefits of arriving at a decision or of structuring a discussion. The indirect benefits are in the form of increased familiarity of the participants with the problem landscape and an increased awareness of the various aspects of the problem. The indirect benefits of this sort may in fact outweigh the direct benefit in some cases. Some writers on scientific and other forms of creativity theorize that the process consists of familiarization, subconscious incubation, and then an unexpected insight. The indirect benefits of the IBIS might aid in the process of familiarization.

A negative critique of the IBIS approach is that it can require so much paperwork and form shuffling that it inhibits the participants from thinking. This problem is related to the critique with regard to thoroughness. The key words relevant to this negative criticism are *ponderousness, clerical tasks,* and *housekeeping.*

A negative critique related to the preceding critique with regard to paperwork is that the prospect of having to go through the paperwork may in itself discourage the use of the technique and/or the initiation of a discussion.

The IBIS is a structure for the step-by-step development of information about a problem. One point of view is that the process of information takes place in the form of vague breakthroughs of awareness rather than in the form of a step-by-step process. A parallel argument in science would be between the pedestrian, step-by-step approach of what Kuhn calls "routine science" and the paradigmatic revolutions that he describes in his book *The Structure of Scientific Revolutions.* Another analogy would be the comparison of a mechanical process with spontaneous combustion.

PARTICIPANTS IN THE IBIS PROCESS

Kunz and Rittel describe the IBIS as a procedure for structuring the problem-solving efforts of decision-making cooperatives into the form of arguments, that is, as a procedure for use by groups.[17] The IBIS can also be used by individuals for a variety of purposes.

The IBIS was developed with the intent of being an argumentative decision-making procedure. However, many issues arise and are argued without reaching decisions. A worthwhile function of the IBIS might be problem exploration and definition, even when no decision is expected to result.

The use of the IBIS in a group situation requires first, of course, that the group accept the notion of using it. The second requirement is agreement on some rules of procedure and, if appropriate, decision rules. Agreed-upon rules of procedure might be a traditional set like *Robert's Rules of Order*, and the agreed-upon decision rule might be a simple majority vote. There are many alternatives to both these decisions.

The comments above imply that a group is using the IBIS as an argumentative procedure in which all sides to the argument participate. An alternative group use is the situation in which one side of an argument uses the IBIS as a means of simulation or role playing without involving the other side in the IBIS procedure, as a means of anticipating positions and arguments that might be set forth by the other side in order to develop counterstrategies in advance. Yet another alternative situation is one in which one or the other side does not yet exist, has not been identified, or is incapable of speaking for itself. Simulating positions and arguments in behalf of future generations is an example of the last-mentioned alternative situation.

The use of the IBIS for either decision-making or problem exploration, described above as pertaining to the use of the IBIS by groups, also applies to the use of the IBIS by individuals. An individual might wish to use the IBIS as a means of thorough problem exploration, as a means of arriving at some decision based upon explicitly stated and retraceable reasons, or as a means of role playing or simulation. The process of listing all the positions and arguments that one can think of with respect to a given issue is one that encourages looking at a situation from multiple points of view, one that encourages thoroughness, exposes underlying values, and one that makes it possible to record the reasons for various possible decisions in an understandable and communicable form.

Mann lists the participants in an IBIS as those directly involved in the debate, expert consultants, judges and critics providing evaluative commentaries, and documentarians.[18]

INPUT AND OUTPUT OF THE PROCESS

The input to a specific debate will vary according to the nature of a topic. Kunz and Rittel refer to a "trigger phase" that evokes an image of the problem area.[19] The output of the procedure in the most direct sense is a plan, policy, or solution to a given problem. The output in a more diffuse sense is a deepened awareness of the problem and of the points of view involved in the problem area, on the part of all the participants.

BASIC ELEMENTS OF THE IBIS

The core components of an IBIS are *issues*, the *positions* that people take on issues, *arguments* offered in support of or in opposition to positions, and *references* cited to support arguments. A secondary set of components includes *questions of fact or consensus, answers*, and *evidence*, which are special cases of *issues, positions*, and *arguments*, respectively. A further set of components includes *topics, forms, tables, matrices*, and *trees*, which are a means of recording and interrelating the core components and secondary components.

THE CORE COMPONENTS

The processes of information and decision are ones in which *factual images* are interpreted in the light of *values* or *deontic images*. While people living in a same culture at the same time often share *factual images* to a very great degree, disagreement about *values* or *deontic images* or *norms* is not uncommon. Thus, while a group of people, all of whom have some interest in a design or planning project, may share their dominant *descriptive-factual, causal-factual,* and *instrumental-factual images* of the situation, they may disagree sharply as to *values* and, thus, as to which decisions are proper or acceptable in the situation. It is for this reason that the abc's of environmental planning are sometimes said to be "argumentation, bargaining, and conflict." The degree to which people share *values* or *norms* or *deontic images* seems to be decreasing over time.[20] The degree to which we are willing to recognize that knowledge of the world is in the form of selectively perceived *images* rather than in the form of absolute or "true" facts seems to be increasing over time.[21]

Churchman views the *clients* of a decision situation as being all those whose interests are affected by the decision and who therefore have a right to be represented in the decision situation.[22,23] The existence of more than one client to a decision makes it highly probable that even if *factual images* are shared, *values* or *deontic images* will not be, and therefore that many questions about the decision will not call forth agreed-upon answers but will instead cause different clients to adopt different positions with regard to which answer is correct. The recent movement for user/client participation in design and planning has been motivated in part by the feeling that since different people have different values, those who are affected by designs and plans should make the value judgments involved, instead of having value judgments made in their behalf by professional designers, planners, or administrators. This point of view is summarized in Rittel's statement, "The best world is one in which nobody plans for, in behalf of, or at anybody else."

This point of view also relates closely to the general swing towards "consumerism" and towards "accountability" and "openness" in government and public institutions.

The IBIS has as its basic component the *issue*, that is, a question upon which people take different *positions* rather than agreeing upon a single answer. The core components of the IBIS are *issues,* the *positions* that people take on the issues, the *arguments* that people offer in support of or in opposition to positions, and the *references* that people cite in support of their arguments.

Issues

The basic component of the IBIS is the issue. An issue is a question for which there is no agreed-upon answer but, rather, a variety of positions adopted by different people. Examples of issues are the following:

Should automobiles be prohibited from coming into the city center?

Should all dwellings be required to be of noncombustible construction?

Should a city have aesthetic requirements in its building code?

On all of the above questions several positions are conceivable and likely to be taken by different people. Among the positions that might be taken are, "Yes," "No," "The question is absurd or irrelevant," "Yes, but . . . ," "No, but . . . ," and so on. It is unlikely that all members of any large group would take the same position on any of these questions.

In constructing an IBIS issues may or may not be classified in terms of the type of knowledge they deal with. Dehlinger and Protzen described several classifications.[24] A basic classification scheme is to divide all issues into those dealing with *factual images* and those dealing with *deontic (value-based) images* and to then subdivide the factual issues into *factual-descriptive, factual-causal,* and *factual-instrumental* issues. This classification scheme yields four types of issues, as shown below. Dehlinger and Protzen (1972) suggest that each type of issue might generate a different type of position-argument structure.[25]

Deontic Issues.   Those dealing with values, norms, or prescriptive images.

Factual-Descriptive Issues. Those deaing with factual images that describe reality.

Factual-Causal Issues. The subset of factual issues dealing with cause and effect (if A, then B).

Factual-Instrumental Issues. The subset of factual-causal issues dealing with causes that the decision-maker can control (if I do A, then B).

Some of the advantages of classifying the issues in an IBIS are the following: Each type corresponds to the type of knowledge involved and perhaps therefore to the groups of people who should be involved in the argumentation; as mentioned above, there might be a different position-argument structure and working routine appropriate to each type; and issues can be grouped by types for access and review.

The latter point is potentially important in long-term deliberations; it might be useful, after several months or years, to be able to retrieve and review all the *deontic issues* that have been discussed, excluding the *factual issues* in order to focus on the basic, value-based agreements and disagreements that have emerged.

Informal discussions with people who have applied IBISs to various problems have revealed some reasons for preferring not to classify issues. One problem with issue classification is that it is sometimes not clear which class an issue belongs to—and in fact a single issue might belong to more than one class—and therefore that ambiguous classification would result in ambiguous retrieval. Another problem with issue classification is that new classes are always conceivable in new problem contexts and using old categories for new problems might result in inappropriate or obstructive classification of issues. The arguments against issue classification are supported by the fact that labels like, "I-001 is a deontic issue," can always be applied to issues for later retrieval if it seems appropriate, even if no comprehensive issue classification scheme is in use.

The *deontic* or seemingly *deontic* issue statement is normally of the form, "Should there be X?" The following are some sample deontic issues:

Should the exterior siding of the house be redwood?

Should a special gasoline tax be collected in order to support the development of public transportation systems?

Should asphaltic roof systems be outlawed in high air pollution areas?

Should provision for the future installation of solar heating apparatus be a requirement of all new construction?

Should construction be forbidden within a specific distance of known earthquake-causing faults?

Should the kitchen and dining room be one continuous space, with no visual isolation between them?

To return to the question of whether or not to classify issues, some of the above might be either ambiguous or mixed issues. The "redwood siding" issue might at first seem to be a deontic or aesthetic issue, but turns out to be a factual issue, depending on material availability or price, or an instrumental issue, depending on questions of heat gain or fire resistance.

An instrumental issue, a causal issue, and a factual-descriptive issue can each be identified quite simply by the way in which it is stated. This can be used as an argument either for or against issue classification. The pro argument is that if the types are so distinct as to automatically generate different statements, then any divisions that are so powerful should be made into distinct classes. The con argument is that if the types are so easy to distinguish that they generate different statements right from the start, then further classification is a needless formality.

The following are typical forms of statements:

1. Deontic: Should there be X?
2. Factual-descriptive: Is there X?
3. Factual-causal: What causes X?
4. Factual-instrumental: How can X be brought about?

All the above types of knowledge are, of course, interrelated. One's values will have much to do with the descriptions, causes, and means that one is willing to accept, and so forth.

Rittel lists a further issue type in addition to the deontic, factual-descriptive, factual-causal, and factual-instrumental types described above.[26] The additional issue type is the conceptual issue, for example, "What is meant by Z?" Rittel also gives examples of the basic four issue-types[27]:

1.	Deontic:	Ought Y to be the case?
		Ought X to become the case?
2.	Factual-descriptive:	Is W the case?
		Can V be expected to become the case?
3.	Factual-explanatory (also referred to as factual-causal):	Why is U the case?
		Why should T be the case?
		Why should S become the case?

4. Factual-instrumental: How can R be made to become the case?
                                    What are the possibilities for making Q
                                       become the case?
                                    What are the consequences of doing P?
                                    Who will be affected by applying O in
                                       situation N, and in what ways?

The issue of whether or not to classify issues does not require a generalized resolution but is a matter to be considered in each specific IBIS application.

Positions and Arguments

A *position* is taken by somebody in response to an issue. Some common positions in response to an issue in the form, "Should there be X?" are the following:

P 0   The question is irrelevant or inadequate or absurd.

P 1   Yes.

P 2   No.

P 3   Yes, but _____ .

P 4   No, but _____ .

If all participants in an IBIS take the same position on an issue, then that issue ceases to be one and becomes a *question of consensus*.
In a group decision-making IBIS the positions listed should be all those taken by participants, and possibly further positions that the participants wished to hypothesize for purposes of exploration. An individual using the IBIS approach to explore a problem might wish to pose all the positions that he could think of, as a means of encouraging the thorough investigation of the problem and all associated arguments and outcomes. A specific issue might call forth two or more positions. Factual (descriptive, causal, and instrumental) issues generate different kinds of positions than do deontic issues. For example, positions on a factual-descriptive issue might be the form, "A, but not B, is the case," "B, but not A, is the case," "Both A and B are the case," and so on. Positions on a factual-causal issue might be, "C causes D," "E causes D," and so on. Similar forms apply to the instrumental issue.
Arguments are set forth to support or oppose *positions* taken on *issues*. A single *position* might be supported by many different *arguments*. A

single argument might be offered in support of more than one position, and in fact the same argument might be used to support different or even opposing positions. In situations in which several arguments support the same position some of the arguments might be quite incompatible with each other. An issue–position–argument–reference tree is illustrated in Figure 1.

References

References are journal articles, books, or other documents cited in support of *arguments*. The place of references in a typical issue–position–argument–reference tree is shown in Figure 1. References are assigned R numbers in the order in which they are cited; thus the first reference that anybody cites is R-0001, the next is R-0002, and so on.

It might be worthwhile in some situations to assign key words or index terms to each reference and compile a keyword index as the list of ref-

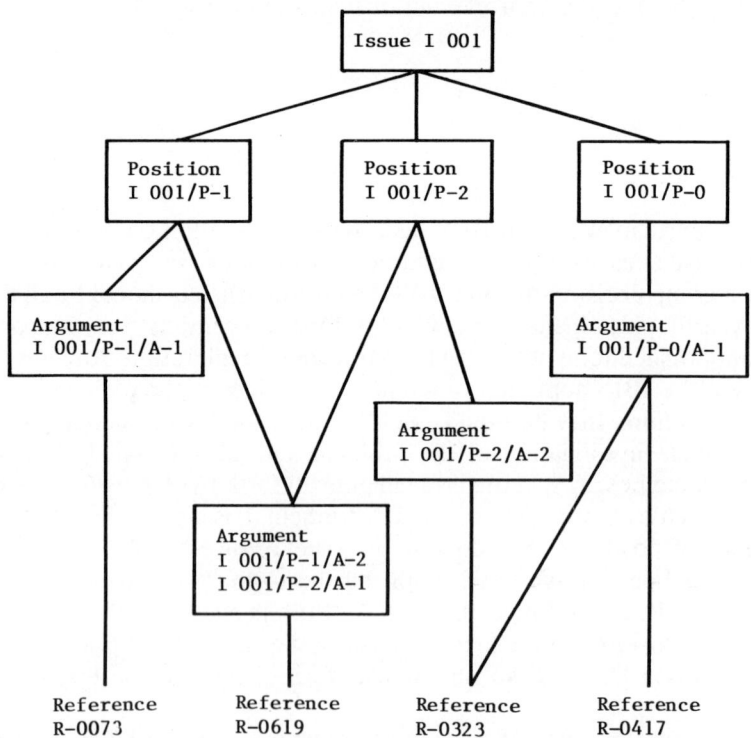

Figure 1   An issue–position–argument–reference tree.

erences grows, in order to allow for future searches of the reference list on specific subjects.

Questions, Answers, and Evidence

Some questions raised will not call forth multiple positions, but only a single position. An issue on which only one position is taken will be referred to as a *question of fact* or *question of consensus*, and the position taken on it will be referred to as an *answer*. Arguments in support of the answer will be referred to as the *evidence* offered. *Questions, answers,* and *evidence* are special cases of *issues, positions,* and *arguments,* respectively.

Some questions do not generate multiple, opposing responses, but only a single, agreed-upon answer. Every *issue* might turn out to be a *question* upon which there is consensus among discussion participants. Conversely, every *question* might turn out to be an *issue* upon which people take different positions, even though the person who raised the question might originally have thought of it as a question of consensus or fact.

The expression *question of consensus* is perhaps preferable to the expression *question of fact*. Historically, many questions for which there were widely agreed-upon answers, and which were thus thought of as questions of "fact," have evolved into issues upon which people took conflicting positions, and then sometimes back into questions with widely agreed upon answers. During the periods of Copernicus and Galileo, the question, "What is the shape of the earth?" had a widely agreed-upon answer: "It's flat." Copernicus, Galileo, and commercial interests based on navigation all helped to turn this into an issue with two positions: "It's flat," versus "It's round." Now the question, "What is the shape of the earth?" again has a widely agreed-upon answer, "It's round," with some dissenters here and there. Similar examples could be constructed around questions about the causes of disease, the nature of matter, and so on. The expression *question of fact* has an implication of absoluteness and truth that the expression *question of consensus* does not have. The example set by intelligent and disinterested persons in the past, who have regarded some questions and answers as matters of fact or "truth," only to have the perceived "truth" very much changed with the passage of time, may lead one to prefer the more modest expression *question of consensus* to the less modest expression *question of fact.*

A *question of consesnsus or fact* is simply a question upon which all discussion participants take the same position. The position that is agreed upon is the *answer*. An *answer to a question* is thus a special case of a *position on an issue.*

Project:		Page of I-	
Preceded by:	Issue Statement	Succeeded by:	Related to:
	Positions/Answers:		
Issues Questions Topics Positions			
Context/Background:			

Date:	Action:	By:	Date:	Action:	By:

Model Issue-Form II: Donald P. Grant, Architect and Planner - 1976
The top section of the sheet with the issue-statement is 1/8 of
the length of the sheet, so that eight can be photocopied onto one
sheet to make up an issue list.

The issue + position section is 1/2 sheet to allow photocopying
two per sheet.

Figure 2   Issue form.

The statements that are offered as evidence in support of an answer are referred to as *evidence*. *Evidence* is thus a special case of *arguments*, that is, the special case in which there is only one position being supported.

References may be cited in support of *statements in evidence* just as for *arguments*. A given answer may be supported by several statements of evidence and each statement of evidence may refer to several references.

Topics

A *topic* is formed by the clustering of several issues and questions around one general subject.

A general *topic* statement may initiate a discussion before any specific issues have been stated. Conversely, a discussion may start with the statement of one or more specific issues, with *topics* emerging only as issues and questions begin to form clusters.

Any issue or question may relate to more than one *topic*.

Forms, Tables, Matrices, and Trees

Forms can aid in the recording of an IBIS discussion and in encouraging thoroughness during the discussion. Forms can also help to keep bookkeeping time and effort at a minimum in order to avoid having paperwork detract from the discussion.

Forms are intended to be used constantly as a discussion proceeds.

Tables, matrices, and trees can serve to interrelate the components of an IBIS for later reference. While forms are intended for the recording of a discussion as it proceeds, tables, matrices, and trees are more likely to be developed as a means of organizing the material generated during a discussion after the discussion has ended and before the next discussion begins.

The basic form for the IBIS is the *issue form*. In some applications of the IBIS the basic *issue form* is the only form used. The basic *issue form* itself is sometimes divided into two parts, one for stating the issue and positions and the second for stating arguments.

The *issue-form* should provide spaces for the following: an issue's issue number, the date the issue was raised, the issue statement, a brief context statement about the issue, and a variety of other data. An issue form is illustrated in Figure 2. An additional argument form is illustrated in Figure 3.

No.:	Argument Statements with Author and Date	Source:	I-
			Page of
			References

Model Argument Form: Donald P. Grant, Architect and Planner - 1976

Figure 3 Argument form.

In some IBIS applications questions of fact or of consensus have been recorded on a special *question form* that was different from the basic *issue form*. One argument for doing this is that questions generate a different structure than do issues and therefore a different form is appropriate. One argument against a separate *question form* is that questions are only special cases of issues and having a question form would only proliferate the number of forms and thus add to paperwork. A related argument against the use of separate question forms is that it is not always known in advance whether an issue will turn into a question, or vice versa, and that having forms only generates needless rewriting when such changes occur.

Other forms generated in IBIS applications have included forms for topics, references, positions, and other things.

No one approach to the question of what forms to use is likely to be appropriate to all IBIS applications. One point emphasized by Dehlinger and Protzen and others who have applied the IBIS approach is that the characteristics of the IBIS should be modified and adapted to each specific situation.[28]

Tables, matrices, and trees are useful for interrelating the components of an IBIS discussion. The interrelation of components aids in searches of past discussion records and in retracing the way in which decisions were reached as well as the arguments and answers that went into the decision-making.

Tables, matrices, and trees are usually constructed after one discussion meeting has concluded and before the next discussion meeting takes place.

Tabular or matrix format is useful for displaying the following sorts of relationships:

1.1.4.2.1. Predecessor/successor relations among issues. The contents of this sort of table or matrix are indications such as, "Issue 001 is the predecessor of Issue 002," and, "Issue 002 is the successor of Issue 001."

1.1.4.2.2. Simple relations among issues. "Issue 001 is related to Issue 003."

1.1.4.2.3. Predecessor/successor and simple relations among questions and among questions and issues.

1.1.4.2.4. Relations of issues and questions to topics.

1.1.4.2.5. Relations of references to issues, questions, and topics.

In the matrix below, three types of relationships are shown. The relationships are indicated by the letters P, S and R inside the matrix.

P indicates that the issue/question in the horizontal row is the PREDECESSOR of the issue/question in the vertical column.

S indicates that the issue/question in the horizontal row is the SUCCESSOR of the issue/question in the vertical column.

R indicates that the two issue/questions are related, but not in a predecessor/successor relationship.

An issue is indicated by its issue number; for example, I 001 or I 053.

A question is indicated by an issue number followed by a Q in parentheses, indicating that what was potentially an issue turned out to be a question; that is, the special case of an issue upon which only one position was taken.

ISSUE/QUESTION x ISSUE/QUESTION MATRIX

Principle Diagonal	I 001	I 002	I 003 (Q)	I 004	I 005 (Q)
I 001		P	R	–	–
I 002	S		P	R	–
I 003 (Q)	R	S		P	P
I 004	–	R	S		R
I 005 (Q)	–	–	S	R	

Principle Diagonal

Since this is a square matrix, with the same list in the columns and in the rows, all the information conveyed in it is contained in the upper-right triangular half of the matrix, above the principle diagonal. The other half of the matrix is complementary to the half above and to the right of the principle diagonal.

Figure 4 Matrix relating issues and questions to issues and questions.

In some cases it may be desirable to relate specific positions on issues to other issues and to questions, topics, references, and so on.

The question of which tables or matrices to construct is an obvious one: "Which relations are discussion participants likely to want to trace?" Constructing elaborate tables, matrices, and trees for which there is no probable use can lead the IBIS to bog down in paperwork and waste a great deal of time.

An *issue/question-to-issue/question matrix* is illustrated in Figure 4.

An *issue/question-to-topic matrix* is illustrated in Figure 5.

An *issue/question/topic-to-reference matrix* is illustrated in Figure 6.

Perhaps the clearest form for displaying the relations among components of an IBIS is a treelike diagram. An issue–position–argument–reference tree is illustrated in Figure 1.

A tree might illustrate relations among issues without showing positions

---

Topics are indicated by their topic numbers; for example, T 001.

Issues are indicated by their issue numbers; for example, I 002.

Questions are indicated by their issue numbers followed by (Q); I 003 (Q).

ISSUE/QUESTION x TOPIC MATRIX

	T 001	T 002	T 003
I 001	x		
I 002	x	x	x
I 003 (Q)			x
I 004	x	x	x
I 005 (Q)			x

This is not a square matrix with the same lists in the columns and rows, so there is no principle diagonal, and the entire matrix is necessary to convey the information shown.

Figure 5   Matrix relating issues and questions to topics.

or arguments or might show positions, arguments, and even references as well as issues and questions.

The actual diagram referred to as a "tree" may have untreelike characteristics, as illustrated in Figure 1, in which some of the branches rejoin each other.

The difference between an *issue* and a *question* is illustrated nicely by comparing the issue–position–argument–reference tree with a question–answer–evidence–reference tree, as in Figure 7. An issue is an issue only because there are two or more positions, and a question is a question only because there is only one answer; thus the question tree cannot branch before the answer level, while the issue tree must branch at the issue level.

References are indicated by their reference numbers; for example, R 001.

Issues are indicated by their issue numbers; for example, I 001.

Questions are indicated by their issue numbers followed by a (Q); for example, I 003 (Q).

	R 001	R 002	R 003	R 004	R 005	R 006	R 007
I 001	x	x					x
I 002			x	x			
I 003 (Q)				x	x	x	
I 004							x
I 005 (Q)		x					
T 001	x	x	x	x		x	x
T 002			x	x			x
T 003		x	x	x	x	x	x

This is not a square matrix with the same lists in the columns and the rows, so there is no principle diagonal and the entire matrix is necessary to convey the information shown.

Figure 6　Matrix relating issues, questions, and topics to references.

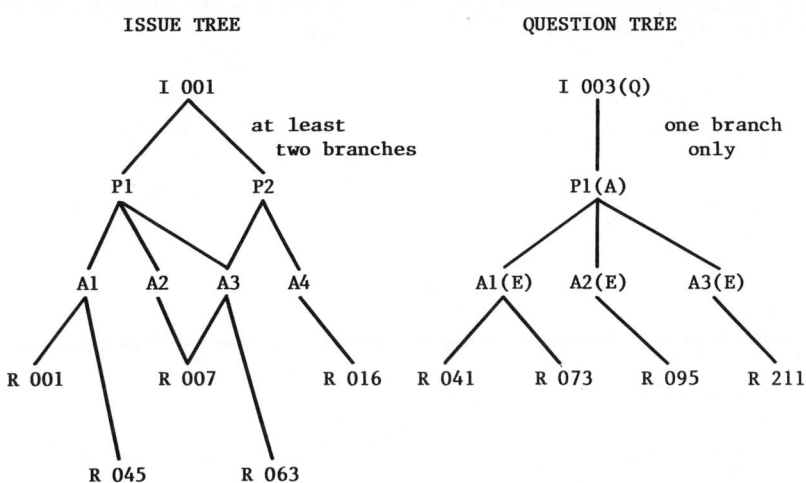

An <u>issue</u> is indicated by its issue number; for example, I 001.

A <u>question</u> is indicated by its issue number followed by a (Q).

A <u>position</u> on an issue is indicated by its position number; for example, P1.

An <u>answer</u> to a question is indicated by its position number followed by (A); for example, P1(A).

An <u>argument</u> for a position (or against a position) is indicated by its argument number followed by (E); for example, A1(E).

A statement of <u>evidence</u> in support of an answer is indicated by its argument number followed by (E); for example, A1(E).

A <u>reference</u> is indicated by its reference number; for example, R R001.

The basic distinction between the tree-representation of an <u>issue</u> and that of a <u>question</u> is the number of branches. Only one branch can leave the question indicator, and at least two branches must leave the issue indicator.

Figure 7   An issue tree compared with a question tree.

A tree relating issues/questions to issues/questions without showing positions and arguments is illustrated in Figure 8.

A tree illustrating the relations among issues/questions and also showing positions and arguments and references is illustrated in Figure 9.

Whether or not to construct a tree representing some aspect of a specific IBIS and what kind of tree to construct are questions that must refer to the information needs of particular IBIS participants. There is no point to constructing trees, tables, or matrices that are not useful to the discussion participants in a specific situation.

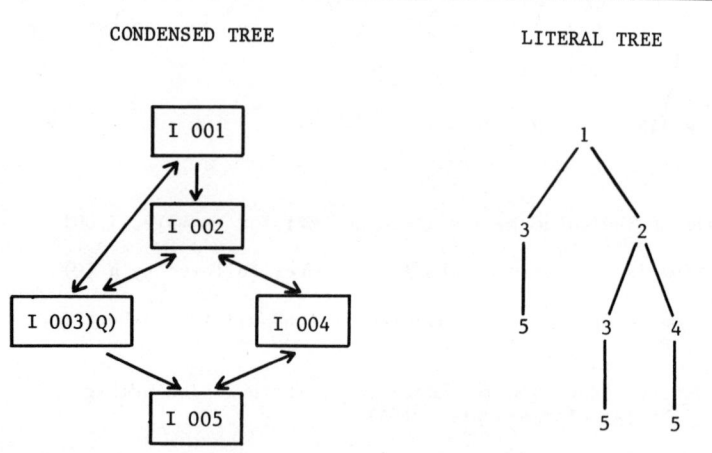

CONDENSED TREE   LITERAL TREE

A single-ended arrow ──→ indicates a predecessor/successor relationship.

A double-ended arrow ←──→ indicates a relation that is not a predecessor/successor relation.

An issue is indicated by its issue number; for example, I 001.

A question is indicated by its issue number followed by a (Q); for example, I 003(Q).

The origin of the term "tree" can be seen in the right-hand figure labelled "Literal Tree." To maintain a branching structure in the graphic representation, some points like "3" and "5" are repeated. In the condensed version, they are not repeated.

Figure 8 Tree showing issues/questions but not positions, arguments, or references.

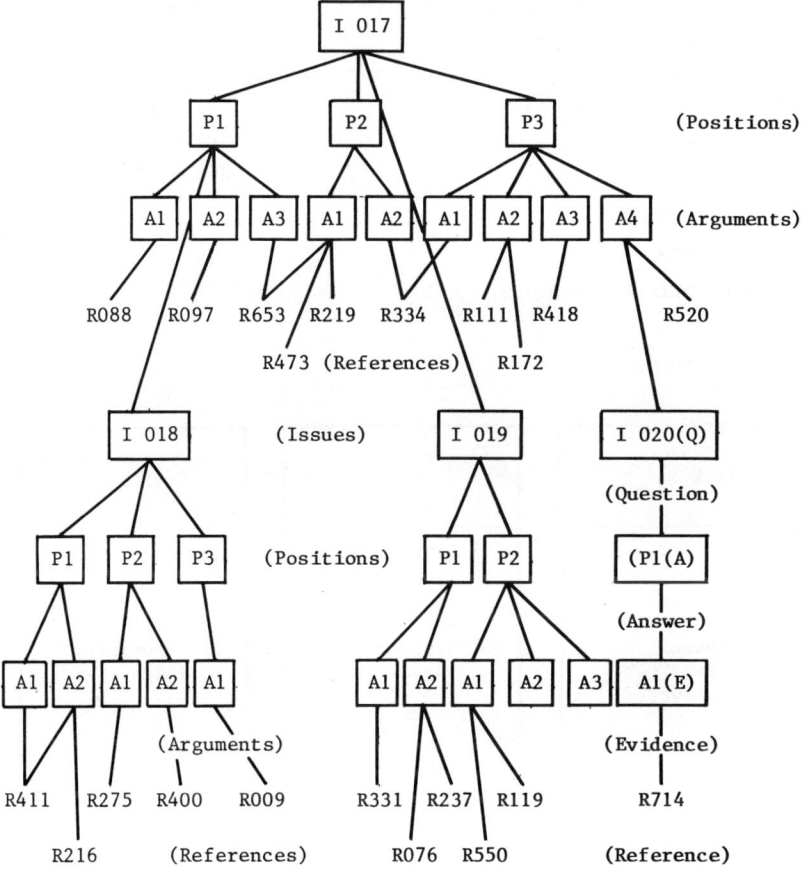

The above tree implies that I 018 is a successor to Position P1 on Issue I 017, and that I 019 is a successor to Issue I 017 regardless of position on I 017.

Another implication is that Question I 020(Q) is raised by Argument A4, Position P3, Issue I 017.

Figure 9 Tree showing issues/questions, positions/answers, arguments/evidence, and references.

Some efforts have been made to conduct IBISs, using the computer to carry out clerical tasks. The construction of trees, tables, and matrices by computer is quite easy once the data from an IBIS is recorded in computer processing format.

Rittel describes an additional tool for maintaining order among IBIS records and making such records readily accessible.[29] The additional tool is an issue map, which is intended to be a graphic display of the state of the argument. The parts of the issue map are the following:

1. Evidence bank.
2. Issue bank.
3. Documentation system.
4. Handbook.
5. List of topics.

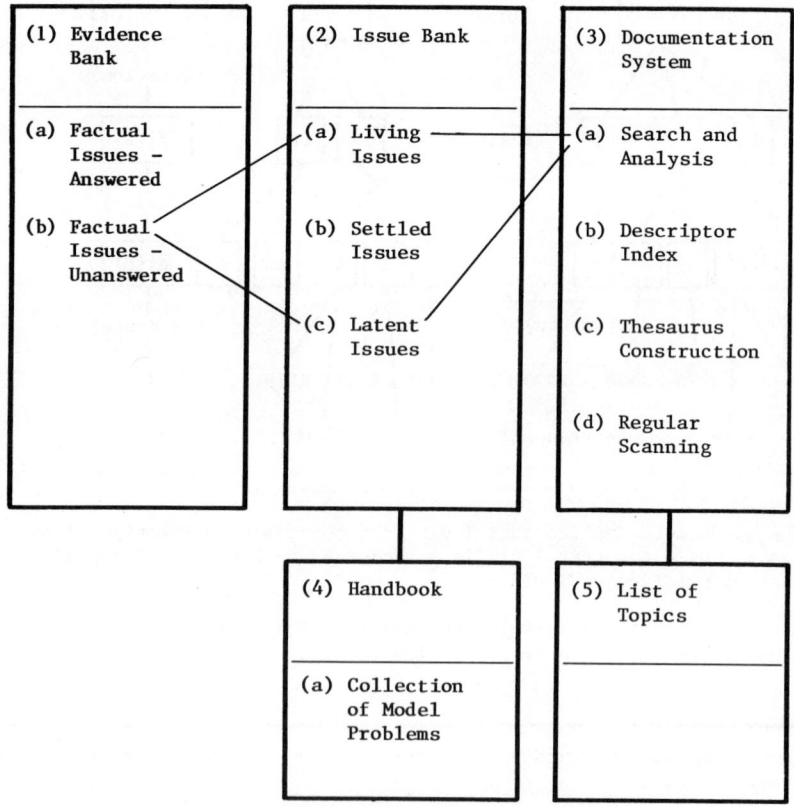

Figure 10 Issue map.

The contents of each part are as shown:

1.	Evidence bank	Factual issues—answered
		Factual issues—not answered
2.	Issue bank	Living issues
		Settled issues
		Latent issues
3.	Documentation system	Search and analysis
		Descriptor index
		Thesaurus construction
		Regular scanning
4.	Handbook	Collection of model problems
5.	List of topics	List

An outline map made up of the above parts is shown in Figure 10.

OPERATIONAL PROCEDURES

The first step in using the IBIS is to decide upon rules of procedure. This applies whether the IBIS is to be a group dicussion or an individual exercise. Subsequent steps must deal with such matters as decision rules, forms, and notation. The following list of steps is not intended as a decisive statement of the "only way" or "correct way" to implement an IBIS, but simply as one way of doing it. Every planning and design situation has its own context and characteristics and is subject to forces that should shape any methods or procedures applied to it.

Procedural rules might be some predetermined set like *Robert's Rules of Order* or an ad hoc set of rules agreed upon on the spot. Some of the things to be dealt with might be the form in which issues, positions, arguments, and references can be entered into the discussion (written or spoken, required format or no required format), the order in which entries can be made, the determination of who has the floor, how long participants may speak, and so on.

The second step is to decide upon a decision rule. One possible decision rule is that no decision be made but, rather, that the purpose of the discussion be to explore the problem and familiarize the participants with various sides of the problem. Other possible decision rules are decision by simply plurality, by simple majority, by some rule applied only when a quorum participates, or some variation thereof, like the weighting of votes according to interest, age, rank, experience, or some other basis. One approach to the weighting of votes was reported in some of the early work on the Delphi approach. Participants were asked to assign a weight to their own votes on the basis of their self-assessed expertise or qualification.[30]

The third step consists of the decision as to which forms to use. This is an important step because the forms themselves can either help or hinder in recording the course of deliberations. The basic form is the issue form. One type of two-part issue form is shown in Figures 2 and 3.

Step four has two aspects: One is the aspect dealing with the form in which the progress of the argument is to be illustrated in graphic or other form; the other aspect is that of how the parts of the IBIS are to be related to each other. The illustrative aspect of this step has been discussed in earlier sections describing tables, matrices, trees, and maps. The clearest form of illustration is in the form of trees, which is also the form requiring the greatest expenditure of time and effort. Matrices are more compact and economical of time and effort, but they are less informative upon casual inspection. A more fundamental aspect of this step is that dealing with how parts of the discussion are to be related to each other. This aspect was discussed among Rittel and colleagues in a project dealing with the analysis of scientific and technical information processes among the European Communities. One method of relating parts of an IBIS to each other is to make a notation on a sheet of paper for each of the parts to be related. Another method is to note relations in a matrix instead of on each individual sheet. For purposes of illustration assume that a discussion has been going on for some time and that issue number I 063 has been reached. It is decided that issue I 063 relates to several previous issues, each of which has been discussed and recorded on its own issue sheet (see Figure 2 for an example of an issue sheet). The sheets for the previous issues have been copied, and one copy of each issue sheet has been distributed to each of five discussion participants. The decisions about the relations of I 063 are the following:

1. Issue I 021 is a predecessor of I 063.
2. Issue I 037 is related to issue I 063.
3. Issue I 051 is related to issue I 063.

If it is decided to note relations by reciprocal marking, that means that for the current issue, I 063 (which has presumably not been photocopied and distributed yet), it is only necessary to list issues I 021, I 037, and I 051 on the original I 063 issue sheet, properly identified as predecessor issues and related issues. However, the reciprocal notations are not so simple: *I 063* must be properly noted and identified on the original issue sheets for I 021, I 037 and I 051, *and* on each of the five copies that have been distributed to the participants. Also, if the IBIS is being recorded on microfilm or microfiche, then the previously filmed sheets are now obsolete because they do not have the newly decided-upon relations entered upon them. There are ways, of course, of entering the necessary additions on the earlier sheets—distributing new sets of copies each month, periodically announcing necessary additions to all the participants so that they can make the alterations on their own copies, and so on. However, a more sensible approach seems to be the use of matrices without reciprocal notation. In this approach the relations to issues I 021, I 037, and I 051 would be entered on the current sheet as they arose; but instead of going back and altering the issue sheets for I 021, I 037, and I 051 to indicate their new decided-upon relations to issue I 063, the relations would be entered in an overall *issue × issue matrix*, as illustrated in Figure 4. The amount of work involved in the two approaches is as follows:

Reciprocal Notation	Matrix Notation
Relations entered on—	Relations entered on—
I 063	I 063
I 021 and copies	Matrix and copies
I 037 and copies	
I 051 and copies	

The approach labeled *matrix notation* above carries the same information content but with less wasteful busy work.

Step five is to decide upon notation. This step deals with the question of which forms of notation to use. One form is illustrated here; several others have been used as well.

A SAMPLE NOTATION SYSTEM

Some prefixes for the various elements of an IBIS are the following:

Issues: I
Positions: P
Arguments: A
References: R

Topics: T

Questions: I (Q)

Answers: P (A)

Evidence: A (E)

Some suffixes for issue classification, if desired, are the following:

Deontic issue: I (d)

Factual-descriptive: I (f-d)

Factual-causal: I (f-c)

Factual-instrumental: (I (f-i)

Conceptual: I (c)

Issue Number 1 in this sample scheme for notation is *I 001*. The notation for Issue Number 273 is *I 273*. The notation for Position 1 on Issue 1 is

I 001-P 1

The notation for Position 11 on Issue 37 is

I 037-P 11

A convention for numbering the positions on issues that call for yes or no-type positions is

P 0: Inadequate, irrelevant, or absurd issue or question
P 1: Yes
P 2: No

Not all issues call forth yes/no-type positions. Especially causal and instrumental issues, but conceivably others as well, might call forth a list of positions, each of which is a described object or event or situation, in which case the positions are simply numbered consecutively as they are entered, starting with *P 1*.

The notation for Argument 1 on Position 1 on Issue 1 is

I 001-P 1/A 1

In the notation scheme being illustrated no note is made of relations of "for," "against," or "related to." Some attempts have been made in past IBIS applications to incorporate these relations. One approach is the following:

I 001-P 1/A 1(+) indicates that A 1 is "for" (that is, in favor of) Position P 1.

I 001-P 1/A 2(−) indicates that A 2 is "against" (that is, in opposition to) Position P 1.

I 001-P 1/A 3(r) indicates that A 3 is "related to," but not necessarily "for" or "against" Position P 1.

A more compact notation for several arguments with respect to the same position is

$$\text{I 001-P 1/A 1(+), A 2(−), A 3(r)}$$

Arguments are numbered starting with number A 1 for an *issue,* regardless of *position.* Thus position P 1 may give rise to arguments A 1, A 2, A 3, and A 4, then position P 2 may give rise to A 5, A 6, and A 7, and position P 3 may relate to the already stated arguments A 2 and A 5 and give rise to new ones A 8 and A 9. A single argument may relate to so large a number of positions that it is desirable to number arguments consecutively by *issue* rather than by *position.* Several earlier IBISs were carried out numbering arguments by position, giving rise to problems of having two or more notations for the same argument, corresponding to the two or more positions to which it related. It seems simpler to number the arguments from A 1 and to not assign several notations to the same argument. Thus, if Arguments A 37 applies to both positions P 1 and P 2 on issue I 009, the notations would be

$$\text{I 009-P 1/A 37}$$
$$\text{I 009-P 2/A 37}$$

Since a question is only a special case of an issue, the notation for a question is

$$\text{I 013(Q)}$$

The notation for the answer to question I 013(Q) is

$$\text{I 013(Q)-P 1(A)}$$

The notation for articles of evidence 1 and 2 supporting the answer to question I 013(Q) is

$$\text{I 013(Q)-P 1(A)/A 1(E), A 2(E)}$$

The notation for Reference Number 273 is R 273. References are numbered consecutively as they set forth and are not related to issues, positions, or arguments in their numbering. The notation for Topic 1 is T 001. Topics are numbered consecutively and not related to issues or other elements in the way in which they are numbered.

There are many other notational schemes that might be used instead of the one described above. The use of an abbreviated notation scheme could in fact be avoided altogether if desired. Instead of using an abbreviation like $I\ 001$-$P\ 1/A\ 1(+)$, for example, one could simply write, "Argument 1 in favor of Position 1 on Issue 1," or some variation thereof; however, over the long haul, the abbreviated notation scheme seems to be worthwhile as a means of saving time and effort.

An IBIS discussion, like any discussion, might start with an overall topic from which specific issues are generated, or it might start with one specific issue before any general topics suggest themselves. Step six is to decide whether or not a general topic statement is to be developed as the initial step in the discussion or whether, instead, the discussion is to begin with the statement of a specific issue from which other issues will arise and from which general topics may or may not later appear as groupings of issues.

The initial issue is stated and then discussed and edited into the form in which it is to be entered. A brief statement of the context in which the issue occurs is made if desired.

Participants state their positions on the initial issue. If only one position is taken among all participants, then the issue becomes a question.

If there are two or more positions, then the participants state arguments in favor of, in opposition to, and related to the positions on the original issue, cite references in support of the arguments on the original issue, and exercise the agreed-upon decision rule on the initial issue. The agreed-upon decision rule may be that no decision is to be made, that decision is not mandatory, or that decision may be postponed or avoided by moving on to successor issues. In some cases a decision may be unavoidable even though the participants might feel that they have insufficient knowledge.

The next step is to construct a tree or matrix, or both, illustrating the structure of the argument over the initial issue. This step is optional but desirable for purposes of clarity. The system then proceeds to the recurring step of starting and editing the additional issues to be discussed, with their positions, arguments, and references, and exercising the decision rule. This is followed by the construction or modification of tables, matrices, maps, and trees, as desired.

A further recurring step is to decide whether or not there should be a time limit on the discussion of each issue as it comes up.

One basis upon which time limits might be set is the judged relative importances of the issues at hand. Relative importance weights might be deliberated in order to allocate time and perhaps even research effort in order to avoid wasting a scarce resource (the group's or individual's time) on relatively unimportant issues.

The most fundamental step of all is to decide who is to participate and/

or be represented in the discussion and who decides, and so on through the parallel-mirrors dilemma. The basic issue of who participates in decisions is far beyond the scope of a "how to do it" description of a structure for argument.

However, it may be worthwhile to suggest that it is an issue that should be discussed by the participants in a planning or design IBIS, however they themselves happened to become participants, and to further suggest that this basic issue should be discussed early in the procedure and then probably discussed over and over again at intervals throughout the duration of the procedure.

The duration of an IBIS may be several hours, several months, or the life of the system may be of indefinite duration. Mann specifies a duration of several hours to several months for the game format IBIS he proposes.[31] The UMPLIS system in West Germany is intended to be a national-scale IBIS of indefinite duration, operated by the Federal Ministry of the Interior (Bundesministerium des Innerns) for both governmental and public use.

Mann provides a very specific set of operational procedures for the IBIS in a gaming format.[32] Mann suggests several alternative ways of initiating discussion, including the introduction of a problem topic, some controversial solution proposals, or a set of issues. Special decision sessions are held to dispose of issues after debate has taken place. Mann's procedure is intended to structure the activities of raising questions and issues, contributing answers and arguments, proposing solutions and plans, deciding, evaluating, and setting priorities.[33] Dehlinger and Protzen state that the operations procedures of an IBIS should encourage and even generate as many conflicts as possible.[34] Swanson, in a development methodology for the IBIS for use in business firms, proposes a direction committee that would develop a list of source issues to seed further debate.[35] The organization implementing the IBIS would then grow to include issue-addressing teams, problem experts, information scouts, chroniclers, a system-building team and other roles.

GUIDELINES FOR THE USE OF THE IBIS

Guidelines and cautions appropriate to implementing an IBIS-structured discussion are contained in this section. Some comments that qualify as guidelines have already been made in the preceding section outlining a sample step-by-step procedure for using the IBIS.

The nature of the IBIS is such that it encourages thoroughness in problem coverage. A danger is that it might encourage being too thorough.

The essence of this guideline is, "Do not get carried away with being thorough for the sake of thoroughness." The IBIS structure points out the fact that there are often many positions, arguments, and references relevant to a specific issue. However, not all such elements are equally relevant or important, and the relevance of one item, like a reference, for example, might depend on what items (other references) have gone before. Often the chief benefit of an IBIS has been gained simply by pointing out that many branches exist, without necessarily exploring every branch in detail. Time, money, and patience are all scarce resources. Some specific implications of this guideline follow.

An issue is under discussion; positions have been taken. Immediately, 47 arguments come to mind, all in support of Position 7. However, everybody participating in the discussion is persuaded by the first of the 47 arguments, and the position is adopted. Do not stop the discussion to continue on through the 46 additional arguments; it might be worthwhile to later record them for future reference, but the group deliberations should continue on to the next issue to be argued and resolved.

An argument is offered in support of (or in opposition to) a position. Two references come to mind, R 1 and R 2. However, Once R 1 is cited, R 2 is more or less redundant, or vice versa. Move on in the discussion and list the redundant one later, if you have time. The same principle applies to arguments, and even to positions in some cases, for example, the case of an instrumental issue in which the positions are a list of possible means, some of which are closely related.

Great arguing would do well to emulate one of the virtues of great art and great engineering: economy of means. Look for brief, decisive arguments, conclusive citations, and concise statements.

An issue has been explored, positions, arguments, and citations have been listed, and a tree has been sketched for the issue and its related elements. It seems like time to take up an urgent successor issue, but the sketch of the issue tree reveals a lot of branches that have not been filled in and some that look as if there might be further branches. Let them hang and move on and fill in the loose ends later for purposes of future reference.

Helmer lists the avoidance of specious or irrelevant persuasion as being one of the key reasons for using the Delphi procedure.[36] The following are some of the specious reasons for which arguments or citations might be persuasive:

1. The person who sets them forth is a good speaker, has charisma, or has authority or prestige.
2. The majority seems to agree, producing a bandwagon effect.

3. The majority seems to agree, producing an antibandwagon effect.
4. Somebody is afraid that if he supports a novel position or argument, he will look like a fool or will drop in the estimation of others.
5. Somebody is afraid to take a position or offer an argument that will contradict his earlier statements.

One way of dealing with the problems of specious persuasion is to make the incoming statements anonymous and to provide feedback, as in the Delphi approach. An interesting prospect is to implement an IBIS-structured discussion in a Delphi-like format.

Some conflict occurs because people have different interests or opinions and some occurs because people seek out and engage in conflict for its own sake.[37,38] The aim of using argumentative approaches to design, planning, and policy making is to derive some of the benefits that the former type of conflict can yield and to avoid the latter, which has a much more dubious potential for useful results. The latter phenomenon might be referred to as the "king of the hill" syndrome. Rubin and Brown describe three aspects of the social structure of bargaining situations that influence the behavior of bargainers in conflict.[39] One social aspect of bargaining and negotiation is the presence or absence of an audience. The audience may be either physically present of present only in the sense that they are informed of the activities that go on in the situation. The audience might be the constituents of the bargainers, for example, a group that is being represented by the bargainers, or disinterested nonconstituents. The mere existence of an audience seems to encourage participants in bargaining conflicts to behave in a way that gains the approval of the audience, and the more important the audience is to the participants, the stronger this effect. The bargainers' desires to look good in the eyes of an audience will not necessarily produce more constructive behavior; it might produce more hostile and aggressive behavior that is not related to the issues being discussed, if hostile and aggressive behavior is what the bargainer thinks will win audience approval.

The second social aspect of bargaining and negotiation described by Rubin and Brown is the presence of third parties.[40] Third parties might be either formal or informal. Examples of formal third parties are ombudsmen, marriage counselors, and judges. Formal third parties tend to exert an influence towards agreement. Some of the useful functions of third parties are to suggest new alternatives and propose compromises and concessions that might be acceptable to the participants but which they would be reluctant to set forth for fear of looking weak. Third party influences might be exerted through their specific acts or simply on the basis of the knowledge that they will be involved, especially in cases in which the third party is possessed of power, prestige, or a charismatic personality. This, of course, gives rise to a dilemma of sorts insofar as it introduces a form of specious persuasion: an influence towards agreement based on the position or personality of a third party that is unrelated to the actual issues and conflicts of interest at hand. This particular aspect of third party influence tends to hold down conflict for its own sake, but for specious reasons, and may also hold down useful or necessary conflict over substantial issues. Rubin and Brown[41] claim that the pressure of third parties is toward the following:

1. Fairness.
2. Social responsibility.
3. Reciprocity.
4. New alternatives.
5. The reduction of irrational behavior.
6. The regulation of outside intervention.
7. The regulation of the costs of conflict.
8. The provision of means of face saving for graceful retreats and concessions.
9. The provision of informal communications channels.

This largely positive view of the reduction of conflict is characteristic of American social and behavioral science. An alternative point of view expressed by Coser[42] and Mills[43] is that conflict is often preferable to the reduction of conflict for specious reasons. Mills and Coser both criticize American Middle-class social science for automatically assuming that conflict is naughty and for failing to study the positive and functional aspects of conflict and the resolution (rather than avoidance) of conflict.

The third social aspect of bargaining and negotiation listed by Rubin and Brown[44] is the number of parties involved. The greater the number of parties, the more complex, harder to coordinate, and prone to multiple conflicts is the situation. There is also a frequent and quite sensible tendency in three party conflicts for the two weaker parties to form a coalition to counterbalance the power of the stronger party. This is a sensible course of action, but it may have little to do with the substance of the issues to be resolved.

A related caution that has already been discussed in the preceding section refers to the work of Simmel,[45] Coser,[46] and Mills.[47]

Rubin and Brown point out several ways in which the physical setting of a bargaining and negotiation situation may influence the behavior of the participants.[48] One example is that bargaining on one's own territory seems to be an important source of strength. Another example is that the physical arrangements may be important to the participants as indicators of relative power or status. Many people probably learned more than they ever wanted to know about this aspect of bargaining during the protracted discussions about physical arrangements prior to peace negotiations at Panmunjon in the 1950s and in Paris in the 1970s.

The physical arrangement of things such as seating influences behavior and the climate of interchanges and might tend to increase or decrease the intensity of conflict for irrelevant reasons. Many studies of such influences of arrangement on behavior have been reported by Festinger, Schachter, and Back,[49] 1950, Caplow and Forman,[50] 1950, Whyte,[51] 1956, Hall,[52,53] 1959 and 1966, and Sommer,[54] 1969.

In his study of the IBIS in a gaming format Mann stresses the need for physical features such as display walls for posting issues or, alternatively, a long table where issues can be laid out.[55] Other needed physical features called for are writing surfaces for all player-participants, prepared forms, display space for rules and the state of the game, and the decision file. Issue maps or networks might also be worthwhile displays.

Rubin and Brown point out that issues themselves are variable components of bargaining and negotiation situations, just as the social and physical characteristics of the situations are variables.[56] Issues can be manipulated toward being more general or more specific, toward being combined with other issues or divided into several smaller issues, or by means of different wordings with different connotations and undertones. Adding new issues to a discussion might open the way to agreements that were impossible before, or, for that matter, do quite the opposite.

Intangible issues may arise during the course of bargaining about tangible issues. Intangible issues are sometimes related to a feeling of loss of face and are sometimes attempts to obstruct agreement and interchange. Tangible issues, with tangible gains and losses associated with them, are likely to be easier bases for negotiation and compromise than intangible issues.

A "zero-sum" situation is one in which no party can gain except as a result of another party's loss. "Non-zero-sum" situations are those in which the process of interaction either increases or decreases the total body of resources distributed over the participants. The production of goods from raw materials is generally viewed as a positive-sum activity, in which the value of the goods produced exceeds the value of the resources consumed in production, including labor. War is generally a negative-sum situation in which the total pool of resources is decreased by the activity. A zero-sum situation is one in which one side can gain only to the degree that they make the other side lose, and it is likely to make agreement difficult and conflict intense.

Issue situations should be made as explicit as possible. The number of issues to be discussed, the sequence of discussion, and the format and presentation of the issues should be explicitly set forth. The alternatives being considered should also be explicitly illustrated as possible.

It is sometimes useful to consider issues or topics within the framework of model problems or model queries. Model queries, expressed perhaps in the form of model positions and arguments imposed on each issue, or even in the form of model issues imposed on each topic, might include questioning in one of the following forms:

1. The Kantian imperative as a query source.
2. Opportunity seeking as a query source.
3. An entropy query.
4. Analysis of the problem landscape in terms of being what Boulding characterizes as conflictual, benign, or malign in nature.
5. An indifference query.

An oversimplified statement of the Kantian imperative is that every action should be scrutinized as though it were to become the universal model for further actions. There are at least two ways in which this attitude might be expressed in the form of model responses in an IBIS. One approach is to develop positions based on the assumption that the positions are to be viewed as universal models. Another approach is to develop arguments based on the consequences that might be expected if each position were taken as a universal model. These two approaches generate two model queries.

MODEL QUERY ONE

Model Query One is, "Is there a position on each issue that might be viewed as a universal model?"

In the case of a deontic issue with the form, "Should there be . . . ?" the positions take the forms, "Yes," "No," and "Inadequate question," plus the possible addition of variations such as "Yes, but . . . ," and "No, but. . . ." Model Query One applied to a deontic issue is a matter of inquiring as to the appropriateness for universal application of one of the small number of positions.

In the case of factual-descriptive, factual-causal, and factual-instrumental issues the positions to be taken are likely to be lists, sometimes long lists, of different answers. Taking an instrumental issue as an example, the form of the issue might be, "How can we achieve situation X?" The positions on such an instrumental issue might be a long list of the form, "P 1: by doing A; P 2: by doing B; P 3: by doing C," and so on. The application of Model Query One to this situation would be to seek out a position that held promise of being a suitable model for universalization.

MODEL QUERY TWO

Model Query Two is, "Is there an argument on each position appropriate to the prospect of universalizing that position?"

The application of Model Query Two is a matter of generating arguments based on the predicted consequences of universalizing each position on each issue during an IBIS. For example, for the position "P 1: We can achieve X by doing A," an argument in support of P 1 might be that if everybody did A, or if A were to become a universal principle, then everybody would benefit, resources would be conserved, or the number of undesirable situations would be reduced. Conversely, an argument against a position might be the form, "If everybody did A, then the following undesirable consequences would follow."

IMPLEMENTATION TECHNIQUES

Pencil-and-paper implementation of an IBIS by a group might begin with an informal argument among the group members. Much of the argument generated might be irrelevant, with many impressionistic comments, leading to a discussion about which aspects of the informal argument should be recorded, that is, which relevance filters to apply. Recording of the selectively filtered material from the argument would be undertaken by individual participants after the group meeting, with the later addition of arguments and references. The key to the success of this approach could be an intelligent reduction of data at the end of each day.

One approach to reducing the burden of paperwork in order to facilitate argument would be to have a stenographer take down everything that was said. The court reporter comes to mind as an example of this approach in action. The recorder or stenographer would then type up the record of the discussion for editing by participants.

A second approach to recording with a minimum burden of paperwork would be to conduct the IBIS through the medium of computer conferencing instead of as a face-to-face meeting in real time. Computer conferencing is described in Linstone and Turoff,[57] Vallee, Johansen, and Spangler,[58] and Turoff.[59]

An approach to indexing the record of an IBIS for future access is automated indexing through the use of machine-readable text and an indexing program. This concept has occupied information scientists in recent years but to date has not been implemented on a large scale. There are some basic problems that must be solved before automated indexing becomes a usable tool. Once the record of an IBIS is indexed in key-word form, there are many approaches to accessing that are usable in currently available forms.

A provocative prospect for recording is the automated transcription of the voices of the participants in an IBIS for full-text printing, for purposes of editing and review by the participants. This is technologically but not economically feasible at this time.

Hans Dehlinger in Stuttgart has worked on computer programs for IBIS implementation.

Kunz and Rittel suggest an issue bank, an evidence bank, a collection of model problems, a topic list, an issue map, and a detailed documentation system as parts of an IBIS, and state that implementation might be computer supported.[60]

RELATED CONCEPTS

Several relations between the IBIS approach and the sociological study of conflict have already been described in the section on guidelines for the use of the IBIS. Of particular interest is the theoretical work of Georg Simmel[61] and the later, related work of Lewis Coser[62] and C. Wright Mills.[63] Work on bargaining and negotiation has been collected and analyzed in Rubin and Brown.[64]

Boulding has developed several aspects of the economic analysis of conflict situations.[65] Some of Boulding's work was described in the preceding section on guidelines for the use of the IBIS.

The Delphi technique has been referred to several times as a concept related to the IBIS and also as a possible format for the implementation of the IBIS. There is a very large body of literature on the Delphi approach. A positive description and a number of applications are contained in Linstone and Turoff.[66] A negative view of the Delphi technique is contained in Sackman.[67] Three substantial bibliographies are contained in Linstone and Turoff,[68] Sackman,[69] and Hudson.[70]

C. West Churchman describes several different inquiring systems, including a basically rationalist approach, a basically empiricist approach, and two synthesizing approaches.[71] These are the Leibnizian, Lockean, Kantian, and Hegelian inquiring systems, respectively. The underlying assumptions of the IBIS relate to the Kantian and Hegelian approaches, especially the latter. There is some justification in regarding the IBIS as a form of a Hegelian, dialectic inquiring system.

Churchman describes a dialectic approach to planning that he refers to as "counterplanning."[72] This is a dialectic approach in which a plan stimulates the generation of a contradictory, opposing approach referred to as the counterplan. The objective is to synthesize a third plan based on what was learned in developing and arguing the plan and counterplan.

Geoffrey Broadbent (informal discussions, 1973) describes an approach to architectural design that relates in part to counterplanning and to the concept of the dialectic or Hegelian inquiring system. It is an application of Popper's "conjecture and refutation" paradigm to architectural design and involves the generation of a design in the spirit of setting forth a controversial and information-laden hypothesis that invites attempts at refutation, perhaps in the form of counterplans, perhaps only in the form of negative arguments directed at the design itself. Partial refutation of the original design would result in modification of the design to eliminate the refuted features. Complete refutation would result in the generation of a wholly new plan (or counterplan, in some instances), which would in turn itself be subjected to attempted refutation. A design that was not successfully refuted in any way would presumably be adopted.

EXAMPLES OF IMPLEMENTATION

The applications of the IBIS approach during the first eight years of its existence as a concept have included educational planning, housing, environmental planning on a national scale, and curriculum planning.

Dehlinger and Protzen describe an educational planning application entitled "University Planning System HKP (Hochschule Kapazitäts Planning)" carried out in the German Federal Republic.[73] Dehlinger and Protzen also describe the application of the system to housing planning in the Martin Luther King Development Corporation Planning Information System in Miami, Florida, and lay out the general guidelines for an environmental planning information system on the IBIS model.

The largest scale application of the system to environmental planning is the UMPLIS (Umweltplanunginformationssystem), which has been under development for several years for the German Federal Ministry of the Interior by the Studiengruppe für Systemforschung in Heidelberg. Descriptive literature on the UMPLIS system and other early IBIS applications is listed in an annotated bibliography by Grant.[74]

The IBIS concept has been applied to the problem of the generation and flow of scientific and technical information among the 10 countries of the European Community in a project of several years' duration commissioned by the European Community and carried out by the Studiengruppe für Systemforschung. The basic model for the European Community STIEC (Scientific and Technical Information in the European Community) system is outlined in Grant.[75]

The IBIS approach was used in curriculum planning sessions for architectural education at the California Polytechnic State University, San Luis Obispo, in an attempt to reconcile widely divergent and intensely conflicting education objectives in a very large faculty, and is under consideration as a long-term planning method for the same campus.

The applications listed here were the source of the limitations of the system defined earlier in this chapter, as well as of the guidelines for implementation that were described here in some detail.

REFERENCES

1. E. Burton Swanson, "A Methodology for IBIS Data Gathering and Development," *Design Methods and Theories,* Vol. 11, No. 4, Oct.–Dec. 1977, pp. 256–261.

2. Olaf Helmer, *Social Technology,* Basic Books, New York, 1966.

3. Harold A. Linstone and Murray Turoff, Eds., *The Delphi Method: Techniques and Applications,* Addison-Wesley, Reading, Mass., 1975.

4. C. West Churchman, *The Design of Inquiring Systems,* Basic Books, New York, 1971.

5. Horst W. J. Rittel, Lectures on the IBIS, Architecture 130, Department of Architecture, University of California, Berkeley, Calif., Spring 1972, from notes by Diana Lee-Smith.

6. Horst W. J. Rittel, "The State of the Art in Design Methods," *DMG–DRS Journal: Design Research and Methods,* Vol. 7, No. 2, Apr.–June 1973, pp. 143–148.

7. Horst W. J. Rittel and Melvin M. Webber, "Dilemmas of a General Theory of Planning," *DMG–DRS Journal: Design Research and Methods,* Vol. 8, No. 1, Jan.–Mar. 1974, pp. 31–39.

8. Swanson, op. cit.

9. Hans Dehlinger and Jean-Pierre Protzen, "Some Considerations for the Design of Issue-Based Information Systems (IBIS)," *DMG–DRS Journal: Design Research and Methods,* Vol. 6, No. 2, Apr.–June 1972, pp. 38–45.

10. Donald P. Grant, "A Compromise-Generating Planning Information System for Public Housing Location Conflicts," *The DMG Fifth Anniversary Report: DMG Occasional Paper Number One,* The Design Methods Group, Department of Architecture, University of California, Berkeley, Calif., January 1972, pp. 23–28.

11. Thorbjoern Mann, "A Design Game Based on the Argumentative Model of Design," *DMG–DRS Journal: Design Research and Methods,* Vol. 6, No. 4, Oct.–Dec. 1974, pp. 161–165.

12. E. Burton Swanson, *A Sketch for an Executive Information System,* Center for Information Studies, Graduate School of Management, University of California at Los Angeles, Los Angeles, Calif. 1975, Information Systems Working Paper 17–75.

13. Donald P. Grant, *The UMPLIS and IBIS Approaches to Environmental Planning*

Information Systems: An Annotated Bibliography, Council of Planning Librarians, Exchange Bibliography Number 1000, 27 pp., POB 229, Monticello, Ill. 61856.

14. Dehlinger and Protzen, op. cit.

15. Werner Kunz and Horst W. J. Rittel, "Issues as Elements of Information Systems," *The DMG Fifth Anniversary Report; DMG Occasional Paper Number One,* The Design Methods Group, Department of Architecture, University of California, Berkeley, Calif., January 1972, pp. 13–15.

16. Donald P. Grant, "How to Use the IBIS as a Procedure for Deliberation and Argument in Environmental Design and Planning," *Design Methods and Theories,* Vol. 11, No. 4, Oct.–Dec. 1977, pp. 185–220.

17. Kunz and Rittel, op. cit.

18. Mann, op. cit.

19. Kunz and Rittel, op. cit.

20. Basil Honikman, "Introduction," *Responding to Social Change,* Dowden, Hutchinson, and Ross, Stroudsberg, Pa., 1975.

21. Kenneth Boulding, *The Image,* University of Michigan Press, Ann Arbor, Mich., 1956.

22. C. West Churchman, *The Systems Approach,* Dell Books, New York, 1969.

23. C. West Churchman, *A Non-Technical Introduction to Operations Research,* Fybates lecture notes, Berkeley, Calif., 1969.

24. Dehlinger and Protzen, op. cit.

25. Dehlinger and Protzen, op. cit.

26. Horst W. J. Rittel, Lectures on IBIS, Architecture 130, Department of Architecture, University of California, Berkeley, Calif., Spring 1971, from notes by Donald Grant.

27. Ibid.

28. Dehlinger and Protzen, op. cit.

29. Horst W. J. Rittel, Lectures on the IBIS, Architecture 130, Department of Architecture, University of California, Berkeley, Calif., Spring 1972, from notes by Diana Lee-Smith.

30. Helmer, op. cit.

31. Mann, op. cit.

32. Mann, op. cit.

33. Mann, op. cit., p. 163.

34. Dehlinger and Protzen, op. cit.

35. E. Burton Swanson, "A Methodology for IBIS Data Gathering and Development," *Design Methods and Theories,* Vol. 11, No. 4, Oct.–Dec. 1977, pp. 256–261.

36. Helmer, op. cit.

37. Georg Simmel, *Conflict,* The Free Press, Glencoe, Ill., 1955 (first published in German, 1908).

38. Lewis Coser, *The Functions of Social Conflict,* The Free Press, New York, 1956.

39. Jeffrey Z. Rubin and Bert R. Brown, *The Social Psychology of Bargaining and Negotiation,* Academic Press, New York, xii.

40. Ibid.

41. Ibid.

42. Coser, op. cit.

43. Mills, op. cit.

44. Rubin and Brown, op. cit.

45. Simmel, op. cit.
46. Coser, op. cit.
47. Mills, op. cit.
48. Rubin and Brown, op. cit.
49. Leon Festinger, Stanley Schachter, and Kurt Back, *Social Pressures in Informal Groups: A Study of Human Factors in Housing,* Stanford University Press, Stanford, Calif., 1950.
50. T. Caplow and R. Forman, "Neighborhood Interaction in a Homogeneous Community," *American Sociological Review,* Vol. 15, 1950, pp. 357–366.
51. William H. Whyte, *The Organization Man,* Simon & Schuster, New York, 1956 (especially Chapter 25).
52. Edward T. Hall, *The Silent Language,* Doubleday, New York, 1959.
53. Edward T. Hall, *The Hidden Dimension,* Doubleday, New York, 1966.
54. Robert Sommer, *Personal Space,* Prentice-Hall, Englewood Cliffs, N.J., 1969.
55. Mann. op. cit.
56. Rubin and Brown, op. cit.
57. Linstone and Turoff, op. cit.
58. Jacques Vallee, Robert Johansen, and Kathleen Spangler, "The Computer Conference," *Futurist,* Vol. 9, No. 3, June 1975, pp. 116–121.
59. Murray Turoff, "The Future of Computer Conferencing," *Futurist,* Vol. 9, No. 4, August 1975, pp. 182–190, 195.
60. Kunz and Rittel, op. cit.
61. Simmel, op. cit.
62. Coser, op. cit.
63. Mills, op. cit.
64. Rubin and Brown, op. cit.
65. Kenneth Boulding, "Conflict Management as a Learning Process," in A. de Reuck and J. Knight, Eds., *Conflict in Society,* Churchill, London, 1966, pp. 236–258.
66. Linstone and Turoff, op. cit.
67. Harold Sackman, *Delphi Critique,* Lexington Books, Lexington, Mass., 1975.
68. Linstone and Turoff, op. cit.
69. Sackman, op. cit.
70. Ivan Hudson, *A Bibliography on the "Delphi Technique,"* Council of Planning Librarians, Exchange Bibliography, Number 652, September 1974, 17 pp., POB 229, Monticello, Ill. 61856.
71. C. West Churchman, *The Design of Inquiring Systems,* Basic Books, New York, 1971.
72. C. West Churchman, A Non-technical Introduction to Operations Research, Fybates lecture notes, Berkeley, California, 1969.
73. Dehlinger and Protzen, op. cit.
74. Donald P. Grant, *The UMPLIS and IBIS Approaches to Environmental Planning Information Systems: An Annotated Bibliography,* Council of Planning Librarians, Exchange Bibliography, Number 1000, POB 229, Monticello, Ill., 61856.
75. Donald P. Grant, "How to Operate an IBIS in Support of a Research and Development Project," *Design Methods and Theories,* Vol. 11, No. 4, Oct.–Dec. 1977, pp. 221–255.

CHAPTER 7

KANE SIMULATION (KSIM)

JULIUS KANE
PAUL FITZGERALD

KSIM (Kane SIMulation) is a language/process that permits nontechnical users to work with sophisticated mathematical tools without the need for becoming technically proficient in prolix methodologies. KSIM (pronounced kay-sim) does for systems analysis what FORTRAN or BASIC does for computer programming. It is far simpler to use than, say, Forrester's DYNAMO. KSIM also has the advantage that it can utilize subjective or emotional data as freely as objective or numerical data. The purpose of KSIM is to give a user group a "hands-on" model of a decision-making situation in which the members themselves have a significant personal stake. For best results KSIM requires deep emotional involvement in the definition, analysis, and conclusion(s) of a problem.

Although a mathematical process, KSIM is based upon linguistic considerations and experiences. Scenarios do not have much meaning unless there is a "willing suspension of disbelief." Participation requires plausibility, and this concept is constructed by an entirely different process than formal validation procedures. A true statement and a plausible statement are not the same thing. While a person may know what is true, he or she generally will *act* upon that which is plausible. KSIM recognizes the difference between truth and plausibility or knowledge and belief, and it tries to converge these differing perspectives into a coherent and consistent structure. Where there are differences, or divergence, this gives the user group a clear signal that all is not well and that either more information or a reevaluation of the formulation of the problem is required.

KSIM has been primarily designed for the pragmatic and skeptical user. Imbedded within its postulates is the assumption that world variables interact with each other with nonlinear feedback coupling and have coupling coefficients (variables) that are often state dependent. KSIM has three stages of complexity. At its simplest it is a signed digraph, emphasizing "level" interactions. At its second stage it recognizes "derivative" coupling. At the third and last stage the couplings are arbitrary functions. At this last stage KSIM is as general a systems language as can be desired.

KSIM is a general cross-impact language that shares many features with other processes described in this volume. Its uniqueness is that it has been designed with primary reference to psychological barriers and the essential constraints of small-group interactions. It assumes that in any group of concerned people there is more information than can be articulated by any one individual. In practice, it sets out a series of procedures that encourage group interaction and communication and greatly enhances such exchanges. To a great degree KSIM is more an encounter session or a form of spontaneous theatre than a mathematical methodology. However, the output of KSIM can be refined to any degree of validity or rigor desired.

At the user's option KSIM can make full use of the largest computer systems, exploiting graphic displays and other presentations. On the other hand, KSIM can be programmed for simple hand-held computers.

What is KSIM? It is a mechanism to make systems analysis a transparent tool, a language that does not impose its own constraints upon the formulation of a problem. However, as any language, it is best learned with an experienced tutor at the outset.

For large computers KSIM program tapes are available without charge, except for duplication and handling costs, from Dr. Julius Kane, whose telephone number and address appear at the end of this chapter.

In what follows we present an overview of KSIM, emphasizing its concepts. This chapter presents basically Stage I KSIM. For its further development the user is referred to the technical literature. In the sequel the first person is used to emphasize those portions for which Julius Kane bears the primary responsibility for design, conception, and philosophy. The third person is used where he has had the most able assistance of Paul Fitzgerald in implementation, particularly with applications involving NASA and the U.S. Navy.

FUNDAMENTALS

In the fields of management and politics, how often do we hear that decision-making is an *art, not a science*? The phrase is so trite that we give it little thought; and yet the cliché does contain truth. Why else would it be so frequently repeated? What is the difference between the art and the science of problem-solving? It might be answered that problems are unique, whereas science deals with reproducible phenomena. But the only difference between a "problem" and a "reproducible phenomenon" is the scale and complexity of unknowns. A box sliding down an inclined plane can be described by a small number of variables. The clash of personalities and issues at a diplomatic debate has a staggering number of imponderables and unknowns. Yet there must be some common rules of logic and behavior. We expect our diplomats to go to school and serve apprenticeships. The education of Soviet, Chinese, and American diplomats or managers has much in common. If there is an essence, there must be a science of that essence—but how to express it?

The problem of problem-solving is articulation. Its practitioners become skilled by *doing*. Expressing is another matter. The abilities acquired by a skilled executive are mostly lost upon his retirement. We pass along management skills in much the same way as ancient folklore and tradition. Cocktails replace campfires, and while the costumes may be different, the rites of passage are similar.

The transmission of folklore or *art of practice* is very much a one-to-one encounter. The success of such encounters is highly contingent upon the personality interface. Facts and information are often resisted (or distorted) if they collide with preconceived notions or puncture the ego of the recipient (see the extensive psychological literature on *cognitive dissonance*). One reason why science makes progress is that it is abstract—it is removed from personalities. Its messages can be distilled and transmitted without the need for a psychological interface.

In this activity science has had the great assistance of that most dispassionate of all endeavors—mathematics. This symbolic language permits the extraordinary compression of relationships into the shorthand of equations. If a picture is worth a thousand words, then surely a formula is worth a million pictures. The information that can be transmitted by a formula is staggering. Once the symbolic language is mastered, the horizons of comprehension are broadened towards infinity.

As an exercise, we can set down the following levels of comprehension or interaction:

Level 0:   *Intuition,* the sensual or inarticulate
Level 1:   *Verbal exchanges,* illiterate encounters
Level 2:   *Written language,* books
Level 3:   *Visual images* (pictures, maps, etc.)
Level 4:   *Symbolic* (formulas, equations, etc.)

Level 5:   ???   (the next level of conscious-
                        ness—a return to the intuitive
                        or spiritual?)

From any level to the next, the complexity and volume of information exchange increases by orders of magnitude. If we examine the above table, we note that managers tend to interact at levels 1 and 2, artists are usually at level 3, and scientists work most often at level 4. There is such a great hurdle of *level of organization* from one plateau to the next that there is a great communications barrier. For example, scientists are simultaneously extraordinarily sophisticated and abysmally naive. While there are exceptions, scientists make terrible managers. A major problem is that they are schizoid in their communications abilities: They have difficulty in shifting from level 4 concentration of information to the subtle and often ill-defined patterns of human interaction at level 0. In the company of a colleague they can exchange information in megabyte bursts; involved with the lay public or a household problem, their narrow concentration of depth often disassociates them from the realities of creature contact. To continue with another cliché, there is a failure of communication.

With such considerations in mind, I put together KSIM with the aim of leaping over the barriers of dialogue and finding a mathematical shorthand that could be used to enhance exchanges between level 0 *intuition* and level 3 *articulation and comprehension* by making use of level 4 techniques.

I was heavily motivated by my experiences in partial differential equations. Crudely put, such equations describe the interaction of infinitely many variables, which must simultaneously satisfy infinitely many internal equations and a multiplicity of boundary conditions. As so stated, the task would appear hopeless. It was observed, however, that certain canonical problems could be solved fully—the vibrating string, the flow of heat in a bar, the potential of a dipole, etc. Each of these problems is solved by making simplifications and assumptions. Once these canonical problems are studied and understood, it becomes possible to say much about more complex, real-world problems.

The equations of mathematical physics tend to sort themselves into one of the three following forms:

Elliptic $$A\,\frac{\partial^2 u}{\partial x^2} + B\,\frac{\partial^2 u}{\partial y^2} + C\,\frac{\partial^2 u}{\partial z^2} = 0 \tag{1}$$

Hyperbolic $$A\,\frac{\partial^2 u}{\partial x^2} + B\,\frac{\partial^2 u}{\partial y^2} - C\,\frac{\partial^2 u}{\partial t^2} = 0 \tag{2}$$

$$\text{Parabolic} \qquad A \frac{\partial^2 u}{\partial x^2} + B \frac{\partial^2 u}{\partial y^2} - C \frac{\partial u}{\partial t} = 0 \qquad (3)$$

In these equations A, B, and C are arbitrary *positive* parameters. It matters very little what the relative magnitudes of A, B, and C are. What is most important is the *signature pattern*. All pluses, as in equation 1, result in an equation that describes *equilibrium* phenomena—the rest position of a stretched membrane for example. The minus sign in equation 2 results in *wave propagation*. All of special relativity emerges from this equation, with $A = B = 1$ and $C = 1/c^2$, where c is the velocity of light. The phenomena of equation 3 are appropriate to the mechanisms of heat flow. As a consequence of the first-order time derivative, there results an "arrow of time." The solution(s) of this equation has entirely different behaviors for future time as compared to past time, as anyone who has worked with the equations of mathematical physics knows. (This is *not* the case for the hyperbolic equation 2.)

KSIM emphasizes the coupling of variables and especially the *parity* of any interaction. Simply put, a positive coupling indicates that variable A benefits B—for a negative coupling, the contrary occurs. However, we distinguish between the action of variable A on B and that of B upon A. Just because I am your friend does not necessarily mean that you are my friend.

IMPACT

We write $A:B = \alpha$ for the impact of A upon B. This impact, *considered in isolation of other effects,* has the form

$$\frac{1}{A} \frac{dA}{dt} = -\alpha(A \ln A)B \qquad (4)$$

We call the parameter α the coupling coefficient of B upon A, and note that the differentiation above is a *logarithmic derivative*. That is, the fundamental changes considered are *percentage* changes rather than *absolute* changes.

In the equation 4 the changes are linearly related to the coupling coefficient α and the level value of B. While B is always positive and in the range [0, 1], the coupling coefficient can, and frequently is, *negative*. A negative value of α means that increasing values of B will diminish A.

If $-\alpha(A \ln A)B$ had been a constant, say, k, then equation 4 would have the familiar form of exponential variation,

$$\frac{dA}{dt} = kA \qquad (5)$$

and little more would need to be said. However, the fact that the KSIM
k has significant functional variation results in many important effects.
The term $A \ln A$ means that the percentage change will be zero whenever
A is either 0 or 1. This is an important difference between equations 4
and 5 above. It means that the growth of A in 4 is severely limited when
it is near *threshold* ($A \sim 0$) or *saturation* ($A \sim 1$).

This is a far more realistic variation than the exponential growth of
equation 5. Equation 4 is of the type considered by Gompertz, who found
that it was a reasonable model of many types of biological, organic,
economic, and historical forms of growth. For positive values of α the
solution of equation 4 has the form shown in Figure 1. In early phases
such growth exhibits exponential characteristics. However, once resource limitations become appreciable, the growth levels off and eventually flattens out.

For real-world systems growth of the exponential type in equation 5
is impossible, for such variations quickly imply unboundedness. Mathematically, however, equations of the type in equation 5 are highly tract-
able. They have easy solutions and highly intriguing characteristics. As
a consequence, they have assumed a role all out of proportion with their
significance. Because such equations admit such an easy conceptuali-
zation, they have had the unfortunate consequence of heavily biasing
many theoretical and empirical undertakings.

Every businessman would like to think that because his growth this
year and last was, say, 6%, that it will be 6% forever. Imbedded within
our culture is the cliché *if present trends continue. . . .* The mathematics

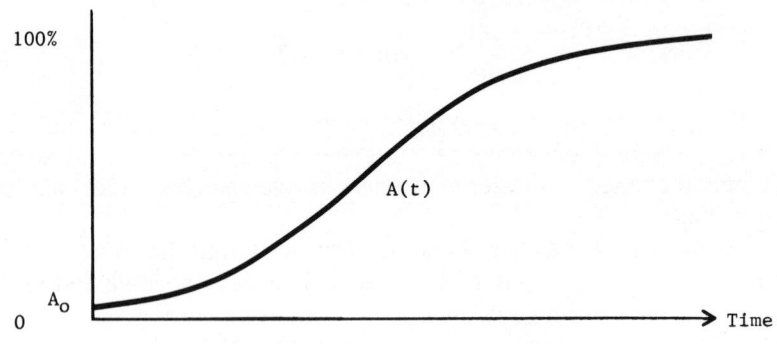

The solution of $\dfrac{dA}{dt} = -\alpha A^2 \ln A$ for positive α

Figure 1 Logistic variation.

of compound interest is intriguing, and all too many people get carried away by the beguiling implication of doubling and redoubling, ad infinitum. It is the sort of fallacy that led educators in the early 1960s to plan for a continuously exploding college population. Who can we trust if the professors do not learn their own lessons?

The logistic in Figure 1 is a far more realistic model of real-world behavior, primarily because it incorporates from the outset the saturation effects of finite resources. Unfortunately, its mathematics is that of nonlinear differential equations, not readily grasped by people without an analytical bent. This was one of the reasons why KSIM was developed. Just as it is possible to speak without being conscious of the mechanics of grammar, syntax, and etymology, so it should be possible to use the language, gestalt, and conceptions of mathematics without being involved in its technicalities. While a knowledge of the mechanics can aid in comprehension, it is not necessary for many important applications. In fact, questions of technique can often obscure or intrude upon the real purposes of a KSIM exercise—a methodology for decision-making. The problem is similar to that of weather prediction. For many applications one only wants to know when to wear a raincoat, and not how to manipulate equations of state.

In systems analysis we seek to make predictions and comparisons of system behavior. Even for deterministic systems this is in a sense impossible, for it is impossible to make a full and complete definition of a system. If we do not know what it is, how then can we make predictions about it?

Yet we also know that systems exhibit attributes—what we call system variables. These seem to have more tangibility and measure. They can be quantified and reasonably described. If ambiguities are accepted, then reasonable definitions—and prediction—can emerge. To make these ideas clearer, ask what is AT&T. No one knows. Every person will give a different answer. One person will say this, another that. Precise description is impossible. Yet it has attributes—cash flow, patents recorded, number of employees, and so forth. These can be measured. These will be weighted differently by different observers. The difference in weightings is simultaneously an expression that each observer has different *knowledge* of AT&T, and a different *personality*.

In systems modeling, the model is not only a picture of the system, but of the analyst as well. In the KSIM process this is very important, for it emphasizes that it is as important to understand the psychology and motives of the people defining the system, as it is to understand the abstract behavior of the system.

KSIM DEFINES VARIABLES BY DYNAMIC ENCOUNTER—NOT FORMAL DEFINITION

With these *caveats* we can set out the following Kantian* program: system, excitation, response. The dynamics of action and reaction behave in an entirely different fashion than formal, static definitions. In the language of a novelist's dictum, "Show, don't tell."

In this fashion, *even if the system is not fully designated or understood, insight can be obtained into its behavior patterns by gaining experience with KSIM analogues.* At the outset it is important that one should not be too formal or rigid in variable definition for depending upon the utilization of the variable; it may exhibit contradictory or perplexing behavior in terms of the effects it produces. For a simple example, consider tax rates. Two extremes are meaningless—0 and 100%. The first is obvious; the second takes only a moment's thought. If taxes are 100%, then there is no incentive for anyone to do any work. It then follows that between 0 and 100% there is some tax rate† that optimizes revenues. If the gross, effective rate is less than this figure, then increasing taxes increases revenues. If this observation is accepted too easily, then one is led to the fallacy that increasing tax rates continuously increases revenues.

One of the features of *structural analysis* is that a concept can be learned and understood within the context of one example and then applied to many other diverse situations.

In a sense, the "Laffer curve" is a rediscovery of the wheel. It has been found many times, in different situations, involving different examples. However, the reasoning is always the same. Consider a lake teeming with fish. We can measure *productivity* as the ratio of the number of fish caught to some measure of maximal measure of sustained yield. Similarly, *fishing effort* is a fraction of the maximum effort wherein the lake is fished by repeatedly dragging a perfect net of extremely fine mesh. Again, a little reflection will reveal that there are two fishing efforts that result in zero productivity—no fishing effort *and maximal fishing effort.* Obviously, if you do not fish, you do not catch any. What should be just as obvious is that *if you catch all the fish, you have destroyed the fishery.* In resource economics and ecology the Laffer curve has long been known as the yield–effort curve (see Figure 2).

Many other situations exhibit the same phenomena. Consider a cor-

---

* Kant put it as: thesis, antithesis, synthesis.
† Implicit in this example is the understanding that we are dealing with a *gross, effective* tax rate and not a *marginal* one.

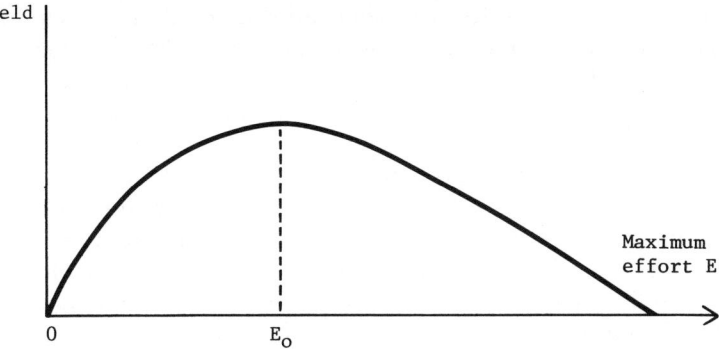

In general, there is no yield if effort is either zero or
fully maximum. There is a crucial or optimal value of
effort E_O. Below this value, increasing effort increases
production or yield; above this value, it is counterproductive.

Figure 2 The Laffer Curve, one of many yield–effort curves.

poration that is concerned about the relationship between *productivity* and *communications*. Clearly, if there is no communication, then everything will be in chaos and nothing will be produced. On the other hand, if the employees do nothing but communicate, then they are spending all their time communicating and not in producing (otherwise known as "lolligagging").

Another such pair of variables is *national security* and *defense budget*. (You have no security because you have nothing to protect when the defense budget goes to 100% of the GNP). As exercises, try coupling the following variables:

Response (Yield)	Effort
Market penetration	Advertising
Satisfaction	Affluence
Health	Medicine
Education	Schooling
Recreation	Sports
Comprehension	Formalism
Being	Having

In the exercises above the reader is directed to the works of Ivan Illych, particularly as regards education and medicine: He makes the point that too much schooling results in less education and that too much "doctoring" results in less health (iatrogenic diseases). For having/being, see the works of Erich Fromm. All this is a highly articulate way of saying simply that in many important respects society is on the wrong side of the yield* curve.

In an important sense, this is where KSIM differs from many abstruse forecasting methodologies. KSIM emphasis *less* formalism and *more* comprehension, whereas most "academic" or econometric procedures tend to reverse KSIM's priorities. Clearly the preceding examples could similarly be used to describe the yield of *comprehension* as a consequence of the effort of *formalism*. Decision-making requires answers *now*, with the data and processing facilities presently available. The ideal predictor is an exercise in wishful thinking and if it could be realized, would probably be unworthy of the effort. KSIM, on the other hand, tries to keep the procedures simple, easily comprehensible, and on the "correct" side of the yield curve. (There is a whimsical problem of semantics here because the "right" side of the yield curve is on the *left* side as we have drawn it.)

STRUCTURALISM

The preceding examples also illustrate another feature. Much insight can be drawn from a simple observation that responses begin at zero when effort is zero and is also zero when effort goes to the other extremity— one, *maximum effort*. Between zero and one the response curve will have a peak. This is a *geometric* observation and independent of the numbers describing the peak. This is a geometric truth that does not depend on arithmetic details. I submit that this is what we mean when we talk of the *art* of problem-solving, the ability to grasp the *geometry* or structure of relationships. What is termed the *science* of problem-solving is an invocation of *arithmetic* processes, finding numbers to quantify relationship(s). Alternatively, and more commonly, these considerations are given the labels *subjective* and *objective*. I find this unfortunate because our culture heavily biases the objective in favor of the subjective. Supposedly, an objective decision is better than a subjective decision: Actions based on feelings are deemed unworthy and irresponsible; there is refuge in number. The bureaucrat occupies himself with census and inventory. It is the line manager who has to make decisions based upon his percep-

---

* With all due respect to Laffer, it is perhaps more honest to call it simply the "yield" curve.

tions of the structure of relationships. This is *not* the same thing as acting on whim. Structures are geometric in nature, and in mathematics geometry and arithmetic share equal prominence, are polar opposites and inseparable. Geometric considerations can be arithmetized and vice versa. Each approach has merit and distinction. For applications the relevance of one does not deny the validity of the other.

In physiology we have found that the brain divides itself into two hemispheres, each specializing itself differently. The left brain is analytical, the right brain holistic. The mechanics of science tend to be left-brain activities; the processes of artists tend to be right-brain activities. Alternatively, there is an extensive literature deeming left-brain activities to be *male* and right-brain activities *female*. I find these labels most unfortunate, tending to inflict cultural or social biases upon symmetric organic phenomena. Would anyone dare suggest that half the brain should be excised to favor the other?

We have yet to decipher the full mechanisms of brain function. My understanding of its functions leads me to conclude that each half of the brain operates as the "transform" of the other, in the same sense that Fourier transforms convert point information into wide-band spectral information and vice versa. The exchanges between transform pairs is a fascinating and informative study that need not occupy us here other than to observe that mathematics does not give priority to either member of a transform pair. If you wish to label high frequencies "female" and low ones "male" in the fashion of soprano/basso, then you are free to do so, but what useful purpose emerges from such an artificial separation?

To summarize, where a parity exists in nature between polar pairs and our culture or technology emphasizes one pole in favor of the other, then this should be a clear signal *of imbalance in our culture or technology.* Whatever their merits, computers do have the unfortunate feature of beguiling us with numbers, *arithmetical* processes. KSIM is a process designed to reverse that trend and to restore the balance by promoting recognition and use of *geometry* by emphasizing the *structure* of relationships.

EXAMPLE OF THE USE OF KSIM

The formalism of KSIM has been extensively described in the literature (see references in Fred Roberts, *Discrete Mathematical Models*, Prentice-Hall, Englewood Cliffs, N.J., 1976). Here we highlight a few of its features and give some examples of its use.

As a simple example, think of FARMERS growing LETTUCE on an island inhabited by RABBITS and FOXES. The capitalized nouns are the variables of the system. On the island there are natural biological limits to the amount of LETTUCE, RABBITS, or FOXES that are set by such factors as sun, rain, soil fertility, disease, insects, reproduction, and so forth.

The matrix of interaction is shown in Figure 3.

Imbedded within the matrix of Figure 3 are a number of assumptions. The island is well suited for the growing of lettuce (LETTUCE : LETTUCE = +3), rabbits breed more rapidly than foxes, and the farmers have stabilized their reproductive level (FARMER : FARMER = 0, but the good crop of lettuce (LETTUCE : FARMER = +3) and the availability of rabbits for stew (RABBIT : FARMER = +1).

For simplicity we can assume that all variables have the initial value of 0.5, or half the maximum possible. If we sum the signs in each row, we obtain the value of the net force acting on that variable. We see the following:

LETTUCE grows (having a net impact of +3).

RABBIT is stabilized by the combined adverse action of FOX and FARMER.

FOX grows because it has no stabilizing negative impact.

The FARMER population will grow because of the good yield of LETTUCE and RABBIT.

Without going to the computer at all, we have a good idea of what will happen. FOX and FARMER will grow, depleting RABBIT. As RABBIT is depleted, LETTUCE will grow more, thus furthering the growth of FARMER.

On the other hand, an entirely different scenario emerges if the farmers aggressively hunt foxes. If the value FARMER : FOX is set as −3, then the

	LETTUCE	RABBIT	FOX	FARMER
LETTUCE	+++	---	0	+++
RABBIT	+	+ +	- -	-
FOX	0	+	+	0
FARMER	+++	+	0	0

Figure 3 Interaction coefficients of the LETTUCE/FARMER model.

fox population will be depleted, negating the predation effect of foxes upon rabbits. As a consequence, rabbits will thrive and ruin the lettuce crop.

DIAGONAL ENTRIES—SELF-INTERACTIONS

One distinguishing feature of KSIM is its emphasis on self-interactions. Many cross-impact methodologies assume or require that diagonal entries be null. Yet things do have an important influence upon themselves, ACTIVITY begets ACTIVITY, EMPLOYMENT encourages EMPLOYMENT, INFLATION induces INFLATION, and so on. It is one thing for a self-interaction to be set equal to zero because it is a null interaction; it is an entirely different thing for a methodology to force such behavior because of the [im]proper assumptions of the methodology.

KSIM was originally developed for ecological applications. In population biology self-interactions are most important. For this reason KSIM had the guidance of the behavior of living systems rather than abstract mathematical assumptions. Yet even with such pragmatic, natural, guidance, important questions of interpretation remain. Consider the effect, for example, of RABBITS upon RABBITS. The obvious leap to the conclusion is that this is, perforce, a positive entry—clearly rabbits beget rabbits. But this is so only if RABBITS is considered to be a census variable. But consider the self-interaction of a rabbit upon a rabbit (here, I use lowercase letters to emphasize that I am using an actual rabbit as an example, rather than a generic system label *rabbit*). It takes work to be a rabbit. Isolated, removed from food, any rabbit would lose weight. Therefore, if RABBIT is considered in metabolic rather than census terms, the self-interaction RABBIT : RABBIT *would be negative.* But how can an interaction be simultaneously positive and negative? What is the answer?

The answer is *both.* Everything depends upon what the user wants RABBIT to mean—is it a metabolic variable or a census variable? This is for the user to decide. KSIM cannot, does not, and should not provide the answer. One of the chief functions of a KSIM exercise is to reveal and demonstrate to users the riddles and inconsistencies of their a priori assumptions. Modelers are apt to have preconceived notions of how their system could and should behave. Unfortunately, often the exercise is not to inquire how the system really works, but to *show* the modeler's ingenuity in presenting a prolix "explanation" for behavior.

I have found this to be true in many field exercises. I have been invited to lead many KSIM workshops for government and business agencies. Generally the purposes are twofold: (1) to obtain an insight into the KSIM methodology and (2) to *demonstrate a conclusion*. The group may want more research funding, a greater share of a development budget, a conclusion emphasizing the importance of fish over oil, regionalism over internationalism, and so on. Management groups that invite me to lead their workshops have vested interests. The vested interests pay for the forum and the forum is expected to produce support.

This is an honest use of KSIM, *if* the game is played honestly. One of the important features of mathematics that is not generally appreciated by the public is that *mathematics does not deal with truth, but with validity*. Mathematics is concerned with only one question: Does a conclusion follow logically from the assumptions? It has no mechanism to determine whether assumptions are true or false. Mathematics is an if–then relationship. *If* this is true, *then* such will happen. KSIM permits the free choice of assumptions and interactions, but it does this in a way such that these assumptions are clearly exhibited. Any protagonist can easily challenge an assumption. If an entry has been *plus*, it can easily be changed to *minus*. The effect can be observed on the system. Thus alternative views can be compared.

For this purpose let us consider the self-interaction EMPLOYMENT:EMPLOYMENT. Is this positive or minus? We can argue first for positive. The more time people spend working, the more inclined they are to hire people to do things for them. If a woman is employed, she is more likely to hire housekeepers, child-care persons, accountants, mechanics, and so on. One person employed in a primary sector easily generates employment for a number of others in service sectors.

Alternatively, it can be argued that a Parkinson's law-type behavior prevails. Any economy, with limited resources, can support only so much activity. If more people are working, it simply means that each, on the average, is doing, *less*.

It will be seen that either argument can be made persuasive. What is happening here is that the label EMPLOYMENT is being used to describe two entirely different variables. In the first case EMPLOYMENT was equated with WAGE EARNING, a largely statistical concept. One becomes employed if one is hired and receives a periodic paycheck. In the second case, EMPLOYMENT was equated with PRODUCING. In this one does not look at the formal details of the activities of the worker but, rather, the *conclusions of the effort*. Is a man a baker because he draws a paycheck as baker, even if he produces no bread?

The guidance KSIM offers in the resolution of such riddles is very subtle; it does not give direct answers. KSIM merely says that either alternative can be chosen, *but once chosen, the interpretation must be held consistent with the original assumptions*. Of course, at any time assumptions can be changed or new ones added. But once this is done, the discipline of consistency must be maintained.

This is one of the major values of KSIM. By keeping the focus of attention on the *parity* of signs rather than magnitude, KSIM keeps the focus of attention on the *meaning* of the variable rather than on the size of the variable. The purpose of KSIM is to come to an understanding of the variables and their interactions as opposed to numerical conclusions. In this, KSIM is entirely the opposite, say, of econometric models. Econometric models presume a mechanistic structure of the economy. KSIM makes no assumptions. KSIM allows for free adaptation—wild postulates of structural change, that can be easily investigated. Econometric models can only accept variations in data that are marginal changes of statistics. For example, no econometric model can handle triple-digit inflation, the abolishment of the income tax, or the doubling (or halving) of the defense budget. Any such "wild" change stresses the econometric model to a point where the essential assumptions have been ruptured. KSIM can easily posit new configurations.

Econometric models have an essential inertia. The vastness of the effort required to produce the model and feed the machine creates a fallacy—the prodigious use of brains, data, computers, and effort must yield correct conclusions. The magnitude of such effort intimidates discussion, contradiction, or alternative views.

This is one reason why I have emphasized pluses and minuses. Any KSIM presentation can be easily challenged. All a party need do is, say, change the signs of these variables or the magnitude of those coefficients. Since KSIM is an extremely low budget process, alternative computer models can be easily run. An adverse party can run a number of models of his or her choosing and possibly find an alternative presentation.

STRUCTURAL INTEGRITY

However, this is more easily said than done. In fact, any system quickly defines itself. Many of the variables and interactions are simply not susceptible to manipulation. Let us go back to the LETTUCE/FARMER model. It is true that we can make varying estimates as to how fast the lettuce might grow. But, short of absurdity, we cannot have foxes eat lettuce, or rabbits, foxes. For any significant model that is constructed with anything resembling honest effort, it is found that *most system interactions are nonnegotiable and not susceptible to manipulation*, at least as regards the *sign* of entry. If there is any debate about the magnitude of an entry, the computer can run the model twice—once for the minimum value of that parameter and once for the maximum. Most often the difference between the two runs will be insignificant. If there is a difference, it will usually be that the curves are slightly perturbed, and there is no real *structural* difference. This is a consequence of the complexity of the real and the richness of their feedback interactions, giving the curves a great deal of internal inertia. Even the "simple" LETTUCE/FARMER model involves the complex interplay of 16 matrix coefficients.

There are a number of ways in which the model can be improved or modified. An economic variable, COST, can be introduced, relating to the cost of land. This will have an inhibiting effect on the introduction of new farmers. The model just presented has included reproductive effects but ignored metabolic factors; if desired, these can be included as well. Similarly, insects can be introduced, as well as a two-crop economy, and so on. Once the basic concepts of KSIM are understood, the variations of its use are endless.

PROCEDURE AND APPLICATIONS

The KSIM workshop starts with a day or so of introductory material, particularly an introduction to systems analysis, a definition of "levels of organization," and a discussion of the general nature of composite variables. After the first day of preliminary material and discussion the user group is ready to begin with their first model. At this time the following features of KSIM are emphasized:

1.   The initial use of signs, plus and minus, rather than numbers. This has an important psychological advantage in that it allows members of a workshop group to participate freely. Even in sophisticated circles, only a small segment is comfortable working with numbers. For a variety of reasons most people have great inhibition in giving numerical answers to questions. Consider the following two questions: What is the coefficient of self-reproduction for rabbits? Are rabbits good for rabbits? The first question is grossly intimidating. The respondant is put on the spot, feeling that he or she will betray ignorance. No such intimidation arises for the second question. The party can answer freely, dipping into his store of experience or common knowledge. The use of plus and minus also tends to eliminate the "bamboozle" of experts. Jargon is discouraged and the real issues can quickly be brought into focus.

2.   The KSIM process encourages diaglogue between participants and mutual understanding. It is difficult to be possessive about a plus or

minus. Conversely, one does not emotionally resist an argument presenting a case for plus or minus. No great issue of ego, status, or position is involved. This lowering of the defense mechanisms greatly facilitate group interaction. It should be emphasized that KSIM is much less a mathematical formalism than a collection of processes encouraging group interaction and dialogue. The fundamental hypothesis of KSIM is that often the real problem faced by the workshop group is not mathematical. The problem is one of structural definition.

3. The best exhibition of the structural definition of KSIM is *variable selection*. The biggest difficulty faced by new users of KSIM is the proper selection and definition of variables and the elimination of extraneous considerations. New users will often introduce the same variable, but under a variety of different labels. The choice of variables exhibits much about the biases and predilections of the group. Often highly significant variables are left out. Several examples may illustrate this.

I have often given environmental problems to my students as KSIM exercises. Invariably, the matrices they produce exhibit a predominance of minuses. A student has a background of being exposed to situations beyond his or her control. As a consequence, there is an inevitable manifestation of pessimism. Their KSIM matrices clearly exhibit this. They have no background in thinking *beyond* the parameters of a question. Their teachers have conditioned them to give responses *within* the context of course preparation. Analysis is emphasized and remphasized to the exclusion of imagination. Very little in a student's regimen equips him for a "can do" attitude.

I experience a totally different environment when I work with engineers or senior managers in government or industry. These are people who have acquired their positions with a proven record of problem-solving: They can do because they have done, and their attitude exhibits this. However, their matrices have an entirely different fault: they are invariably loaded with pluses. To hear them explain it, their only limitation is that of budget: Give us the money and we'll do the job. Several examples will illustrate this.

THREE EXAMPLES OF FIELD USE

FOREIGN UTILITY TAKEOVER

In my first example I was called in by a major utility in a foreign country. They were a subsidiary of an American corporation and were worried about being expropriated by a newly elected socialist government. KSIM had come to their attention, and they thought it could be helpful in preparing a variety of strategies to resist takeover. They showed me a number of models they had devised. One emphasized improved CUSTOMER SERVICE on the theory that better service would give them public support. I pointed out, however, that their model failed to include a factor that better service would give increased utilization and hence better profits. This would probably result in increased pressure for expropriation. They showed me a variety of other models. All had a serious defect and had a premise that was never stated explicitly, and this was JOB SECURITY. The models that were being built were not honest and had the nature of self-fulfilling prophecies. The premise was that utility (and the jobs of the managers) was indispensable and that therefore private control (and the jobs of the managers) was indispensable. I suggested that they consider an alternate possibility, *the desirability of being expropriated*. Since this took place in a Commonwealth nation, there was no question of being compensated for expropriation. Indeed, the utility would be in the position of receiving several billion dollars for what was essentially nineteenth century technology. This money could be reinvested in twenty-first century technology with a far greater return. However this suggestion fell flat, for the panel members were intimidated by the idea of learning new jobs and entering the twenty-first century.

LONG-RANGE PLANNING, *NASA*

A second example of KSIM use was with the laboratory and headquarters managers of NASA. The problem was this: NASA was facing lean budget years and flight programs are very expensive. The question was could NASA make future development programs even less expensive than the reusable Space Shuttle by the expenditure of more funds on the class of endeavor known as "Advanced Development"? In effect, a portion of "flight" funds would be internally "taxed," or diverted to a research program, "Advanced Development." Various KSIM models were run, for different levels of diversion or taxation. This presumed that the prototyping and testing of advanced technology projects very early in the development cycle would save substantial funds later by uncovering design and systems operational problems on the ground early and without the press of flight schedules later in the cycle. The KSIM models built by the working group were employed to test this hypothesis. As might be expected, the KSIM models did show this; however, they did not resolve the more basic problem of *responsibility*. The flight manager has to show his results now or in the near future. In essence, current flight programs in development and advanced development for future flight programs are competitive in any given fiscal year's budget formulation. In an ideal world flight and research should cooperate, but with limited resources managers will pursue the success of their objectives in the short term, even if these are self-defeating in the long run.

Unfortunately, in most government planning the administrative and congressional requirements are to keep near-term budgets as low as possible. This requirement coupled with the requirement of the line managers for short-term planning demonstrated results working to the disadvantage of long-term cost efficiencies. The KSIM model showed what "should be" for long-term cost efficiencies and even pointed out in quantifiable terms the budgetary shifts required to optimize long-term cost savings, but unfortunately the model cannot make it happen within the bureaucracy, and robbing the "long-term Peter" to pay the "short-term Paul" remains a fact of life.

But there was another very revealing feature of the managers' KSIM models. They were all highly optomistic. This stemmed from a variety of reasons. By training and inclination the managers were engineers, technocrats. They have a tendency to view the world in terms amenable to precise definition and control. The variables they introduced were highly biased, emphasizing those under their technical control or manipulation. Since they were all competent, experienced, and highly motivated, they presumed *that given the resources*, they could do the job. Their models failed to include competition for their resources from national programs that were not technically biased. It was pointed out to them that the ultimate source of their money was Congress or the public, and that despite internal competition between differing technical programs, all technical programs would have to compete with social alternatives—health, education, welfare, and so on.

As in the prior example, it was most revealing to see that the user group tends to look at the problems from their biases. There is nothing internal in KSIM that can prevent this. However, it is highly recommended in initial experiments with KSIM that involve the disposition of significant sums of money that either an experienced KSIM tutor be available or that the user group make every effort to solicit panel members that are likely to challenge existing biases. This is where the real utility of KSIM begins to manifest itself.

The NASA group used KSIM-III, which is beyond the scope of this introductory paper. However, it might be useful to examine some of the graphical output produced by this group once the need for including "social alternatives" in their model was appreciated. Figure 4 shows output presuming that NASA is not successful in competing with "social alternatives."

Both Figures 4a and 4b illustrate a field run wherein SOCIAL ALTERNATIVES compete agressively with NASA's objectives. In the sequel the labels are the following: NASA CAP—NASA capability (i.e., the ability to successfully conclude congressionally authorized missions on time, within budget, and safely); SOC ALTS—funding of competing social programs; EX MIS—existing (authorized) missions for NASA; MIS APRV—congressional approval of new missions; AD BEN—mission definition (the ability of NASA to fully, conceptually, and philosophically understand their own motives and objectives); and the other variables represent internal prorations of NASA's budget between internal competing programs.

Figures 5a and 5b are reruns of the field runs shown in Figure 4. In this instance a clearer idea of MISSION DEFINITION results in better competition with SOCIAL ALTERNATIVES. In this instance, it was not really assumed that NASA was competitive with SOCIAL ALTERNATIVES but, rather, that with a clearer idea of its philosophy and purpose, NASA's objectives would become consistent with social programs and public benefit.

A significant question that was only briefly touched upon at that workshop was, "Is there really competition between NASA and social alternatives?" Are not the objectives of NASA truly the dreams of society and an achievement of culture and civilization? How can the realization of human purpose be less than the primary social goal? Pursuing this particular question may provide another fruitful application of KSIM.

U.S. NAVY—RESEARCH VERSUS FLEET SUPPORT

In a somewhat similar vein, a related problem was considered by a R&D laboratory of the U.S. Navy. The laboratory had sufficient funding for fleet support, but restricted funds for fundamental R&D. The laboratory managers raised the question that their future technical competence was being erroded by the heavy emphasis on fleet support. How best to rebuild their basic R&D program was the question addressed.

A KSIM exercise was run, investigating the interrelationship of the competing variables. In this instance the interactions matrix partitioned itself into a fleet matrix and one for the laboratory, together with their interaction terms.

This particular group problem-solving experiment pointed out another feature of the KSIM modeling process—the uncovering of counterintui-

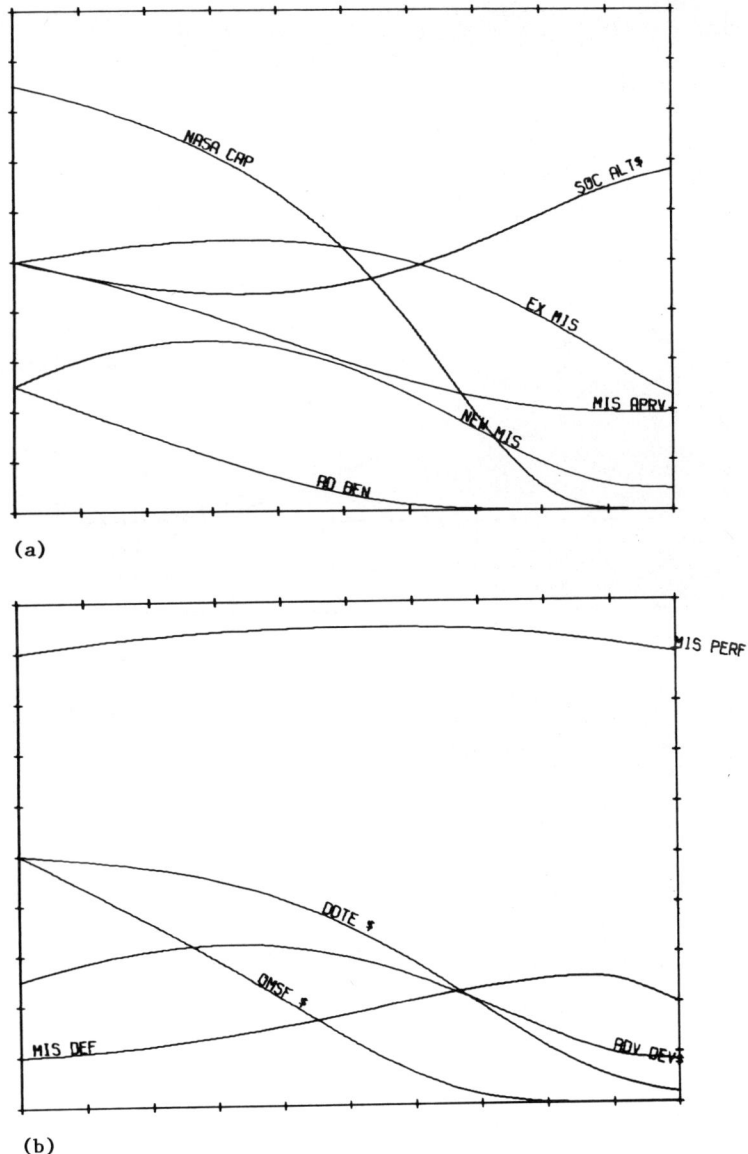

(a)

(b)

Figure 4 KSIM runs performed by NASA Headquarters.

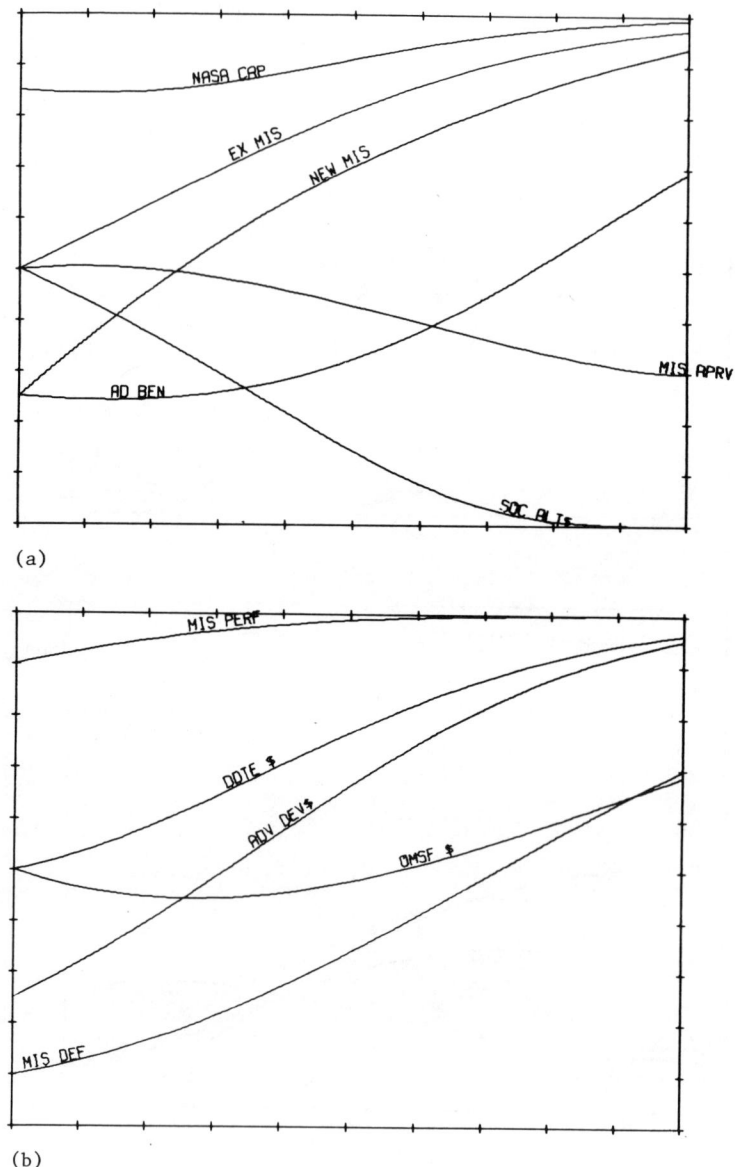

(a)

(b)

Figure 5 Reruns of runs in Figure 4.

tive results. Government (Congress, the administration, the courts) often takes a specific action (the passage of a law, the issuing of an executive order, a court ruling), expecting certain results when their action is implemented, only to find out that its implementation produces something entirely different and sometimes counterproductive to what was intended and expected. Many government actions are taken to solve problems by simply adding more money or people to a program to work on its implementation. The particular model built by this Navy R&D laboratory working group showed that the addition of funds and manpower alone to the basic R&D program did not have the desired effect on the laboratory's future, and only when these two factors were accompanied by a strong factor of "leadership" (with its attendant decision authority) did the model produce the desired results.

The ability of the KSIM model to reveal the existence of counterintuitive results and to explore their implications is but one of its several advantages when employed in an open group problem-solving and decision-making process.

FURTHER INFORMATION

In this brief introduction it is not possible to give more than an overview of KSIM, its uses, and applications. The authors invite telephone queries, user's comments and reactions, complaints, and perhaps praise. KSIM programs are available from Dr. Kane, who can be reached at either (206) 734-9125 or (206) 671-1285. Paul Fitzgerald can be contacted at MARTIN MARIETTA AEROSPACE. Their addresses are as follows:

Dr. Julius Kane   Paul E. Fitzgerald, Jr.
P.O. Box 100   MARTIN MARIETTA AEROSPACE
Bellingham, Washington   P.O. Box 29304
98227   New Orleans, LA 70189

The best library introduction to KSIM and related procedures is *Discrete Mathematical Models* by Professor Fred Roberts (Prentice-Hall). This book also contains an extensive bibliography, including references to more advanced versions of KSIM.

BIBLIOGRAPHY

Boroush, M. A., K. Chen, and A. N. Christakis, Eds., *Technology Assessment: Creative Futures*, North Holland, New York, 1980.

Coady, S. K., et al., "Effectively Conveying Results: A Key to the Usefulness of Technology Assessment," First International Conference on Technology Assessment, The Hague, 1973.

Kane, J., "A Primer for a New Cross-Impact Language—KSIM," *Technological Forecasting and Social Change*, Vol. 4, 1972, pp. 129–142.

Kane, J., et al., "KSIM: A Methodology for Interactive Resource Policy Simulation," *Water Resources Research*, Vol. 9, No. 1, 1973.

Kane, J. and W. Thompson, "Environmental Simulation and Policy Formulation—Methodology and Example (Water Policy for British Columbia)," *Proceedings of the International Symposium on Mathematical Modeling in Water Resource Systems*, (A. K. Biswas, Ed.), Vol. 1, 1972, pp. 39–55.

Kane, J. and W. Thompson, "Health Care Delivery: A Policy Simulator," *Socio-Economic Planning Sciences*, Vol. 1, 1972, pp. 283–293.

Kruzic, P. G., *Cross-Impact Simulation in Water Resources Planning*, Report 74-12 to U.S. Army Engineer Institute for Water Resources, Fort Belvoir Va.

Lipinski, H. and J. Tydeman, "Cross Impact Analysis: Extended KSIM," *Futures*, Vol. 12, 1979, pp. 151–154.

Moll, R. H. H. and C. M. Woodside, "Augmentation of Cross-Impact Analysis by Interactive Systems Tools," M. Eng. thesis of R. H. H. Moll, Carleton University, Ottawa, 1975.

Porter, A. L. and F. A. Rossini, *A Guidebook for Technology Assessment and Impact Analysis*, North Holland, New York, 1980.

Roberts, Fred, *Discrete Mathematical Models*, Prentice-Hall, Englewood Cliffs, N.J., 1976.

Sage, A. P., *Methodology for Large Scale Systems*, McGraw-Hill, New York, 1977.

White, K. P. Jr., "The Extremal Behavior of KSIM Models: Existence, Stability and Implications," *IEEE Transactions on Systems, Man, and Cybernetics*, at press.

White, K. P. et al., "On Counterintuitive Behavior and the Role of System Dynamics Models in Policy Planning," *Large Engineering Systems*, (G. J. Savage et al., Eds.), Sanford Press, Waterloo, Ontario, 1978.

CHAPTER 8

NOMINAL GROUP TECHNIQUE (NGT)

SANDRA L. GILL
ANDRÉ L. DELBECQ

Nominal Group Technique (NGT) is one group process skill that has proved very useful where complex problems need to be delineated or where solution strategies to complex problems demand multiple perspectives. It was designed in 1968 by André Delbecq and Andrew Van de Ven to respond to the increasing need for practicing administrators and professionals to increase their rationality, creativity, and participation in problem-solving meetings associated with program planning. It is particularly well suited for problem identification, the group analysis of problem situations, idea generation, and group selection and the consideration of alternatives. Most users find it helpful in whole or in part in a wide variety of decision contexts. When properly used, NGT can be a powerful tool for increasing a group's creative capacity to generate critical ideas and understand problems and the component parts of their solutions. Thus participants can aggregate (pool) individual judgments and arrive at desirable group decisions.

Where there is no "one best way," where there is controversy and numerous alternatives, there is a need to obtain judgmental decisions. These are different from the noncontroversial demands for expertise or the typical leader-centered tasks of coordination and information exchange. When judgmental decisions are obtained as though there were no difference of opinion, the result is exacerbated conflict, resistance, and decreased motivation to support the result arrived at. Judgmental

271

decisions demand judgmental decision-making techniques. NGT is one of these techniques.

NGT is a structured group meeting for the consideration of a single critical issue, for example, "What do you think are the most important elements of a successful planning endeavor?" The process of NGT occurs through four basic steps:

1. Silent generation (independent thinking and listening) of ideas in writing.
2. Round robin listing from group members to record each idea in a terse phrase on a flip chart.
3. Serial discussion of each recorded idea for clarification and evaluation.
4. Individual voting on priority ideas, with the group decision being mathematically derived through rank ordering or rating.

Each of these steps will be illustrated in the next few pages. It is useful here to point out that this series of independent steps maintains two unique components of judgmental problem-solving identified in small group research: (1) the idea-generating phase, in which a careful search for ideas and facts is generated and (2) the evaluation phase, in which individual information items are evaluated. NGT makes use of each of these, but in a fashion supported by small group research. Whereas idea generation and creativity operate best in the absence of criticism, idea clarification and evaluation are improved when interaction among group members is available. Another benefit to the NGT format is its repetitive opportunity to carefully consider each alternative. The typical group meeting generally succumbs to a focus on a few ideas promoted by a few key individuals owing to their status, personality, and so on. In NGT attention will be given at least three times to each idea generated before the final decision is made. This provides full consideration to the group's entire range of thought as well as to the meaning and merits of each alternative.

Finally, by evaluating ideas mathematically, judgmental errors are reduced in determining priorities. Disagreement is acknowledged where it occurs, yet the full scope of individual preferences is recorded anonymously for future consideration or reference. In short, NGT is a systematic group process that maintains sensitivity to individuals' contributions and simultaneously provides an objective method for making decisions. Both are needed for efficient problem-solving.

THEORETICAL BACKGROUND

Before we present the group process guidelines for NGT we will present
the theoretical background from which this method was developed.

In the 1950s and 1960s a wellspring of research on small group interaction and management practices arose. One of the primary areas of
investigation was that of individual and group creativity. This research
is particularly useful today, since so many problems are complex, demanding multiple professional input and participation for ultimate commitment and action. How does a leader successfully create the conditions
for creativity?

First, the composition of the problem-solving group must itself include
the common elements of creativity. Individuals must be of diverse characters, that is, multiple chronological age and age with the organization,
scientific and professional diversity or specialization, and diverse perspectives in terms of organizational and departmental representation.
However, this creative set of individuals must also share a common interest in the problem, flexibility in terms of tolerating alternate and diverse
points of view, and must sufficiently trust one another to allow differences
of opinion and perspectives to be expressed. Finally, each person must
have sufficient personal intelligence and breadth of knowledge to be a
useful respondent to the issue at hand.

We find, however, that in most problem-solving groups almost the
opposite set of conditions prevails. Typically, administrative problem-
solving groups are selected from individuals who tend to be physically
proximate to the group leader; the selection and interaction with colleagues who are nearby is a simple habit in most administrative units.
Very often these individuals who share office area are often from similar
administrative levels and may, in fact, not be terribly variant in terms of
age, attitudes, or even prior activities in work experience.

Conditions for group creativity, then, demand a variety of components;
each person must have equal participation; there must be a sense of group
centeredness around a common issue or shared problem; the group must
avoid premature evaluation and provide adequate time to explore each
idea presented. The group must share an exploratory attitude and must
ultimately share judgments and come to a common decision.

In fact, what we find typically happening in administrative problem-
solving groups is that a few "dominant" individuals command most of
the "air time"; consequently, there is a narrow focus on a limited number
of ideas. The status differences and competition among group members
is a common characteristic, and very often such groups are characterized

by an early sense of pressure or need for conformity, with tension to adopt the most quickly expressed, seemingly reasonable proposition. What seems reasonable, however, is not necessarily the most inclusive or creative idea!

In summary, then, creative problem-solving groups need to have three major characteristics: (1) the group composition must be heterogeneous enough to bring wisdom and alternatives to the fore; (2) participation among members must balance to allow for clarification and insight; and finally (3) members must share a collegial and exploratory rather than a status-oriented style.

GROUP COMPOSITION

    Diversity

Multiple ages		Physically proximate
Multiple specialties		Similar administrative
-Scientific	VS.	level
-Professional		Similar age
Multiple organiza-		Similar attitudes
tions		Shared prior activities

    Homogeneity

        Interest in problem
        Flexibility
        Trust

    Personal Characteristics

        Intelligence
        Breadth of knowledge

GROUP PROCESSES CHARACTERIZED BY:

Equal participation		Dominant individuals
Group centeredness		Focus on limited number
Avoidance of prema-		of ideas
ture evaluation		Status inequalities and
Problem centeredness	VS.	competitiveness
which provides		Pressures for conformity
adequate time for		Strain toward early
each idea		solution
Exploratory attitude		
Shared judgments		

Figure 1 Summary of conditions for creativity.

NGT was developed from studies that examined the character of communication between two different groups: (1) a "nominal group," that is, one in which members interacted through written communication, and (2) the interacting group, where members exchanged ideas through spontaneous verbal communication. Research showed that groups that wrote prior to verbal interaction generated a higher number of ideas on a given problem, a greater quality of ideas, as determined by external judges, and a set of ideas that were more inclusive than those that were generated by spontaneously interacting groups. However, interacting groups were better at the evaluation issues of weighing alternatives and selecting alternatives when they were able to clarify through interaction.

▷ This set of individual and group characteristics for conditions demanding creativity then led Delbecq and Van de Ven to the development of NGT (see Figure 1).

PROCESS STEPS FOR NGT

STEP 1: SILENT GENERATION OF IDEAS IN WRITING

The first sequence of steps among group members in a nominal group is to sit down at a table, focus on a question or issue that is presented in writing on a flip chart or individually provided work sheets, and to silently think, prior to writing the responses, in short phrases on a work sheet or tablet. This step of silent thinking and independent ideation provides a number of benefits. First, it provides adequate thinking time for each individual to independently search his or her own experience and thoughts. Since it occurs in the context of a larger group in which each individual is working simultaneously, it provides an example of problem focus and obvious effort. Because this step is done silently, individuals are prevented from the mutual interruptions of thoughts and efforts that occur in interacting groups. Since there is no discussion, each individual can fully search and explore his or her own mind and be prevented from the typical group focus on a limited number of ideas. The silent ideation process prevents giving greater importance to the ideas of higher-status individuals in the group, and, finally, because this step is undertaken by all the group members simultaneously, the group remains problem focused and centered.

STEP 2: ROUND ROBIN RECORDING

While Step 1 is normally completed within five to seven minutes, the second step is the process by which the output of these first five minutes of the meeting is recorded. Consequently, it ranges from 30–45 minutes, depending on the size of the group. Step 2 is where each of the individual thoughts of the group members is recorded in a serial list. Typically, the recorder will ask each person by name to read a single idea from his or her list, write that down verbatim, abbreviating where it will be easily understood, and then proceed to the next person. Each idea is taken as an individual thought, and the process of taking only one idea from a single person until all ideas from all participants are recorded separates the idea from the individual personality who contributed it. In this fashion, ideas are seen for their intrinsic content rather than as an output from a particular personality.

The benefits of this second step, that is, round robin recording, are that, first of all, it literally guarantees equal participation from each member. The process of taking only one idea per person and proceeding in a round robin fashion until all ideas are recorded literally provides each individual equal opportunity to participate. Since only one idea is taken per person on each round, the process of serial recording eventually begins to separate ideas from their contributor's personality. Perhaps most important, however, is the third benefit: This recording process provides a written record and guide that captures the aggregate of group thought. Since ideas are written on a flip chart (verbatim), the entire cognitive map of the group is recorded. This process of actually writing each idea also provides a moment of additional focus and opportunity for reflection as participants observe the ideas being written. It also provides an opportunity to compare the newly recorded item with those that have been previously recorded. Since every idea is captured, including new ideas that occur as members see their colleagues' thoughts being recorded, no particular notions are lost. Only participants have the "authority" of deciding whether previously recorded thoughts capture the essence of their own contribution; no group editing or dominance by a higher-status individual can thus take place. Further, the process of recording each idea in writing provides a total collection of the clues and thoughts the particular group has to offer. Finally, the process of recording ideas frequently triggers related but previously unthought of notions; this "hitchhiking" is both promoted and captured.

The process of literal recording also allows a nonconflicting opportunity for alternative perspectives to be presented. Since there is no discussion at this particular stage, rather a simple reading and recording process, the aggregation of ideas will undoubtedly include some that are in direct opposition to others. However, the absence of discussion at this point allows that conflict to be recorded and for differences to be explored in silence. Those differences will be discussed in a semistructured fashion in the next step.

Finally, this process of individual solicitation and aggregate recording does provide a valuable process step of focusing each member of the group on the ideas so that prior to discussion the thought is at least visually recorded and preserved. This process of group focus does promote problem and issue centeredness, in contrast to personality conflicts ◁ or the usual group tendency to rush to judgment about the first few ideas presented. Depending again on the size of the group and the particular interest the given issue has for members, anywhere from 25 to 50 or 60 ideas may be recorded. This process generally takes between 45 minutes to an hour, which results in a visual, written record of the aggregate thoughts of the group.

STEP 3: DISCUSSION FOR CLARIFICATION

The next step is the first opportunity individuals in the group have to discuss their ideas. By this time they have experienced almost an hour of reflective thought and reading and most groups are very ready for discussion. Notice that Step 3 is to discuss each idea in a serial fashion for clarification rather than debate. The benefits of going back to the list, recorded in the preceding step, and reading each item verbatim provides a second opportunity to record that idea in the group's mind. Since the purpose of this particular step is to clarify the terse phrases that have been previously recorded, differences can certainly be discussed, but the recorder monitors group discussion so that the focus is on adding information. ▽All questions and comments that serve to clarify the phrase are encouraged and supported. Very often it is useful to have group members provide an illustration so that the entire group gets a sense of shared understanding rather than having a single individual provide a single illustration of a particular concept or phrase. An additional benefit, however, of this particular step is that each idea is given its own time for presentation and discussion, hence there is a norm to move on to the next item as soon as the preceding item has been clarified. This provides an opportunity for a full discussion on a full range of ideas rather than the typical group tendency to focus on the first few items and neglect all those that follow. This is also a procedure that begins to eliminate any unconscious misunderstandings or deliberate manipulations. The process of offering illustrations and clarifying comments serves to weed out any particular misinterpretations, while at the same time allowing for differences of opinion to exist. Hence members can disagree with the value or feasibility of a particular notion but not feel compelled to eliminate that idea from the written record. Again, it does depend upon the controversy of the issue and the number of people in the group, but typically this third process step of discussing each item in order generally takes between 45 to 60 minutes to complete.

STEP 4: PRELIMINARY VOTE

Very often the final step in a nominal group can be the fourth process step, the preliminary vote on item importance. The benefits of this voting procedure are many. First, it allows members to select from a total list of ideas a limited set of phrases that they, as individuals, feel contribute most to the issue at hand. This opportunity to independently and silently select one's own priorities guarantees equal participation in the selection process. Members are prevented from having to conform to the perspectives or attitudes of superiors or the higher-status individuals who may be present in the group. Since the second and third steps have provided an opportunity to focus on the idea, the unconscious or conscious process

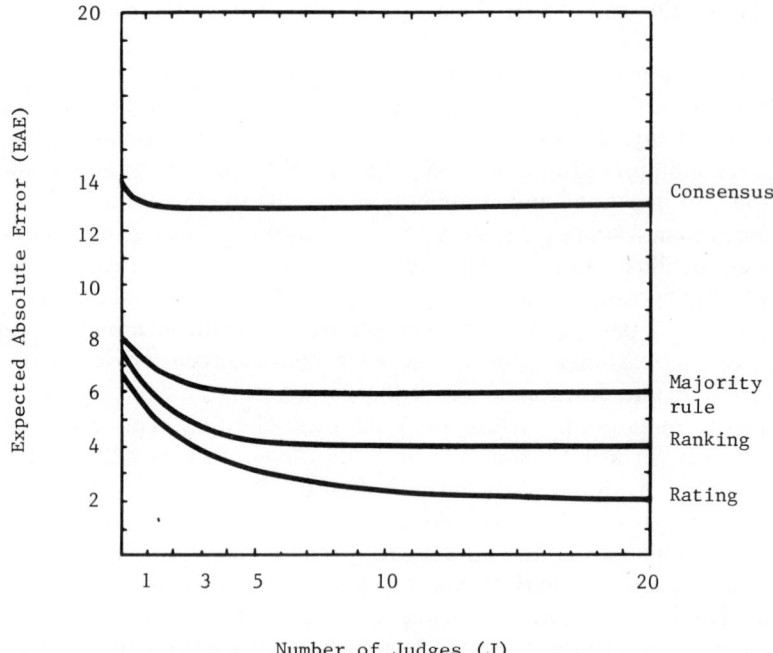

Number of Judges (J)

Figure 2 Impact of decision-making alternatives and number of judges on error reduction.

WHAT DO YOU THINK ARE THE MOST IMPORTANT BARRIERS TO MORE
CREATIVE STAFF MEETINGS?

		Votes		Σ
1.	Inter professional jealousy	3, 1	=	4
2.	Boss-subordinate relations	4, 1	=	5
3.	Negative attitudes	3, 3, 5	=	11
4.	Politics	1, 2, 5	=	8
5.	Different agendas	5	=	5
6.	Lack of information	5	=	5
7.	Personality conflicts	2, 3	=	5
8.	Fear of being wrong	1, 2	=	3
9.	Confusion of purpose	5, 5, 4, 5	=	19
10.	Lack of leadership	3, 4, 3	=	10
11.	Too many people	3, 4, 2, 4	=	13
12.	Lack of common approach	--	=	0

Figure 3   Illustration of nominal group results and posted votes on flip chart.

of selecting an idea on the basis of the author or contributor is generally eliminated at this point. So independent judgments are obtained through the individual, anonymous process of selection as well as enhanced by the previous steps in which ideas are separated from their contributor's personality. Since the selection process is done by individuals, using 3" × 5" cards, the social pressures of conformity are reduced and group accuracy is increased because each individual can fully concentrate and contribute what he or she thinks are the optimal choices.

Small group research has been quite clear about the relative benefits of alternative group processes for selecting priorities. For example, it has been established that the process of public consensus in a group, for example, a straw vote or a public showing of hands, leads to judgment by (conscious or unconscious) conformity rather than a true expression of one's independent thoughts. The process of majority rule, in which individual thoughts are allowed and then aggregated by voting or posting of preferences, significantly reduces judgmental errors as compared to consensus procedures. Even more error reduction occurs through the process of ranking or rating (see Figure 2).

The posting of the individually selected and anonymously recorded ballots provides a very important form of feedback to the group (see Figure 3). First, each individual's vote and selection is duly recorded. Second, the relative agreement and disagreement on those items that group members feel would be important becomes obvious as the various weights are posted. Finally, those arenas in which there was diverse opinion are highlighted, that is, where on a single item an individual gave a very high or a very low ranking compared to the vote of other individuals (note Item 4 in Figure 3). On such diversely valued items further group discussion may be necessary in order to eliminate misunderstanding or to validate the presence of a clear difference of opinion.

The last two steps of NGT are a simple repetition of the previous steps of discussion and voting; that is, following the posting of the preliminary vote, those items that received diverse weights or that seemed to be quite similar with other items could be discussed (Step 5, discussion of the preliminary vote). Then, a final vote, Step 6, may be taken to provide closure and a second final ballot.

A CASE ILLUSTRATION

It is much easier to observe a nominal group than it is to describe one, and the following case illustration is provided to highlight leadership behavior as well as the actual group behavior. Let us imagine that a group of staff members wishes to identify the barriers to their own creativity. The group leader decides to use NGT. Prior to aggregating the staff members the leader very carefully selects a meeting room that will allow the participants to sit around a table (rectangular) and have the use of a flip chart, butcher paper, blackboard, or other posting instrument. Further, since the staff members total between 15 and 20, the room is large enough to accommodate at least two tables so that no more than nine individuals are seated in a single group. These tables are sufficiently far apart to allow the individual subgroups to have their discussion in recording steps without interfering or overpowering one another's conversations. Finally, there is a wall or surface against which the flip chart sheets can be posted as they are filled during the recording process.

The leader assembles all the supplies prior to the meeting. These include the posting paper, masking tape, broad-tipped felt marker pens (preferably water color rather than permanent ink!), and half dozen or so 3" × 5" cards on which each participant will record his priority items. Work sheets or tablets and pencils are provided so that each person has paper and pencil to record his or her thoughts during the first step. Refreshments are situated in an area away from the work tables but proximate enough to allow for a comfortable break at the end of the voting process.

LEADER'S INTRODUCTION (10 MINUTES)

The leader welcomes the members to the group and establishes a sincere rapport with each of the group members. He or she explicitly requests the participants' complete cooperation and commitment to the seriousness of the issues. The leader informs individuals that they have been particularly selected because of their experience and wisdom on the issue at hand and makes it very clear that the purpose of this particular meeting is to direct attention to the problem, not solutions, gossip, or pat ideas that any one of them may have brought with them. He or she informs each participant that the quality and output of the meeting is clearly dependent on their full participation and asks that each of them participate actively during each step of the structured process.

Since the leader has previously pretested the nominal question, he or she is confident that the question will provoke appropriate information. Consequently, as the leader passes out the nominal work sheets, he or she points out that the question is written exactly the same at the top of the work sheets as it appears at the top of the flip chart. The leader then reads the question, "What do you think are the most important barriers to more creative staff meetings?" A group member asks what kind of response the leader is looking for, and he or she is careful to provide an example that clarifies the question but does not unnecessarily lead the group into providing only a certain set of responses: "For example, when I read this question, I can think of a wide variety of barriers that inhibit me, but the whole purpose of our using this process today is to make sure that each of us fully contributes our own ideas; what this question provokes in me may be similar or could be very different from what you see as the barriers to our more creative staff meetings. What I am asking each of you to do is to think carefully and then write down as candidly and as briefly as possible those ideas and phrases that deter our creativity as you see it."

The leader then explains the purpose of Step 1: "Now that each of you have your own work sheets and can clearly see the question, I am going to ask that we sit quietly for a full five minutes to give each of us an opportunity to do our homework. In response to this particular question, it will be most helpful if you write your ideas down in short, brief phrases rather than in complete sentences, because after we finish our first step I will then be recording each of your ideas on the flip chart paper in front of us. The briefer you are in your phrases, the more quickly the second step of recording will proceed. Please do not concern yourself with proper spelling or punctuation, because we simply will not worry about those things at this time. The quality of this step will depend on each of us fully and silently participating, so I hope that you will use these next five minutes to think and work quietly and with full enthusiasm.

"Some of you may find that it takes a little bit of time to warm up, while others of us will be able to list a wide variety of thoughts and then finish ahead of the others. Whether you start slowly or finish early, it is important that you maintain silence so that each person has an equal amount of time to think and reflect and write. At any time during the process you will be able to add additional ideas as they occur to you, but the importance of this step is to try to generate as many different ideas as you can. Are there any questions on what you are to do at this particular stage?"

The leader then sits down and quietly models appropriate behavior. During this step when members finish early or begin to interrupt their peers, the leader monitors such behavior with a quick, indirect statement.

At the end of five minutes the leader stands up and says, "Some people may still be writing. Please feel free to continue. In the interest of time I am going to start recording. What this means is that I will ask each of you to look at the list you have just finished and give me one short, terse phrase, which I will then write verbatim, exactly the way you have stated it, on the flip chart. I will number each phrase and leave a column on the right-hand side of the paper, where after discussion we will record our priorities and our voting results. But for the second step I am going to try to write as quickly as possible each of your phrases as you give them to me. Now, notice that I am only going to ask you for a single item from your particular list and will then go on to the person sitting next to you. In this round robin fashion I will be able to record each of your ideas, one idea and one person at a time. Therefore if you see that your colleagues have already posted an idea that you feel is identical to your own, please give me a new item; or if you feel as though all of your ideas have already been recorded and you have not yet thought of additional items, simply say, 'Pass.' As we progress through this recording process you may feel free to reenter at any time a new idea that occurs to you when it is your turn. In other words, if you have exhausted your idea list, have said 'Pass,' and then have seen something that provokes a new set of ideas, when it is next your turn please feel free to reenter the recording process.

"It will be very important during this process that we do not criticize or try to restate any of our ideas. The entire purpose of this second step, round robin recording, is simply to accomplish just that, a written list of each of our nonduplicative items. I am sure you will see that there will be many similar items and variations on common themes. We can deal with the issue of categorization at a later point in time should you wish to do so; however, for this process the only conversation should be from an individual who is giving their brief phrase and my recording of it. I will probably restate the item as I am writing to make sure that I have exactly the words as you have given them to me. Again, I am not going to concern myself with misspellings and I will probably abbreviate common terms to hasten this recording process. Are there any questions about the purpose or procedure for this second step?''

The recorder then proceeds as previously described and generally about 30 to 45 minutes later has achieved all of the recording. A volunteer at the table usually assists the recorder in posting the filled flip chart sheets so that the recorder can continue writing as quickly as possible and so that each of the previously recorded items remains posted in front of the group. This posting, as items are recorded, allows each group member to review the list to make sure that he or she is not duplicating previously recorded items.

At the end of the recording process the leader comes back to the group and says, "Now that we have all of our thoughts and ideas recorded, it is important that we go back and reread each of these items to clarify their meaning. The way in which I will do this is to go back to the first item on the chart, read it just as it appears, and ask you if you have any questions. It is important in this discussion step that we limit our discussion to the purpose of clarification. It is not important to try to persuade or convince one another of the relative validity or importance of these items; rather, the purpose here is to make sure that each of us as individuals fully understands each of these short phrases, which, as you can see, are fairly brief. Each of us will have an opportunity to select those that we feel are most important and useful in the next step, Step 4, preliminary vote. For the next half-hour or hour, however, let us go back to clarify each of these items to make sure that when we do cast our vote, it is with full understanding.

"If you have an illustration to offer in terms of what this particular phrase meant to you, please do that. We would like not to force the author of each item to defend or define for the group; rather, this is a group process in which we need to make sure that we share our understandings. I do expect that we will have some disagreement over the importance of these items, but the opportunity to express our agreements and preferences will come in the following steps. Any questions here about how we shall proceed?''

A member of the group asks, "What if I feel that my idea is being misrepresented by the group's discussion of it?" The leader responds, "Well, in that case I would encourage you to suggest that an additional item be added to the list or that you offer an illustration, or we can always add clarifying words or comments. But it is important that I not eliminate any of these ideas, because, after all, we all are considered equally expert on this particular issue. Does that answer your question?'' The member nods his head.

The leader then begins by reading each phrase and asking for questions or clarifying examples. He or she is careful to pace and time the group so that each item receives approximately equal time in its consideration and clarification. Nonetheless, the group needs to be reminded periodically of the remaining time left so as to move on quickly as soon as clarification is achieved. This particular step is very lively and enthusiastic because group members are able to generate a variety of illustrations from almost each phrase. They also frequently become involved in evaluating the merits of various phrases, but with leader's facilitation and cooperation from individual group members they are able to move on to the remaining tasks.

At the end of this step, which takes approximately one hour (30–40 seconds per item), the leader introduces Step 4, preliminary vote. "Now we are at the point where we can select a subset of priority items from our common list and then, as a group, rank order our individual preferences. For this process, I am going to distribute these 3" × 5" cards. Pleast count them to make sure you have seven per person. Once you get your cards, I would like you to get into a position where you can clearly see all of the completed chart pad sheets so that you can select seven priority items from our entire list. This means that you will probably have to get up and walk around to make sure that you can see each item clearly. Please feel free to do so if necessary.

"The important point here is to very carefully select those seven items that you feel are the most useful and most important issues for us to be concerned with. As you make your selection write both the number of the item as it appears on your list and at least a few key words from that phrase if you do not wish to write the entire phrase itself. You will see, as we go about recording our voting values, that it will be most important to distinguish the number of the phrase from the value we will give it as part of the ranking procedure. So the reason we ask that you list a few words or the complete phrase in writing is to make sure that we know precisely what phrases you are casting a preference for. Are there any questions about how you proceed?''

A group member may say, "Let me see if I understand what we are supposed to do. First, you want us to make sure we can see everything so that we do not inadvertently overlook an item of importance. Then we are to take these 3" × 5" cards and as we pick out an item, write it down, both the number and the phrase on the card. And you want us to write only one item per card, I gather. Then, do you want us to put them in priority?'' ''No, we will do that as an entire group,'' the leader responds, ''Some of you will probably take more time than the others to complete your selection process. If you do finish ahead of your colleagues, simply double-check your cards to make sure that you have the number and the phrase accurately recorded, then sit quietly and after a few minutes I will call time and we will do our rank ordering as a group. The reason I ask you to rank order your cards with me is that I will be using a particular numerical procedure for doing so. Thank you for asking that question. That was very helpful.''

The group members then begin to make their individual selections, and after approximately five minutes they have completed writing one item per card.

The leader then gives the following instructions: ''Now I would like you to spread out all of your individual cards in front of you so that you can see them all at one time. Looking at your entire set of seven, now select that item that you feel is the most important phrase you have selected. Give that item a value of seven; we are putting seven cards in rank order and the highest value for your highest or most important card is the value of seven. Make sure you draw a circle around that number or somehow distinguish it from any item numbers that may appear on your card. Again, give your highest or most important card a value of seven.

''Now, turn that card over and you have six more items remaining. Now select the least important item and give that card a value of one—least important item gets a value of one. Please stay with me, because there is a variation in this procedure.

Now, you have five remaining cards; looking at the most important of the remaining items, give that card a value of six: The most important value of your remaining items gets a value of six. Okay, turn that card over and you should have four remaining ballots. Select the least important and give it a value of two; look at your remaining cards, give the most important item a value of five; now you have two cards left, give the least important card a value of three and the last remaining card automatically gets a value of four. Now turn your cards over and make sure that you have no values that are greater than seven.'' The group may chuckle and laugh as one person makes a correction, having confused his voting value with his item number. The leader continues, ''Now, making sure that you have rank ordered each of your phrases and item cards, pass those cards over to the person sitting next to me, who will then shuffle them and post your vote.''

The assistant reads first each item number, then the voting value, while the recorder posts these numbers on the original chart pads. At the end of this recording process the entire group is able to see the relative ranks they have assigned to each of the items that have been selected. They are also able to see that many of the items have not been selected at all. In this way the breadth of the items of concern, as well as their relative rank order, is established. Each of the voting values is summed per item so that the entire list is then completed, and the various ranks are identified by Roman numerals; that is, the item receiving the highest voting value becomes I; the next highest sum becomes II, and so on. In situations in which items receive an equal value or sum of votes, they each are given the same rank order, that is, Roman numeral, and are considered a tie.

The group is usually very interested in seeing these numerical results through their list and may want to discuss several of the items that get similar votes or when a single item gets several high values and several low values. They may decide at this point that it is not important to recombine similar items into general categories, however; they think that should they wish further details they may come back to this preliminary set of results, recombine similar items under general categories, and then consider a second final vote. The leader encourages them to nominate two or three members of the group to create category titles and decide which phrases would be combined under which categories. He or she cautions that even though categorization will occur, it is not "legitimate" to alter any of the original phrases; that is, each phrase should be rewritten in its original form, with its original group number under the newly created categories. Furthermore, should individuals wish to cast a second vote, they have the option of selecting a single item or a categorical vote; in either way the individual's preference or priority will be expressed.

After a brief break the leader asks the group to reconvene for a few minutes to discuss their reaction to this process. Group members may be surprised at the length of time they have actually been interacting. Others may wonder how useful this process may be with more complex problems. The leader responds that NGT is clearly a single-purpose or single-question group process. It provides a full range of participant interaction but is limited to focusing on a single issue. It also demands that participants be knowledgeable prior to participating in this particular group process. A discussion may follow in which common problems in using NGT are identified. These include the fact that some participants who are happier with greater prominence or status feel restricted. A member may bring up the fact that certainly this process does not guarantee that the higher-level decision-makers will necessarily accept their input, although it does allow for the full participation of each member. It is clear that this procedure demands adequate physical facilities and space and time. The skills of the leader and/or recorder are very important so that that person does not consciously or unconsciously manipulate the group. Perhaps the most important point is the need to be very careful to frame and pretest the nominal question; a poorly written question may generate more useless information than anyone would ever wish to summarize! One advantage to using this procedure is the fact that even with a large number of participants, smaller groups can be determined and a large number of people can simultaneously provide input, the difficulty, of course, being the aggregation and summarization of that input at a later point in time.

SUMMARY

In summary, NGT is a very useful and successful group process technique for problem identification or designing solution components. It does demand appropriate group composition, physical facilities, time, and follow-up expertise. In general, members using this technique have found it as successful and as edifying as other group meeting processes.

CHAPTER 9

PROGRAM PLANNING METHOD (PPM)

SANDRA L. GILL
ANDRÉ L. DELBECQ

The program planning method (PPM) was developed in 1971 by André L. Delbecq and Andrew H. Van de Ven. PPM offers an explicit process for structuring the character of participation within each phase of planning. It assumes that the cornerstone of successful planning is accurate, valid problem identification and that the best-qualified people to identify problems are the groups affected by the potential program or product for which planning takes place.

The model was originally distilled from studies of technological and product innovation and was then empirically tested in the sociopolitical context of local agency and community planning endeavors.[1] As a planning method, it is both systematic and quasistructured, in which particular individuals are incorporated at specific stages in a seven-step sequence.

The benefits of PPM have been empirically validated elsewhere.[2] In general, PPM provides a process by which clients and consumers can be usefully incorporated in the overall planning process. This generally results in an increased legitimacy of the program or product in the view of the recipients or users. It also decreases their potential resistance to the program implementation phase and typically increases the overall program or product effectiveness because of careful incorporation of the concerns of all parties during the design phase.

AN OVERVIEW OF PPM

In this article we wish to identify tasks associated with the major planning phases and offer a variety of strategies by which each of these subtasks can be accomplished. A complete bibliography is provided for further evidence of research results and application of PPM.

An overview of the PPM is provided in Figure 1. In general, PPM is a process for organizational innovation, that is, the management of large-scale organizational change. The use of PPM is appropriate in situations in which major behavioral, cognitive, or attitudinal changes are to result

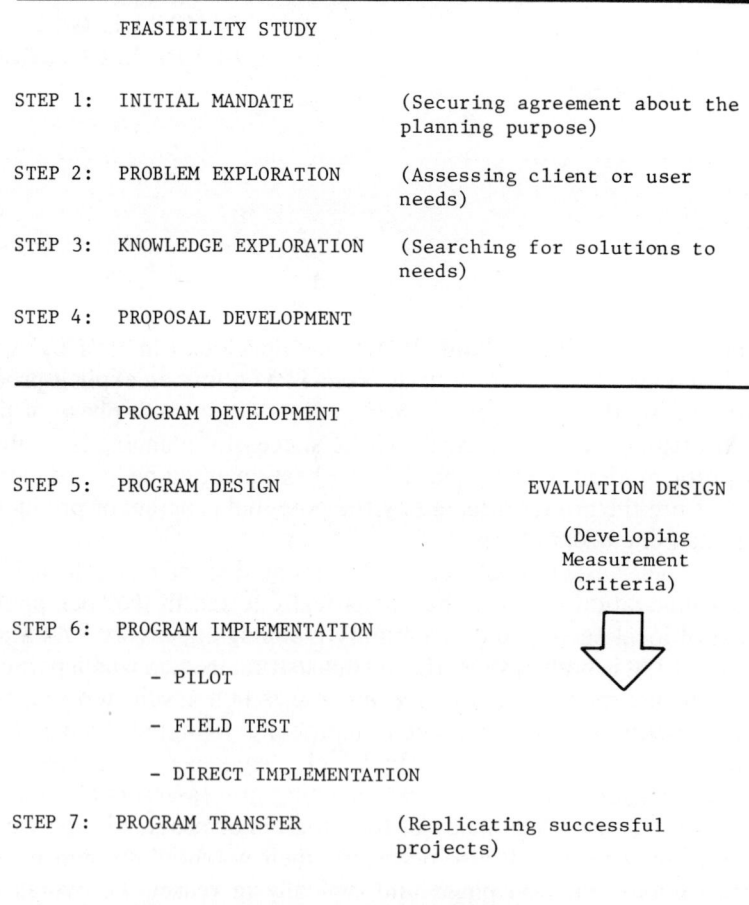

Figure 1 Overview of PPM.

from the outcome of the planning endeavor. Research on product and technological innovations shows that innovative changes require much greater effort than a brief, seemingly inconsequential temporary change in the behavior of a small group or a single individual. Rather, here we are looking at planning for a program, project, or product that will require major reorientations in thinking and behavior across a large group of individuals.

PPM incorporates seven major phases: The first phase, the *initial mandate*, focuses on securing agreement about the planning purpose from a representative group of decision-makers. The second phase, *problem exploration*, is concerned with client or user needs assessment. The accurate identification of the problem or basic conditions for which an innovation is required is then followed by the third phase, *knowledge exploration*, in which a creative search for solution or design components specific to the problem previously identified is conducted. Following these first three phases, a *preliminary proposal*, phase four, is prepared. The preliminary proposal is submitted to a structured review process and revised into a *formal proposal*, phase five, on which the program design is then based. The sixth phase, *program implementation*, may take one of three major forms. We will describe the characteristics of each of these forms for program implementation in detail. Finally, and hopefully, there is that last state of transferring an innovative program to a broad scale replication in which the successful program or project elements are duplicated in a wide variety of settings. The program or project evaluation is incorporated during the last three phases of design implementation and, ultimately, transfer.

ANALYZING PLANNING COMPLEXITY AND RESOURCES

The recent research of John Bryson (1978, 1979)[3,4] demonstrated that when given a choice, "wise and experienced planners" do plan contingent to the planning circumstance; that is, as resources become more constrained and/or as the planning conditions become more complex, wise and experienced planners will choose a greater number of strategic planning alternatives.

Bryson's research demonstrated that successful planners engage in a preliminary assessment of the planning circumstance. This assessment includes a ball park estimate of the planning resources: available time for planning, staff, and available budget resources. In addition, wise and experienced planners are careful to assess the complexity of the planning project: How many groups must be involved in the planning endeavor

VARIABLE	EASY TO DEAL WITH	DIFFICULT TO DEAL WITH
Number of groups involved	Single	Multiple
Degree of value agreement		
Awareness of problem	High awareness	Low awareness
Priority given to problem	High priority	Low priority
Intensity of concern	High intensity	Low intensity
Technical difficulty		
Understanding causation	Causation understood	Causation not understood
Technological sophistication	Simple, routine technology	Highly sophisticated technology
Time available	Ample time available	Severe time constraints
Money available	Ample money available	Severe money constraints
Impact on organization structure	No reorganization required	Significant reorganization required
Impact on resource allocation	No change in resource allocation pattern required	Significant change in resource allocation pattern required
Coalition development	Established coalitions and stability in the organization network	Non-existent coalitions and turbulence in the organizational network
Character of planning staff	Large and skilled	Small and unskilled
Technical quality of proposal	High technical quality	Low technical quality
Environmental stability	Stable	Turbulent

Table 1 Conditions Regarding the Complexity of Program Planning Efforts

(for political or economic purposes)? How large should they be? What is the extent of the relative agreement between these groups on basic values and key issues? What is the degree of organizational impact that a potential planning effort and outcome will have on the existing organization form or structure? To what extent are there "slack resources"— time, staff, budget staff, support services, and so on—that can be called upon if needed during the planning effort? To what extent is the actual technology of the proposed planning effort understood? Is the proposed project something that has been clearly tested in previous situations, or is it experimental or pilot in nature? To what extent are the communication and administrative networks already established—or will this project demand a coalition of fragmented, diverse individuals who do not share a common organizational membership? These elements are illustrated in Table 1. In Bryson's simulated contingent planning model the experienced planners carefully analyze these conditions and then choose their planning strategies in such a way as to maximize their access to needed resources. In some cases one would imagine that a planning situation would be so complex and the resources so limited that the chances for success would be minimal, and that such a project might even be terminated or refused. Many of these judgments will demand intuition and personal experience and are too complex to specifically analyze here. Instead, what we propose is a series of tasks and strategies, each linked to the successful completion of a series of planning phases, which, when achieved in a linear fashion, do promote the probability of successful planning for innovation.

PHASE 1: INITIAL MANDATE (SECURING AGREEMENT ABOUT THE PLANNING PURPOSE)

TASK 1: DEVELOP AN UNDERSTANDING WITH KEY DECISION-MAKERS

The first key task in securing the initial mandate is to develop an understanding with key decision-makers about the overall planning mission and major planning steps. This would include a discussion of the value and timeliness of planning for a particular purpose at this point in time. What is it that makes this particular issue salient or critical to our organization? Why is it important that we look at this issue now rather than at a later point in time? Who are the key groups, representatives, or constituents who should be involved at each stage of the planning effort? What is the form and frequency of reports that key sponsors and decision-makers wish to have as a form of feedback information? What kind of resources are likely to be necessary during the initial planning stages?

These and related issues are key components for testing and securing the initial mandate. If there is very little agreement or resistance to the discussion and contracting of these initial items, then it is perhaps a signal that further resistance will occur, especially if planning efforts become more expensive and demand more resources. Hence the major suggestion is that if the planner cannot engage in discussion and secure agreement on these preliminary issues, then contingency planning is a significant risk.

TASK 2: FORM A COORDINATING COMMITTEE

Assuming that the planner can engage in discussion and reach agreement on these previously mentioned issues, then the next task is to form a coordinating committee, either formal or informal, in order to link the planning effort to three critical perspectives: the administrators', the resource controllers', and the key client's or provider groups'. The responsibilities of this formal or informal coordinating committee are several: (1) It provides the planner a support base through which issues and planning endeavors can be discussed and ironed out; (2) it provides an additional resource link for the planner so that he or she does not become perceived as the single advocate or representative for the particular project or endeavor; and (3) it enlarges the communication base or network; presumably, each member of this coordinating committee would assist in clarifying and communicating within their own network the various procedures and progress levels that are being achieved throughout the course of the planning project.

Achieving an understanding about the initial mandate and the value and timeliness of the planning effort can be accomplished in a variety of ways. The most useful clue is to remember that organizations, as collections of individuals and groups, respond most to those issues that potentially bring fruit or present a threat. When an issue threatens the organizational domain or provides an opportunity for organizational success and prestige, the probability of sponsoring a full planning effort is increased. If, in addition, that portrait of both the benefits and costs of the current planning effort can be supplemented by endorsements from key prestigious or influential colleagues, and supporting documentary information and alliances from influential locals, then the strength of the mandate and the resource base of the initial coordinating committee is enlarged. This provides a basic framework against which questions and threats to the purpose and legitimacy of the planning effort are judged.

Clearly, the absence of clarification at this level can be very dangerous when planning proceeds to more advanced phases.

PHASE 2: PROBLEM EXPLORATION (ASSESSING CLIENT OR USER NEED)

Having achieved the initial mandate, the second major phase in PPM is to conduct a very thorough problem exploration. The purposes or tasks of the problem exploration stage are (1) to identify and document both the amount and character of client or user need, (2) to provide unmanipulated, valid evidence regarding these needs and the resulting problems, and (3) to identify the potential threat or opportunity of response to these needs at this time.

Needs assessments that pertain to the amount and the character of the problem situation can be very powerful. For example, it is persuasive to be able to demonstrate that 80% of a given target population are in need of a given service or program. It is even more persuasive, however, to be able to portray what this need actually feels and looks like and how it impacts on an average client or user. Further, such needs should be demonstrated through valid, objective evidence. The true nature of the problem is far more persuasive than rhetorical. For example, does the needs assessment strategy identify the range of difficulty clients or users are experiencing? Not all users or clients will have the same perceived degree of need or range of difficulty. What are the differences between the client groups who share this need? Do different subgroups within the target population demand different kinds of services? Does the identification of several differences within the target population provide any predictive evidence or suggestions about future needs? What will happen if current needs are not responded to in a timely and professional fashion? What are the costs or benefits of responding to these needs at this time?

Bryson's research showed very clearly that when resources allowed, planners consistently chose to conduct site visits with clients. Field observations and site visits were deemed invaluable in capturing the character and differential range of need among clients or users. In addition, a variety of other strategies can be useful. Conducting client or user problem identification sessions, using NGT, force field analyses, brainstorming, and so forth can provide a wealth of objective and subjective information. These group processes are very useful when a variety of subgroupings among the target population are present, that is, different cultural, professional, religious, or economic concentrations.

Providers of similar services or respondents to similar needs can be an invaluable use of problem identification in terms of identifying the barriers or difficulties associated with a response to such needs. Decision-makers and resource controllers also need to be incorporated in the problem identification arena so that their perspectives can be integrated along with client and provider opinions. In some situations the use of hearings or media can be valuable as a method to analyze public sentiments. Existing information sources, through literature searches, data banks, prior surveys, and so on, can very often provide basic information that need not be duplicated in the current problem identification phase. New distributive information can be obtained through formal or informal research.

In sum, a variety of strategies are useful and necessary for the problem identification phase. To provide a persuasive portrait of the problem such efforts should combine objective needs assessment with an accurate portrayal of client/user concerns.

PHASE 3: KNOWLEDGE EXPLORATION (SEARCHING FOR SOLUTIONS TO NEEDS AND PROBLEMS)

The primary task of knowledge exploration is to engage appropriate experts in the identification of critical solution components. There are several characteristics of this particular search. First, it is important to assess the different perspectives that a variety of experts have on the problem portrait previously identified in phase 2. In addition to the alternative perspectives or schools of thought, it is most important to identify which, if any, of those perspectives have been operationalized; that is, what approaches have already been tried and what are the results of those trials? The essential task is to survey a wide variety of experts in the identification of the essential criteria for a successful solution so that a proposal can be based on a range of theory and experience, parts of which are determined to be essential for program or project success in the current plan.

There are a variety of methods for obtaining such information. Resources permitting, the use of internal and external consultants and experts is highly recommended. The research on technological innovations demonstrates that the majority of innovative ideas has been adopted and pretested in other professional and technical arenas; these ideas constitute an innovation simply because they have not yet been tried in any particular organization or professional group. The cost of duplicating a project that has been pretested elsewhere is far less than the cost of absolute, creative innovation. Literature searches through professional and trade journals, the utilization of resource panels, internal and/or external experts, or consultants, and staff investigations are but three approaches to the incorporation of expert perspectives and opinions.

Another recommended strategy is to use the telephone survey. Essentially, the interviewer prepares a protocol or a checklist of key questions for probes that he or she thinks will be useful in the course of a phone conversation. A paragraph summary statement of the problem is useful so that a specific issue can be clarified immediately with the expert who has been contacted. Using an initial list of rcommendations provided by colleagues, staff, administrators, and so on, the interviewer then telephones each of these experts and also asks each of them to identify other colleagues or individuals who would be useful on that particular issue. This "snowball" method of obtaining nominated experts will eventually identify several common names who then become the key resource individuals to contact.

Regardless of the strategy by which experts are involved, it is important to include a variety of expert perspectives: funders, providers, prior users, and professional technical perspectives.

The outcome of knowledge exploration is a rich, generally diverse set of clues that provide a wealth of information for designing an appropriate solution to the problem previously identified. The remaining task is to assemble that set of clues and suggestions and enter a proposal, which is then subjected to a preliminary review procedure.

PHASE 4: PRELIMINARY PROPOSAL REVIEW

The purpose of the preliminary proposal review is to secure constructive criticism by the various elites as well as to bring them up-to-date on the planning effort progress and secure their commitment and support for further progress.

The procedure for conducting proposal review sessions has been detailed elsewhere.[5] We will summarize it here. Essentially, the preliminary proposal should look like an early draft rather than a complete document. A two- or three-page outline is quite sufficient for this informal preliminary review. This outline should include a summary of the planning steps thus far, that is, the fact that an initial mandate or charter was obtained with the involvement of key decision-makers, that a systematic needs assessment or problem identification process was conducted, that subsequently to the identification of the essential problem elements a search for solutions was conducted (the list of the experts and their titles and affiliations can be attached), and that the purpose of this preliminary review is to incorporate constructive criticism and exchange information on the progress of the planning effort thus far.

By preparing a work sheet so that strengths are identified and "modifications that would improve the proposal," rather than weaknesses, are elicited, a constructive critical review process is more likely. By focusing on strengths, the planning staff or chief planner is reinforced and encouraged on important perceptions of his or her progress so far. The use of the term "modifications that would improve the proposal" also puts the burden of constructive criticism on the reviewers; this alleviates the translation and demoralization of dealing with a laundry list of negative comments at the end of a difficult planning effort. Using NGT, the various strengths and modifications followed by discussion and possibly even a rank ordering of the various strengths and modifications can be very useful if the proposal tends to be controversial. Regardless, the outcome of the preliminary proposal review is to identify the perceived strengths, incorporate suggested modifications, educate and communicate the outcomes of each of the planning phases to date to those reviewers, obtain a commitment to resources and support needed for proposal implementation, and generate a sense of enthusiasm among the various reviewers and stafff.

PHASE 5: THE FORMAL PROPOSAL

This phase produces modifications of phase 4. Research on proposal adoption[6] demonstrates that proposals with high rates of adoption or acceptance have several key characteristics. First, they present clear alternatives or a single alternative that has been endorsed by prior reviewers and has high perceived quality and validity for the problem. In addition, a flexible budget is attached so that the reviewers still feel some sense of partnership in the proposal design. The relative advantage of budget components is clearly presented. Identification of the implications for administration and monitoring of the proposed project are also incorporated in the proposal document. Finally, a valid and clear evaluation plan is incorporated.

Depending on the funding source, of course, the actual format for a given proposal will vary. Nonetheless, these characteristics of successful proposals need to be attended to in either the verbal or narrative proposal form.

PHASE 6: PROGRAM IMPLEMENTATION

The alternative strategies for actual program implementation essentially focus on three increasingly mature forms, namely, (1) a pilot test, (2) a demonstration, and (3) direct implementation. Each of these will be detailed.

THE PILOT TEST FORMAT

The pilot test is an implementation alternative that is very important when the actual effects of the proposed program cannot be determined ahead of time. A pilot test to determine effects and impacts is desirable in terms of justifying additional resource allocations, site location, staffing combinations, and so on. Also, in a situation in which resistance among clients, staff, or administrators is high, a small-scale pilot project is much more feasible than a large-scale, expensive demonstration or implementation.

THE DEMONSTRATION FORMAT

Assuming that the cause and effect relationships are known, that resistance has been overcome, and that the basic implementation requirements are specified, a larger-scale demonstration project is in order. The purpose of the demonstration project is to evaluate a field test of the proposed project and to elicit enthusiasm in ownership among future implementors. Strategies for increasing the sense of enthusiasm in ownership include the participation of a coordinating committee and selected observations and visitations by key potential adoptors. These visitations should be budgeted into the program proposal.

The validity of the demonstration project is that the project can operate and is evaluated in a variety of field situations, rather than in the secure, nurtured environment of a pilot test. In a demonstration one would hope to have at least three different alternative site locations so that the relative effects of the program or project can be evaluated against different circumstances. During a demonstration project it is important that there be significant resources allocated for technical assistance, troubleshooting, the development of implementation, and staffing guides. Further, at least two rounds or sequences of the program are helpful so that debugging and learning can take place during the first round and stabilized effects can be recorded during the second round. A careful evaluation of the core program and its administrative and organizational impacts and implications is essential for demonstration purposes.

THE DIRECT IMPLEMENTATION FORMAT

Assuming that both the pilot and the demonstration are successful, a program is then ready for direct implementation across a range of implementation circumstances. Direct implementation is advocated only when support from both the professional and administrative components is high, when the cause and'effect relationship of the program to its client or target population is understood, when the implementation processes have been simplified and are supported through operational guides and technical assistance, when careful orientation and training of staff and administrative personnel is included in the budget, when there is time for an ample technical learning curve, and when the project is carefully controlled and monitored to make sure that it is implemented according to its original design.

PHASE 7: PROGRAM TRANSFER

Finally, assuming that the program has progressed through these basic stages, from pilot to direct implementation, then the innovation is ready

---

```
Step 1. Initial Mandate

            Task 1:   Develop an Understanding with Key
                      Decision-Makers

            Task 2:   Form a Coordinating Committee

Step 2. Problem Exploration

Step 3. Knowledge Exploration

Step 4. Preliminary Proposal Review

Step 5. The Formal Proposal

Step 6. Program Implementation

             • Pilot Test Format

             • Demonstration Format

             • Direct Implementation Format

Step 7. Program Transfer
```

Figure 2   Summary of steps and tasks for PPM.

for successful replication in a variety of sites. This, then, becomes the final stage, program transfer.

A summary of PPM is shown in Figure 2. In general, such a process takes between 12 and 18 months under benign circumstances. When the political context or technical complexity of the planning endeavor is high, then the planning steps necessarily become more complex and demand more time and resources for successful completion. Both the sequence and the effectiveness of using PPM in contrast to other planning methods have been demonstrated. Finally, this procedure is advocated for innovative planning and is not necessarily assumed to be an essential planning sequence for routine program or project administration.

REFERENCES

1. Andrew H. Van de Ven and Richard Koenig, Jr., "A Process Model for Program Planning and Evaluation," *Journal of Economics and Business*, Vol. 28, 1976, pp. 170–171.

2. Andrew H. Van de Ven, "Problem Solving, Planning and Innovation, Part I: Test of the Program Planning Method, Part II: Speculations for Theory and Practice," *Human Relations Journal*, Vol. 33, Nov.–Dec. 1980, pp. 10–11.

3. John M. Bryson, "A Contingent Approach to Program Planning," Ph.D. dissertation, University of Wisconsin, Wisc., 1978.

4. John M. Bryson and André L. Delbecq, "A Contingent Approach to Strategy and Tactics in Project Planning," *Journal of the American Planning Association*, Vol. 45, No. 2, April 1979, pp. 167–179.

5. André L. Delbecq, Andrew H. Van de Ven, and David H. Gustafson, *Group Techniques for Program Planning: A Guide to Nominal Group and Delphi Processes*, Scott-Foresman, Glenview, Ill. 1975.

6. Arthur P. Brief, André L. Delbecq, Alan C. Filley, and George P. Huber, "Elite Structure and Attitudes: An Empirical Analysis of Adoption Behavior," *Administration and Society*, Vol. 8, No. 2, 1976–1977, pp. 229–48.

CHAPTER 10

A ROLE-ORIENTED APPROACH TO PROBLEM-SOLVING

CHARLES H. BURNETTE

The Role-Oriented Approach to Problem-Solving is a technique for organizing and managing both problem-solving activity and the information involved. It may be applied by individuals or groups, formally through an institutionalized system or informally as a reference for thought and communication. It provides a framework to facilitate problem-solving thought, communication, and behavior, a framework that compliments and supports open-ended techniques of idea generation such as brainstorming, free association, and analogous thinking by providing an operational context in which to assimilate their products.

As problems become more complex and difficult to solve, problem-solving requires the contribution of knowledge and ideas from various disciplines and from people of diverse backgrounds. A way to structure the effort of a group that promotes the solution of complex problems is of increasing importance. Such a process should marshal a comprehensive range of perspectives, information, and potential actions relative to a problem, encourage the free flow of ideas, and structure the efforts of several people into a cohesive, cooperative, and efficient team. In addition, to cope with the growing amount of information a problem-solving method should have the capacity to facilitate both natural and computer-aided communication. Unfortunately, most efforts to assist group problem-solving have focused on the facilitation of the psychological component of the experience without adequately dealing with the information and communication problems involved.

The role structure of the method to be presented addresses these needs by assuring that a balanced set of viewpoints is represented, that a comprehensive range of information is brought into play, and that a common framework for information handling and intergroup communication exists. Each role in the system, whether assumed by an individual or developed by the group as a whole, represents a specialization of information and expertise that is commonly understood and used as a basis for organizing information, communication, and the activity of problem-solving. This use of the framework as a reference by the group is augmented by various techniques of idea generation to assist creative thinking and by principles drawn from the study of group dynamics to facilitate productive behavior. The result is a very effective technique for managing problem-solving activity, for structuring and communicating information during problem-solving, and for analyzing the process of problem-solving and its results.

BACKGROUND

The Role-Oriented Approach to Problem-Solving did not originate as a method in group dynamics but developed as a byproduct of the design of a comprehensive system for computer-aided communication in architecture.[1] Developed with careful attention to the structure of thought and expression, the system was designed to accommodate the normal communications between specialized professionals in the building industry. Given the rich variety of problem-solving communication in architecture, it was evident that the system could be applied to any problem-solving activity, independently of any particular media or application.

The theory behind the process was developed by the author between 1962 and 1969, while obtaining Master's and Ph.D. degrees in Architecture at the University of Pennsylvania. During that time the development of the theory benefited from research on an information system for hospital planning supported by the U.S. Public Health Service[2] and on a classification system for the building industry supported by the Center for Building Technology, National Bureau of Standards.[3] The theory was strongly influenced by linguistics (Chomsky,[4] Fries[5]), cognitive psychology (Bruner Goodnow, and Austin,[6] Miller,[7] Piaget[8] Feigenbaum and Feldman[9]), library and computer sciences (Ranganathan, [10] Vickery,[11] Knowlton[12]), and the practice of designing. The distinctions of the system resulted from the correlation of a vast number of linguistic forms, process models, and content classifications held by their authors to be comprehensive regarding the subject matter represented. Among the most powerful of these models are the conventions of scientific disclosure,[13] the structure of computer languages,[14] and the gestalt principles,[15] each of which can be interpreted in a manner consistent with the distinctions of the system. The system of distinctions became so useful in group problem-solving that the computational form of the system became secondary and, although of great potential utility, has not yet been implemented.

The first use of the scheme as a generalized method for problem-solving occurred at the Philadelphia College of Art, where it became a regular part of the professional practice course for industrial design students and where much of the insight regarding its use as a group problem-solving method was developed. During this period one use of the system was to help the class organize itself into a design firm with a well-defined marketing approach consistent with its members' aspirations and the realities of the marketplace. From this experience it became evident that the method could be used to assess personal aptitudes and affinities for the various roles, to characterize and facilitate interpersonal behavior, to build a powerful sense of group identity and motivation, and, in general, to produce design consensus quickly and effectively. Similarly, it was clear that the system did not require rigid interpretation—anyone could play any role, everyone could share all roles, or the assumption of the roles could vary at will, as long as others in the group understood the orientation employed.

Presentations of the method at an art school, a vocationally oriented university, and an academically distinguished university also revealed the effect of intellectual and cultural bias in the acceptance of the technique by different users: The art school students entered into the experience without inhibitions and adapted it freely to their purposes; the vocationally oriented students worked through the method literally and sequentially, seeking the direction of the instructor; and the academicians dissected the method intellectually before reluctantly using it, forming opinions in the process that inhibited their ability to enter into the experience. While all three groups were ultimately successful in their use of the system, this comparison underlined the need for a behavioral, experiential approach when introducing the system for the first time. Users appear to be most comfortable with the system when it is presented as the framework for a cooperative learning experience rather than as a system to be fully understood before use. Although possible, it is ill advised to lead a group to discover the distinctions of the system in its own behavior. In one experiment members of a departmental faculty engaged in planning course assignments were asked to describe what they would most like to teach; this was then paraphrased in the "language" of the method and happily accepted by each person. When the underlying structure of distinctions and each person's self-specified role within it were revealed as a comprehensive and coordinated whole, the entire system of courses was rejected as too constraining on the invididuals involved. While such reactions might be avoided by acknowledging the distinctions of the method when they occur, it is much more effective to introduce the system as a tool for cooperative use and to let success in using it become its justification.

The method received its first major public exposure at the 1971 Conference of Environmental Design Educators in Key Biscayne, Florida, where it was applied to the problem of developing public awareness of the Martin Luther King Boulevard Project, a major renewal effort then underway in Miami. At this very successful overnight session it was demonstrated that the group facilitator could withdraw once the method was understood and that the system could be used to generate and document concepts with great efficiency; under extreme time pressure the group began to serially process ideas by passing a problem statement card down the "roles" to be filled out appropriately by the specialist in each role. An entire library of proposals fully specified in the comprehensive categories of the system, with a method for evaluating them by the community, was developed within a two-hour period for presentation to the community. As an outcome of this experience, many participants introduced the method into their schools the following year. The author subsequently led a session at Yale focusing on the development of a design program for a community center in an existing railroad station. Multiple actors played the users' role in order to simulate the experience that would result.

At the University of Washington, Seattle, the method was used to solve the problem of determining the population mix and thus the economic basis for the ecologically sound and socially equitable development of a site with excellent water recreation potentials. The solution focused on the communicator's role. The approach was to advertise the way of life in the media appropriate to each social class in order to determine the housing mix and financing plan according to the responses obtained. Further insight into the manner in which each role can contribute to the solution of a problem was illustrated in an application of the method at the Philadelphia College of Art, where the problem was to conceive of a city hall that would directly manifest representative democracy. One student, in the normally passive "resource-specifier" role, brilliantly pre-empted all other roles by specifying that the only construction materials that could be used were those which were brought by hand from the neighborhoods of the city.

The method can also be used to organize the problem-solving efforts of large groups. In a session at the Nova Scotia Technical College over 50 students worked in groups on a single problem. Other demonstrations of the method at the University of Detroit, Louisiana State University, and elsewhere typically led to closure on a solution and consensus regarding the solution among those involved (although the strength of the consensus varied between groups).

THE ROLE STRUCTURE

The basis for the technique is a structured relationship between seven functional "roles" whose utility in problem-solving can be demonstrated (see Figure 1). While the choice of seven roles is arbitrary, it corresponds to the limits of short-term memory and reflects other findings from the study of linguistic forms, gestalt principles, cognitive behavior, and group dynamics, which suggests that $7 + 2$ is the optimum number of distinctions to be considered in a single context.[16] This seven-part framework for the method functions in a manner that is somewhat analogous to the syntax of a sentence in that the "roles" have a functional relationship to one another that is similar to that linking the elements of a sentence. Each depends on the others for subject matter, meaning, and context. As with a sentence, this "role structure" is content free and may be applied at any level of discourse and to any subject, with the degree of license appropriate to the circumstances. Like the sentence form, it provides the organizing framework for a "complete" unit of expression, the elements of which can be easily related to those of other such expressions or used as a reference standard through which to assess and explore the structure and meaning of alternative "expressions" (as in poetry).

Within the closed context provided by this sentencelike structure the individual "role elements" partition the subject matter of the problem or subproblem to which it is applied according to the seven functional categories of the system. Like the noun and verb forms in a sentence, these seven "role categories" can have many interpretations without losing their operational meaning. For example, both the following interpretations represent the distinctions of the system:

1.	Directions	1.	Goals
2.	Elements	2.	Resources
3.	Combinations	3.	Organizations
4.	Forms	4.	Programs
5.	Processes	5.	Implementation techniques
6.	Products	6.	Experiences
7.	Measures	7.	Evaluations

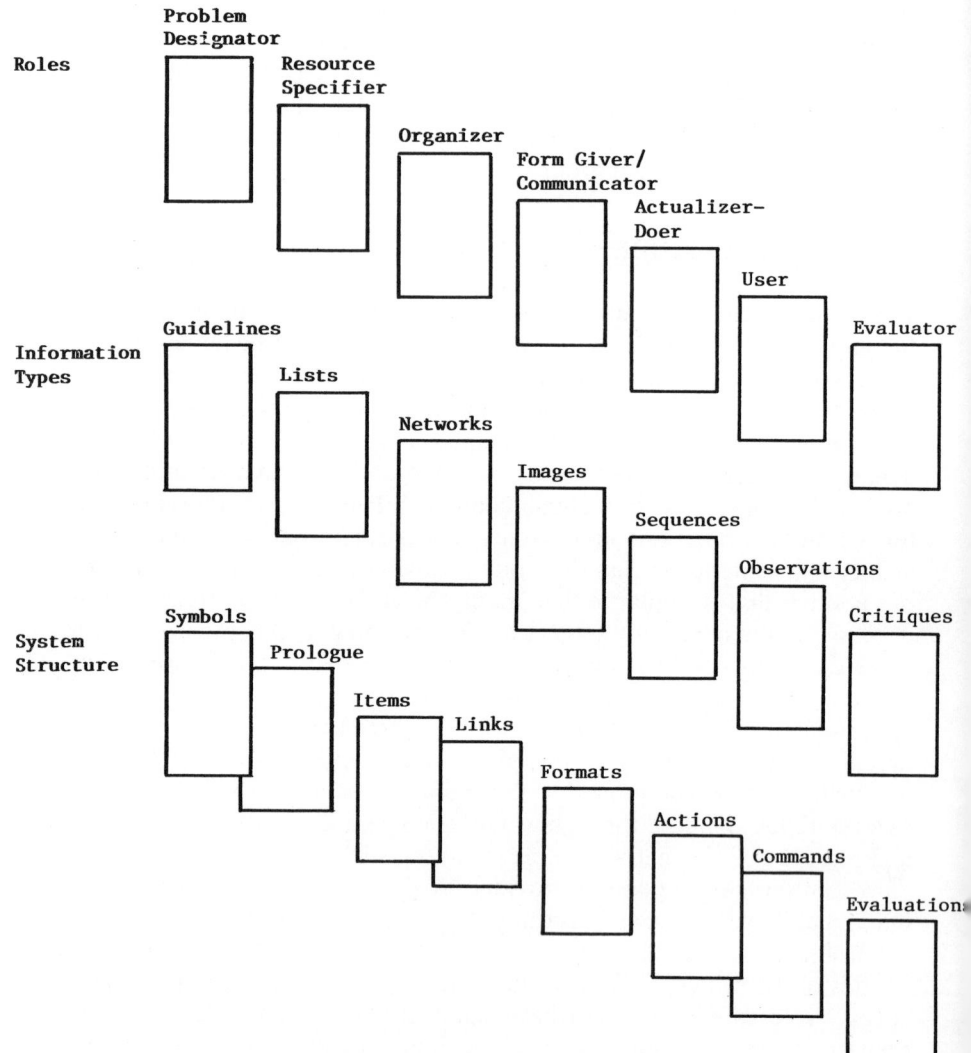

Figure 1 System roles and information types.

In each example the first "role category" is concerned with interpretive information, the second with substantive specification, the third with substantive relationships, the fourth with communicable representations, the fifth with procedural operations, the sixth with empirical description, and the seventh with analysis. As with any linguistic system, the particular labels selected to represent each categorical distinction depend on the subject, the level of abstraction at which it is being considered, and the interpretation of the distinctions of the "language" by those who use it. For example, an architect and a structural engineer may communicate with one another about the design/build process by agreeing to represent it as consisting of seven stages: client interviews, programming, schematic design, final design, working drawings, construction, and post-occupancy evaluation.

If the focus of their consideration changed from the overall process to their particular professional roles, they might choose to interpret the distinctions of the method otherwise: for example, the seven stages for the architect might be project requirements, spatial elements, spatial organization, building form, building construction, functional use, and cost effectiveness. For the engineer, the seven stages might be loading conditions, structural elements, structural principles, structural systems, erecting techniques, behavior under load, and mathematical analysis.

Both perspectives are valid and by no means exclusive or absolute interpretations of the "role categorizations" of the method. They illustrate the fact that each "role" represents a functional orientation that can itself be characterized in terms of particular goals, elements, organizations, forms, processes, products, and measures and may therefore be a complete unit of expression. Thus each role categorization constitutes a complete subsystem expressing one element of the structure of the general system.

More generally, each role represents a specialization regarding the type of information involved, the conceptual patterns that apply, the forms of representation to be used, the processes and skills required, the consequences that can be anticipated, and the values and measures that are appropriate. Each represents one aspect of the overall scheme, expressing that focus in terms of the other aspects.

People easily comprehend and accept the distinctions of the method when they are expressed in terms of the type of information associated with each role, and authority over that type of information is vested in the role. The authority to introduce, deny, or change information being dealt with relative to a given role provides its operational aspect, while the type of information associated with the role defines its substantive aspect.

THE METHOD AND ITS USE

A workshop format centering around a problem brought by one or more of the workshop participants is typically used for teaching the basics of the method. At the beginning of the workshop the structure of the method is briefly described, the roles explained, the group dynamics policies discussed, and the specific Synectics techniques illustrated by brief exercises.

It is desirable, at least in the beginning stages of a group's learning of this method, to have a facilitator whose task it is, after explaining the roles and policies and answering questions regarding the process, to set time limits, answer questions as they arise, call for a switch in or between roles in the event of confusion or blockages, and, generally, to give structured feedback (itemized responses) to improve the internalization of the roles and the dynamics of the group.

Each role is assigned to or selected by one person who is allowed to exercise absolute authority (presumed expertise) over the type of information associated with the role. These roles and the type of information they control are the following.

Problem Designator.   The problem designator presents the problem and controls all directive and interpretive information (should not be played by the facilitator). Typically the form of this information is the narrative goal statement.

Resource Specifier.   The resource specifier identifies, itemizes, and describes resources that can be marshaled in response to the problem statement and controls all descriptive information. Typically the form of this information is an itemized list, with attributes or descriptions for each item.

Resource Organizer.   The resource organizer organizes the specified resources in response to the problem statements and controls all relational information. Typically the form of this information is the linked-node, "bubble", or network diagram.

Form Giver.   The form giver provides physical or communicable forms to manifest the organization of information and controls all representational information. This information is often pictorial, but can take the form appropriate to any media or communication channel.

Actualizer-Doer.   The actualizer-doer realizes and implements the form or plan and controls all procedural information. This information is usually in the form of a program, process diagram, or time line chart.

User.   The user responds to all preceeding roles in terms of perceived needs, desires, and reactions. This role controls all behavioral information. This information typically takes the form of anecdotal reporting, containing a description of the circumstances.

Evaluator.   The evaluator evaluates and measures progress in each or all roles and controls analytic information and formal feedback. This information is typically abstract, often mathematical, or stated as a criterion or value.

Generally these roles, forms, and constraints have proven to be recognizable, easily learned, and quickly internalized by ordinary people, particularly if they correspond to natural aptitudes and are self-selected. Roles are learned through experience and, once internalized by members of the group, become the normal basis for its communication. (Distinctions are made in terms of the roles and addressed to the person playing the role affected.) Roles appear to be readily transferred to any problem type, level of analysis, or particular situation.

GROUP DYNAMICS POLICIES

Three group dynamics policies, based on the Synectics method, are used to foster the positive cooperation of the group. They describe attitudes and preferred responses that are cultivated in all members of the group initially by the facilitator and later by the group itself.

The Spectrum Policy. Participants are encouraged to select the best in each idea, to *itemize* specifically those things being responded to, and to give credit for contributions made.

The Synthesis Policy. Participants are encouraged to build on the ideas of others and to integrate and correlate the various contributions.

The Empathetic Listening Policy. Participants are encouraged to listen, not interrupt, and paraphrase what they hear in order to encourage all participants to contribute to the group.

These policies, while simple and positive in spirit, are difficult to maintain in the face of the normal differences and ego drives that human beings develop, especially in task-oriented groups. Structured feedback by the facilitator to reinforce these policies, with an ongoing review of the group's behavior, helps their internalization.

IDEA-GENERATING EXCURSIONS

Fantasy and metaphorical thinking are also employed as a means to stimulate imagination and involvement to generate ideas and gain insight into the structure of the problem and its potential solution. These idea-generation techniques, most of which derive from the Synectics method, include the following.

Brief Vacation.   Everyone simply takes a rest to release the tension of work. (Let's forget about the problem for a while.)

Brainstorming.   Any idea is encouraged and accepted to quickly produce a list of alternatives to consider. (What items can we list?)

Free Associations.   Connections, however tenuous, with other ideas within the minds of the participants are brought out. (What linkages can we recognize?)

Morphological Synthesis.   The formal properties of things are explored relative to one another with a view to their integration. (What happens when we manipulate the images?)

Direct Analogy.   An example, usually functional in nature, from another context that has characteristics corresponding to the problem is examined. (How does something similar work?)

Empathetic Analogy.   The participant identifies with the object of consideration in order to "get inside" and feel its relationship to the problem from an intrinsic point of view. (What will it be like?)

Essential Paradox.   A search for the nexus of the problem is undertaken by examining its contradiction. (Let's reverse the evaluation criteria.)

These idea-generating techniques provide seven modes of escape related to the distinctions of the system. They are used by themselves or in combination, as appropriate to the situation and the needs of the group. While they do manifest the role orientations of the system, these excursion techniques need not be applied exclusively to the role they represent.

Initially the facilitator introduces the excursions whenever the progress of the group would seem to benefit, doing so until the participants catch on to the procedure and can conduct excursions themselves. Each excursion technique offers a particular kind of license to the imagination to escape from the immediate constraints of the problem and results in a typical kind of behavior/information product, a list of alternative ideas from brainstorming, a useful insight from analogy, and so on. These products are usually not total solutions and must therefore be fitted back into the context of the problem and evaluated. This process is one of the weakest aspects of traditional Synectics theory. Usually its success depends on the sensitive guidance of the group facilitator.

The role-oriented approach offers a variety of "handles" with which to work this return to the realities of the problem. Not only is each role player disposed to assimilate the product of the excursion into his/her own role view, but particular products of the excursion may be returned to the most appropriate role. Whether all or a few roles are invoked, the products of the excursion can be immediately returned to a full framework of evaluation understood by the entire group.

PROBLEM-SOLVING PROCEDURE

After an introduction to the group situation and the approach, each person elects to take or is assigned one of the seven roles. (An "alter ego" back-up person or team can be associated with each role.)

There should ideally be a large circular table next to a blackboard or tablet display. The problem designator should sit facing the blackboard, and the form giver, who will represent the work of the group, next to the board. The resource specifier sits to the left of the problem designator, and so on through the organizer, form giver, actualizer, user, and evaluator. (It is possible, but not preferable, for the problem designator, user, and evaluator to be played by a single person.) Arrangement of participants according to roles is shown in Figure 2.

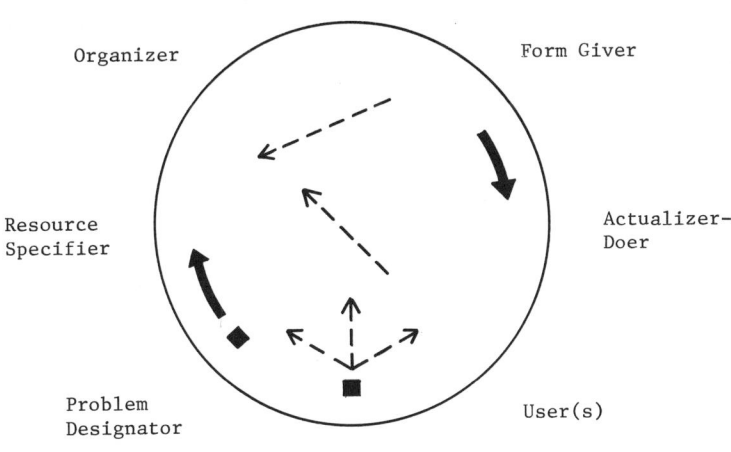

Blackboard or display of writing tablets

Organizer   Form Giver

Resource   Actualizer-
Specifier   Doer

Problem   User(s)
Designator

Evaluator

Figure 2 Seating positions of participants.

Since a logical dependency exists between one role and the next, it is convenient to approach the problem sequentially at first. This may develop from the user to the problem designator or vice versa. Once the first cycle of all seven roles has been completed, the procedure may be opened to questions in order to clarify roles and involve the group as a whole. Usually this discussion results in role focusing, that is, in the selection, more, or less by apparent consensus, of one role/information type as the one requiring the attention of the group. The dialogue between roles evolves as the problem is addressed. Constraints and potentials arise as the needs and contributions of each role become apparent. Said otherwise, as people recognize the constraints on their own solution space or as they see opportunities for others, exchange develops, focusing, broadening, or elaborating the information established in the various roles.

In the basic sequence the problem designator first briefly states the problem, paying special attention to the crux of the problem, its symptoms, the boundaries or criteria for the solution, examples of suggested solutions, itemized reasons why they are inadequate, and what he or she as "client" wishes to take from the session. Usually the other participants are then asked by the facilitator to generate goals based on this statement of the problem. The designator then selects one or two highly interrelated goals for initial consideration. Each role player then responds to the problem as given, to the responses of those in preceding roles, and to the selected goal. For one or two cycles each person is encouraged to respond quickly and in order. Soon, however, there is dialogue, suggestions toward solution, and bartering, as the participants recognize the implications of their roles and of the information being established by others. When anyone (especially the evaluator) feels a blockage in the progress of the group, they may call for an excursion. One person (initially the facilitator) conducts the excursion, typically asking for several goals related to the needs of the situation, then suggesting an appropriate idea-generation technique. For example, the facilitator may select a key word from a goal statement (say, "closure" in the goal of finding a closure for a thermos bottle that does not take up space). He or she may then ask for several examples of closure from a different conceptual universe (say, optics), perhaps calling for the organizing principle, form, or operation of the examples (say, an eyelid or camera shutter) and finally for interpretations related to the original goal (a closing device made out of pliable material that will contract like the eyelid into the neck of the thermos). During such an excursion participants abandon their roles and the free generationn of ideas is encouraged. Only after such ideas have surfaced does the facilitator guide the group back into the role-oriented framework, where the ideas are developed and interpreted in the context it provides.

The procedure continues until the problem designator and users are satisfied with one or more possible solutions or until the time limit for the session is up. In the experience to date all but two sessions have led to solutions that all persons in the group could accept. In almost every case these solutions were obtained within three hours from initial exposure to the method, in circumstances that were less than ideal, and with problems that were very complex and difficult to apprehend. Not surprisingly, an observer at one of the sessions noted that enthusiasm for the process closely modeled the levels of frustration and satisfaction of the participants. Success and insight break rather swiftly, often on the heels of a period of deep frustration and usually from an individual insight.

Variations in the process may be introduced as needed in a particular situation. For example, in many community or advocacy situations there is no problem designator independent from the users, and these roles are collapsed even as they become more differentiated. When there are several interest groups, each briefly states its perspective on the problem and provides information and response to the evolving situation as needed. It is particularly important that problem designators and users itemize their responses to the group, as these roles tend to generate diffused and relatively unstructured information.

The evaluator and facilitator roles are also sometimes collapsed, as both tend to reflect how well the process is fulfilling its potential. The distinction should be preserved, permitting the evaluator to focus on the substantive approach to the problem and the facilitator to focus on the interactions of the group.

Although roles usually remain constant for any one problem session, all participants are exposed to every role and often assist the person in another role. Cooperation typically develops as a consequence of a person in one role realizing the potential or the constraints emerging through the playing out of another. Like chess, the role-defined approach tends to involve its participants in thinking out and evaluating implications and strategies by providing the framework of authority and action into which to project. Synectics ideas and techniques, on the other hand, offer an effective counterpoint to the logical and constraining pattern of the roles by offering escape from them through fantasy and metaphor.

Together, both methods complement and produce the subject matter for each other. Although the value of the metaphorical excursion for improving the creative performance of groups is recognized, it does not have a logically consistent framework to aid in evaluating the idea generated. The structure of the role-defined approach, however, provides such a framework, one that appears to have many applications in analysis, management, and communications.

REFERENCES

1. Charles H. Burnette, "A Linguistic Structure for Architectonic Communication," *Publication of the School of Design*, North Carolina State University, Raleigh, N.C., 1969.

2. Charles H. Burnette, *An Information System for Hospital Planning*, USPHS HM00420-01, Institute for Environmental Studies, University of Pennsylvania, Philadelphia Pa., 1966.

3. Charles H. Burnette, *The ARC System, A Functional Organization for Building Information*, PB 177, 829, Clearinghouse, National Bureau of Standards, Springfield, Va., Nov. 1967.

4. Noam Chomsky, *Syntactic Structures*, Mouton, The Hague, 1957.

5. C. C. Fries, *The Structure of English*, University of Michigan Press, Ann Arbor, Mich., 1952.

6. J. S. Bruner, J. L. Goodnow, and G. A. Austin, *A Study of Thinking*, Wiley, New York, 1965.

7. G. A. Miller, E. Galanter, and K. H. Pribram, *Plans and the Structure of Behavior*, Holt, Reinhart, and Winston, New York, 1960.

8. Jean Piaget, *The Child's Construction of Reality*, Basic Books, New York, 1954.

9. E. A. Feigenbaum and J. Feldman, *Computers and Thought*, McGraw-Hill, New York, 1963.

10. S. R. Ranganathan, *The Colon Classification*, Rutgers University, New Brunswick, N.J., 1965.

11. B. C. Vickery, *Faceted Classification—A Guide to Construction and Special Schemes*, ASLIB, London, 1960.

12. K. C. Knowlton, "A Programmer's Description of L6, Bell Telephone Laboratories Low Level Linked List Language," Bell Telephone Laboratories, Murray Hill, N.J., 1966.

13. A. Tarski, *An Introduction to Logic*, Oxford University Press, New York, 1965.

14. J. E. Sammett, *Programming Languages*, Prentice-Hall, Englewood Cliffs, N.J., 1967.

15. Julius Hochberg, *Perception*, Prentice-Hall, Englewood Cliffs, N.J., 1964.

16. G. A. Miller, "The Magical Number Seven ±2: Some Limits on Our Capacity for Processing Information," *Psychological Review*, Vol. 63, 1956, pp. 81–97.

CHAPTER 11

SYNECTICS

GEORGE M. PRINCE

The word *synectics* is from the Greek *syn* 'together', as in *synchronous*; and *ectics* which has been arbitrarily selected. It is an attempt to capture the essence of an invention where elements that previously seemed to have no relationship are brought together. The same thought is appropriate for group work. Often the members of a group do not have any relationship and find it difficult to cooperate.

Synectics® is used as the name of a company and also to represent a body of knowledge about the creative process, the dynamics of groups, and theories about individual difficulties with speculative as opposed to routine thinking. Synectics® differs from many other systems in that the process and theories were developed by studying tapes of people in groups working on real problems. The process is continually modified as we learn more about how people work more effectively. We have worked with about 10,000 people over the past 20 years.

BASIC PREMISES

Few of us use more than a fraction of the potential we have for good thinking. By systematic and knowledgeable manipulation of myself and the climate around me, I can substantially increase the amount of my potential that I use.

By modifying the way in which I deal with others in a group, I can greatly increase the probability of the group's success.

A REVEALING EXPERIMENT

Many of the problems that make group work difficult and less productive than it might be are not easily visible to the untrained eye. We are so accustomed to the mix of actions that we do not differentiate between those that encourage speculative thinking and those that discourage it. The children in the experiment that follows have not yet developed adult subtlety in their dealings with each other. You will be able to see their destructive as well as their supportive actions; these are the same as those found within an adult group, but more visible. We will later discuss the actions and appropriate measures for handling them. This case history will give you a concrete reference.

In the experiment we gave a three-day course in creative problem-solving to 12 gifted children, ages 9, 10, and 11, from the Needham, Massachusetts, school district. We treated them just as though they were adults. We applied our usual practice of considering our participants as equals and avoiding, so far as possible, any punishment. Our definition of equality was as follows: as soon as a person is able to take responsibility for him or herself, we are equal. That person knows things I do not, so in those areas I can learn from him; I know things that he or she does not, so he can learn from me in those areas. Neither of us has any right to punish the other. We avoid entering into win–lose discussions, although we are available to discuss and resolve differences. We also avoid the stance of being either one up or one down.

After brief introductions we split the group into two teams of six and asked them to take five minutes to organize themselves to work on a problem together. We then gave them an invention problem to work on.

This is an experiment we have run and videotaped thousands of times, so we know what to expect. One team did create an organization new to us. In the session rooms there are five large newsprint pads mounted on one wall. This team split and each of them pulled up a chair to a pad and went to work by himself. The leftover person, a rather shy little girl, worked on her own, using an $8\frac{1}{2}'' \times 11''$ pad.

We gave each person a copy of the specifications of the problem, told them that they had 15 minutes, and left. Like most adult groups, they had agreed to cooperate in several ways: Each would develop some ideas on his or her own; one person was selected to be chairwoman; when everyone was ready, she would guide the group while each person reviewed or presented his or her ideas; she would keep notes about the ideas.

As with adults, these agreements were substantially ignored. As some discovered that they did not have an idea, they needed to discuss the specifications with others. As one would get an idea, he or she needed to tell everyone about it. No one paid any attention to the chairwoman who attempted to keep order. As an idea was explained, the two people who paused to listen first told the idea conceiver why his idea would not meet the specifications and then derided him for thinking such a thing. Often a half-listener would use the idea to trigger an idea of his or her own. Occasionally there was loud appreciation of an idea and everyone would pay attention, and most would leap on the bandwagon for a half-minute or so. Then the dissenter would demonstrate that this idea would not work. The difference was that the children were merciless and loud. Most of the 15 minutes was pandemonium.

We had planned to spend the next hour and a half presenting and discussing some of our findings about the blocks and hindrances to creative thinking. For example, we would say, "Let's look at the differences between routine thinking and speculative thinking. What are some of the characteristics you associate with routine thinking?"

Or we would say, "Self-censoring is a habit that all of us develop and it is probably necessary. How might it keep us from being speculative even when we want to be?"

This particular group of children was unwilling to participate in this kind of a discussion. Based on later observations, I believe that this group was *so* competitive that individuals did not dare expose themselves to certain ridicule if they were even slightly off target.

Early in this discussion I told the children some of my theories about left and right hemispheres of the brain and invited them to doodle using the colored pens that were furnished. They accepted the invitation and gave up the pretense that I had their full attention. Their nonverbal signals seemed to be telling me that they were bored and disinterested in what I was telling them. One boy got on his hands and knees with his back to me and concentrated on his doodling.

I had to keep reminding myself that these children were capable of thinking at a speed of 900 words a minute while I was talking at the rate of 150 words per minute, at most. So they needed to give me only a small amount of their attention.

We had a break and there was another marked contrast to adults. The action was fast and noisy. They made cocoa, opened soft drinks, and created quite a mess with spilled milk, sugar, and cocoa. I mention this because it has significance when compared to later behavior.

We then debriefed the videotapes of their meetings in the usual way: we played a few moments of their tape, asked them to describe what was going on, took note of their observations, and then evaluated what they described. For example, one girl observed, "We were not listening to each other."

After writing this on a newsprint pad, I asked, "Will this increase or decrease the probability of success?"

"It will decrease it because we are not using each other's ideas to build on," said Mary-Alice, displaying a thumbs down signal.

"It will increase it because we are all thinking of ideas so we will have more of them," said Danny.

"It will decrease it because we are not using other's ideas to stimulate us," said Abby.

We spent the rest of the day observing the tape, evaluating actions, and inventing. We concluded by asking the participants to evaluate the activities of the day: "What did you like and find interesting about the day and what did you not like?" The two outstanding likes were, "It was great fun and not like school." They had not known that learning could be fun; it felt very good to be treated like adults and not like two-year-old morons. Their only dislike was that it ended too soon. My colleague and I were surprised. Their constant apparent inattention and occasional rudeness had suggested to us that they were bored and wishing they were somewhere else.

The following day we took them through the Synectics® process, modeled the various roles, and gave them some learning experiments in listening for ideas, for tolerance of the absurd, and other skills that I will discuss later in the chapter. We ended the day by repeating the starting experiment. This time we asked them to organize themselves to work on a problem together and run a meeting on one of their own problems. Each group did well in organizing, using what they had learned about the process.

When they started their meeting, I could not believe what I saw. It seemed that now that they knew all about destructive behaviors, they were skillful in applying them. The meeting was a disaster. I started down the hall to see if the other team was acting the same way. My colleague was starting up the hall to check with me. We were appalled. We quickly decided that it would be heavy punishment to debrief the tape as is usual. We would simply play the tape without comment.

As I watched the tape with my team I gradually became aware that there was a pattern in the destructive behavior. Each child had devised a strategy to compete with the facilitator. One boy would periodically look at one of the television cameras and shout, "It's moving!"

He would then go to the middle of the room and wave his arms at the camera. A girl had a Coke bottle that she used to blow into to make a tune. Another boy folded sheets of paper into hollow balls that he would then burst. No one paid any attention to the girl facilitating except the girl whose problem was being discussed. Now and then one of the distractors would pause long enough to offer an idea, which the facilitator dutifully recorded.

The children's evaluation of the day was much the same as that of the first day. Our evaluation was too punishing to tell them.

The next morning we decided that the level of competition within the group was so high that most of the children were unwilling to let a peer take control. It felt to them like losing in a win–lose situation. We opened the day by explaining that in many situations people play small games of win–lose. When one person loses, he or she makes sure there is *another* small game in which *he* or *she* wins at the expense of the other person. In groups there is really no such thing as win–lose—it always becomes lose–lose as the first loser arranges to get even. We then explained the phenomenon of one up and one down. One boy asked for an example and another boy said, "Sometimes I am sitting in my father's chair and he comes in and says, 'Get out of my chair.' That is a put down."

Every hand in the room went up. Each child had an example that he or she wanted to tell. After that my colleague asked the participants to invent a signal that they could use to tell one another that he or she was getting into a win–lose position or doing a put down. They finally agreed that they would use the V for victory sign. My colleague suddenly said, "You know, we do not want to do that! We do not want to police each other. We want to take responsibility for ourselves. Let us forget the signal."

All agreed and we formed new teams. The assignment was to organize and then run another meeting, taking turns as facilitator while working on one of the participant's problems. We assured them that everyone would have a chance to be facilitator.

Both teams and each facilitator ran perfect Synectics® meetings. Individuals supported one another and there was no destructive behavior. At break time not only was the noise level down, but there were no spills or messes as they made cocoa and had soft drinks. In the large meetings most continued to doodle, but there was a vast difference in their nonverbal signals. I no longer had to comfort myself that they were attending. Their signals of boredom and inattention had disappeared.

After studying thousands of meetings I have become convinced that most of the destructive actions of a participant are grounded in a need to *apparently* win. I use the word *apparently* because in reality the kind of action I am discussing is not an achievement. No one really wins anything. For example, a group member offers an idea. It is not a perfect idea, but it is a good beginning. A second member instantly points out the flaws in the idea and it is dropped. The action of the second member is often justified by, "We do not want to waste time on an idea that will not work." Groups that follow this policy are not nearly as productive and creative as a group that takes every idea as a beginning and attempts to build on overcoming the flaws.

An organic outline of some actions that we find discourage speculation and creativity is shown in Figure 1. When people see this outline they often say such things as, "Challenge! How can you list that as destructive? I use it very effectively with my subordinates."

It is true that challenging sometimes works as a stimulus. Perhaps 2 out of 10 people respond to it well; the other 8 tend to stop using their creative resources when the climate is challenging. They become quite safe in their thinking. Even with the two who deal well with challenge, it is more productive to support and cooperate than to challenge.

The usefulness of this outline lies in raising questions about practices that we have assumed to be harmless or even valuable. I have minutely studied hundreds of instances of these actions and their consequences. I have also experimented extensively with preventing the discouraging actions and stimulating the encouraging actions, and observed the con-

---

ACTIONS THAT <u>DISCOURAGE</u> SPECULATION/CREATIVITY

Be pessimistic	Nitpick	Correct
Preach/moralize	Interrupt	Name call
Be judgmental	Be bored	Blame
Assume no value	Misunderstand	Set up win/lose
Make no connections	Be inattentive	Be competitive
Put the burden of	Act distant	Make fun of
proof on other	Pull rank	Be dominant
person	Get angry	Command
Take ball away from	Disagree	Order
Ask questions	Argue	Direct
Cross examine	Challenge	Threaten/warn
Give no feedback	React negatively	Demand
Be noncommittal	Discount/put down	Do not listen
Put on a stone-face	Be cynical/skeptical	Do not join
Be critical	Insist on early	Use silence against
Disapprove	precision	Scare
Be impatient	Point out flaws	

Figure 1   Actions that discourage speculation/creativity.

sequences of this. There is no doubt that the discouraging actions work against speculation and creativity. This is not to say that it is easy to operate without some of these actions.

I believe that all or nearly all the destructive actions are expressions of competitiveness—an attempt to achieve one-up positions and put my "adversary" in a one-down position, to score a win and cause him or her to lose.

In the case of the gifted children their nonverbal signals of inattention and boredom, and their rudeness, were subtle attempts to put me, the teacher, one down in a way that was safe for them. Children spend most of their growing-up lives in a one-down position. It obviously feels bad and subtracts from their self-esteem. In order to generate *some* feelings of worth, they must win a few. So they devise competitive strategies that they can use to win or be one up without bringing about a real test of strength—which they would inevitably lose. When the gifted children emotionally accepted what they had intellectually observed, that my colleague and I were dealing with them as equals, then it was no longer necessary for them to play one-up games with us.

When they realized that there were ways in which they could operate with each other that were cooperative rather than competitive—equal rather than one up/one down—they adopted them. The acceptance of equality proved to be a powerful grounding for change.

PURPOSE OR GOALS OF SYNECTICS

The individual need to compete appears to be the largest deterrent to good group work. One of the objectives of Synectics® processes and skills is to turn competitive energy into cooperative energy and thus make available more of the creative potential that resides in each of us.

Another purpose is to generate insightful and inventive solutions to a problem. A third objective is to obtain commitment to a line of action. This can be particularly useful when some members of a group hold adversary positions.

Just as the demonstration and protection of equality enabled the children to shift energy from competing to supporting, the climate of a Synectics® meeting can help adults free their energy to make better use of themselves and their teammates.

In addition, Synectics® has evolved some thinking strategies that put people back in touch with imaginative speculative capacities that have inadvertently been repressed.

HISTORY

In the mid-1950s I was in charge of the creative department of an advertising agency. I became interested in how ideas happened and whether they could be made to happen more reliably. I spent time experimenting with artists and copywriters, exploring how they got ideas. I hired one psychologist to work with us full-time on learning more about idea getting. We hired another psychologist as a consultant because he had been doing some independent work in the area of creativity. We were using variations of the popular brainstorming method, with mixed results. It appeared to be better than no procedure, but arriving at a conclusion or solution that struck us as creative was an elusive achievement. But I was and remain greatly impressed with the achievements of Osborne's discovery that he could increase creative productivity by banning evaluation from the idea-getting phase—a major breakthrough in thinking about creativity. In addition, he was the first person to demonstrate that creativity could be deliberately stimulated.

In 1957 I read an article entitled "Operational Creativity" by William J. J. Gordon.[1] Mr. Gordon claimed to have devised a new systematic way to produce good ideas. It was the first new system I had heard of since brainstorming and I was fascinated. We experimented with the process but could not make it operational for us. Nevertheless, it was stimulating and much of what he said struck me as promising.

I visited Mr. Gordon soon after reading the article. At our first meeting we decided to pool our resources to discover everything about the creative process. I joined The Invention Design Group at Arthur D. Little in 1958. The mission of this group was to invent new products and processes for the clients of Arthur D. Little. We were a group of eight people and we prided ourselves on not writing reports but building working models. Sometimes we did well and had reason to be proud of our accomplishments; at other times we failed. We were interested in finding the causes for the failures. We hit on the notion of tape-recording all our conversations and meetings. The thought was that if the meeting or conversation produced a good idea, we could review the tape to learn what we were doing to make that idea happen. This proved to be an extraordinary research tool. We began to discover that successful problem-solving and invention sessions had some patterns. For example, jokes and laughter had a powerful stimulating effect. Quite often when we were stuck on a problem and low in energy and high in frustration, someone would re-energize and stimulate the group by suggesting an analogy that we could explore. In one session we were working on some problems of fish farming and harvesting. One member said, "You know, if we had someone like cowboys in the old West, we could have our herds graze in the open ocean. We would not have to feed our stock—they would feed themselves." This stimulated a barrage of ideas. One was the thought of training porpoises to be our cowboys. It was through the examination of such things that we developed deliberate strategies for stimulating ideas. I will discuss these later, but first I want to describe more of our history.

In 1960 four of the members of The Invention Design Group decided to start their own invention company. We left Arthur D. Little and founded Synectics,® Inc. It was our intention to use our own process to invent products for other companies and also for ourselves. We would use our own inventions to start new companies. These were ambitious goals. Only one of our inventions grew to be a company—a foamed packaging system that came to be known as Instapak. We were more successful in our inventions for others. In the development of that service we gradually shifted from presenting finished inventions to involving the client company in the inventing.

There was an early recognition that when we did the inventing, we were putting the creative people in the client company in a put-down position. They felt competitive enough to make sure that our inventions were rejected after being accepted by the people who had hired us. We did not think of it as clearly as we can now, but we modified the service to solve that problem. The client brings the appropriate people to our laboratory in Cambridge. We are facilitators and keep people cooperating rather than competing, and they do their own inventing and problem-solving. This has been a successful, productive service. A second service we offered was selecting and training small invention groups to operate within a company. This is fully detailed in *Synectics* by Mr. Gordon.[2] This was the first report on our research and was based on our experiences with our own group and perhaps 40 or 50 trainees.

A third service we offered was a course to train facilitators. One effect of these three services was to provide us with a steady stream of people with real problems. We could use them as guinea pigs. We would tape and later videotape these meetings and use them to increase our understanding of the dynamics of the creative process and also of the interactions in the group. By 1970 we had worked with and studied 3000 or 4000 people. We had developed some effective procedures for stimulating creativity and learned a great deal about how to facilitate good meetings. I reported on this in *The Practice of Creativity* in 1970.[3] Mr. Gordon severed his ties with Synectics® in 1967.

In our courses and Problem Laboratories (the service that facilitates problem-solving and invention by the clients) we were able to help people learn to use more of their creative potential, but the increase was in most cases temporary.

As participants returned to the competitive climates of home, they quickly slipped into their competitive ways. It has been my mission, with the help of my colleagues, to invent ways that will help people understand and deal with the forces that work against using the vast potential each of us has. These forces are both inside each of us and surrounding us. In 1970 I hired a therapist trained in transactional analysis (TA). TA is a form of therapy originated by Dr. Eric Berne.[4] The developers of TA had been studying the transactions between people to learn how they effect emotional health. I thought we could learn from their work. For the next few years we had five therapists of various schools of thought working with us. It was not my intent to go into the therapy business, and after we had absorbed as much as we could of their practices to help people think more clearly and effectively the therapists left us to go into business for themselves.

In the decade from 1970 I also experimented with courses for older people—over 65—to see what effect learning problem-solving techniques and cooperation would have on them. I have reported on this research in *Mindspring!*[5] At the present writing I have worked with and observed many thousands of problem-solvers. In addition, I have availed myself of some of the wonderful research that has been done on brain function. Whereas in the 1960s we used procedures and strategies because they worked, in the late 1970s we were better able to understand *why* they worked.

In the rest of this chapter I shall discuss some of the problems that reduce group and individual effectiveness, the Synectics'® processes and skills that are designed to help, and why I believe they work.

THINKING IN GROUPS

Each group depends for its effectiveness upon the capacities of the individuals who make it up, multiplied by the willingness of each person to cooperate and support every other person in the group. If this willingness is zero or low, the group will be ineffective. If it is high, the group will be synergistic—more productive and able than one would expect. If willingness and skill in cooperating is low, then group output may not be as great as any of the individuals could produce on his own.

This is one of those statements I can demonstrate but cannot prove. I can show tapes of a group with little willingness to cooperate and they will demonstrate low output. I can show them after training and they will exhibit a greatly increased capability to accomplish. I can do this repeatedly and predictably, as can any of my colleagues, but this does not represent scientific proof of my statement.

The interrelationship of the individual and the group has a critical effect upon the productivity of the individual and therefore of the group. I want to examine some of the problems we have as individual thinkers and then relate this to group effectiveness.

GOOD THINKING

Good thinking is that type of thinking that is appropriate to the situation. It will range from routine at one end of the spectrum to speculative or creative at the other. Situations in which routine thinking is good thinking are those in which I have tested answers that work. There are many, many times when routine thinking is good thinking, for example, when my pilot is making a landing or takeoff, when I am filing something, or when a surgeon is operating.

On the other hand, there are situations in which I need to think speculatively—when tested approaches are not working in getting a raise, when my child is not doing well in school and nothing seems to help, or when I have a flat tire and my jack is far way propping up one end of my boat.

Nearly everyone I have worked with is quite good at routine thinking. It is when we need to speculate that most of us have trouble. I have repeatedly observed groups that have met to speculate and invent that were unable to go beyond routine thinking. Now I would like to examine the problems that tend to keep us thinking routinely.

THE THINKING OPERATIONS

The individual's thinking operations are the key to his or her performance. I have hypothesized that there are six distinct thinking operations that we use when faced with a problem. The easiest way to make these clear to you is to ask you to experiment with yourself. Pretend that you have been hired by the Thermos Company to invent for them a new stopper for their wide-mouth thermos bottle. The problem you must solve is that of losing stoppers. Mothers complain that when the stopper is lost, the bottle is no good. The company has tried using strings, chains, and hinges, but for some reason mothers do not find these to be satisfactory solutions.

The stopper must somehow be built in. Any solution must retain the wide mouth, be easily cleaned, and be thermally effective. Take a minute or two and develop a beginning idea. A beginning idea is one that does not need to work.

Here is what I believe happens: I wish for a new stopper. *Wishing* is the first thinking operation. Next, I retrieve from my stored experiences something I believe will help me. Let us say I retrieve a spice can that has a built-in sliding "stopper." *Retrieving* is the second thinking operation. Then I image my retrieval—I see it in my mind's eye. *Imaging* is the third thinking operation and it is used with all the operations. Many good thinkers use a split screen in their imaging. In my case I image the thermos on the left screen and the spice can on the right. Then I compare what I have with what I need. *Comparing* is the next operation. I see that the spice can is rectangular so I transform the slide to come all the way out to give a wide mouth. I add some thickness to it for insulation, and so on. *Transforming* is the fifth thinking operation. I recycle comparing, transforming, retrieving, and wishing until I have something that fits my needs and then I do the final thinking operation. I *store* this new concept.

THINKING OPERATIONS AND
ROUTINE/SPECULATIVE THINKING

Both routine and speculative thinking involve these same operations. The only difference is that when the problem is routine—such as tying my shoe—my retrieval is usually close to a precise fit. I have little or no transforming to do, and since I do not develop any new connections, there is little or no storing or learning.

WHY SPECULATIVE THINKING IS DIFFICULT

Since the thinking operations are roughly the same for both ends of the spectrum, it should follow that if I am a good thinker in routine matters, I should be good in speculative situations. I believe there are four general reasons why this is not the case:

1. *My Divided Self.* There are two ways in which my self is divided so that I sometimes work against myself. One is physiological—the two hemispheres of my brain provide me with quite different processing functions. The other is cultural—as I interact with my parents, teachers, and other people in my life, I form two ways of operating, two "selves" that take on different views of the world.

2. *Criteria.*   From childhood I develop standards or criteria with which to evaluate my own performance. Without intending to do so, my criteria devalue speculative/creative thinking.
3. *Habits.*   As I learn to get along with people, I form habits of dealing with my own thinking that are socially necessary, but they inhibit or even prevent speculative/creative thinking.
4. *Climate.*   The climate around me and inside me—the signals that are sent to me, those I send to others, and those I send myself—have a profound effect on my willingness to use my potential for speculative/creative thinking. I want to examine each of these in more detail. The reason that each is important is suggested in Figure 2.

Optimists say that we use about 20% of our potential. Pessimists say it is more like 5%. Warren McCullough[6] suggests that it is even less than 5%. In any case, there is wide agreement that we do not use a large percentage of our potential thinking power. *The basic premise upon which*

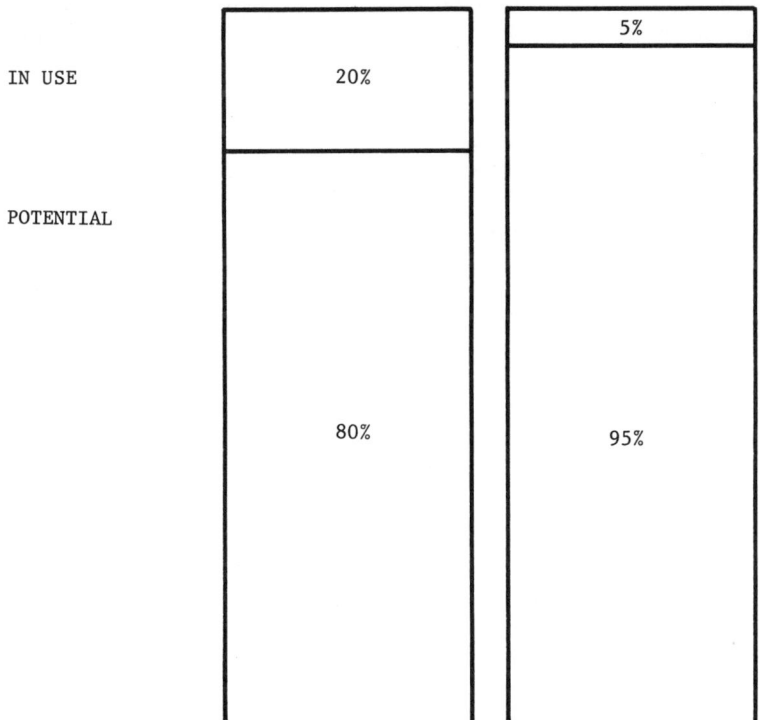

Figure 2   Percentage of potential for speculative/creative thinking.

Synectics is based is that by systematic manipulation of myself and the climate around me I can substantially increase the amount of my potential that I use.

The exciting promise here is that if every individual in a group learns to make greater use of his or her own potential and, in addition, is willing and skillfull at cooperating, the synergy will produce outstanding results. Let us now look at some of the opportunities.

MY DIVIDED SELF—THE EFFECTS OF PHYSIOLOGY

For almost a century we have known that human powers of speech reside primarily in the left hemisphere. Injuries on the left side cause speech damage, while right-hemisphere injuries leave speech intact. In spite of this understanding, we have only recently begun to appreciate how functions of the brain are actually divided between right and left hemispheres.

The real breakthrough in this understanding came in the 1960s when Dr. Roger Sperry and his students Michael Gazzaniga and Jerry Levy began their historic split-brain experiments. . . . They were able to test separately the thinking abilities of the two surgically separated halves of the human brain. They found that *each half of the brain has its own separate train of conscious thought and its own memories.* Even more important, they found that *the two sides of the brain think in fundamentally different ways* (italics Blakeslee's): While the left brain tends to think in words, the right brain thinks directly in sensory images.[7]

THE PHYSIOLOGICAL DIVISION OF SELF

LEFT HEMISPHERE

The left hemisphere controls the right side of my body. Its thinking is linear and logical. Its job is to interpet or translate observations into orderly and precise formulations.

It abhors confusion and ambiguity. It is in charge of language and mathematics. It is literal. If I ask a person who has lost the use of his right hemisphere, "How do you feel?" he answers, "With my hands." This hemisphere is in charge of reading, sequencing, putting in order. This is the boss hemisphere.

RIGHT HEMISPHERE

The right hemisphere controls the left side of my body. Its thinking is holistic—it sees in whole images. It has little language ability and processes observations in images and feelings. It is reached by such communications as poems, rhymes, figurative language, parables, incompletenesses like the few lines of a caricature, scent, and aphorisms.[8]

This hemisphere processes music, is nonlinear, sees simultaneous images: It thrives on ambiguity and confusion. It recognizes patterns and jumps to unlikely, nonlogical connections. It is thought to operate like a hologram (a three-dimensional photograph produced with laser beams). It can take any small piece of a three-dimensional image and from it reconstruct the whole. For example, I can see a part of your face and mentally see the whole.

This brain reads nonverbals signals and interprets tone of voice. It is the main seat of my feelings. It is the subordinate hemisphere.

A SPECULATION

We all know quite a lot about the operations of our left hemispheres because that is our conscious way of thinking. I want to speculate about how the right hemisphere operates. Imagine a large cauldron of clear fluid: Mixed throughout it are millions of tiny electric points of light; drifting through it are clouds of mist. At any instant a few hundred or a few thousand of the points of light form a three-dimensional picture like a well-decorated Christmas tree, as I recall a past experience. A single light point may be part of several such pictures at the same time. In fact, part of one three-dimensional picture may make up a section of another quite different picture.

My right hemisphere takes nothing seriously. The grouping of lights is stimulated by left-hemisphere concerns or observations or its own observations and feelings. The grouping of lights is also for entertainment. The playing around with patterns is to enjoy as well as to help the left hemisphere. When a pattern pleases, or seems to have some relevance to something the left hemisphere is toiling with, the right hemisphere tosses it to the left. It comes to the left hemisphere without prior evaluation or refinement. The right hemisphere simply has the feeling that it is funny, pleasing, useful, or just interesting. It may be that the right side has taken a small portion of a pattern and jumped to a large pattern of conclusions. Easy confusion and a state of disorganized scanning, stirring up, and experimental rearranging—these are the right hemisphere's modus operandi. Ambiguity and confusion are essential to its efficient operation. It rearranges and rolls over many meanings from the same set of data and entertains simultaneously several mutually exclusive possibilities.

My left heimsphere is born knowing how to handle this barrage of insanities. Without anxiety, it speedily sorts, analyzes, rationalizes, and modifies until it uses what it needs to learn the new thing coming at it.

Much of the time it modifies an input and then sends it back for further disorganizing and reorganizing by the right hemisphere. As a result of this early, equal use, the child does whole-brain thinking and so learns and problem-solves with exquisite skill.

To a child of two, three, or four the world is a giant and fascinating unknown. In his wonder and openness he does not attempt to impose order or precision too early. He is comfortable with not understanding instantly. He has a drive to learn and one of the important sources of satisfaction and self-esteem is the accomplishment of learning itself.

As children are socialized, they use their learning skills to learn how to act like grown-ups. In this role there is a need for rules, and much attention is paid to precision, logic, analysis, right and wrong, and so on. The importance of left-hemisphere functions is continually underscored. They learn that honoring right-hemisphere inputs often means trouble and anxiety for them—they begin to employ a device known as selective inattention. Sullivan[9] describes this as the ability to exclude and not notice phenomena that cause anxiety.[10] They tend to become less and less aware of right-hemisphere contributions.

A LEFT/RIGHT BRAIN EXPERIMENT

You can quickly get a feel for the difference between left- and right-brain-dominated thinking. First, using the left side of the brain, think of ways that an automobile is like a child's wagon. After thinking of a few, switch to the right side of the brain and ask, "How is an automobile like a totem pole?"

Most people do the left-dominated comparison easily and comfortably—they come up with such likenesses as both have wheels, both steer, each carries things, and so on.

When people switch to likening a car to a totem pole, they tend to draw a blank. Often they laugh, as though it is a joke. Then likenesses begin to come: Both are worshipped, they are works of art, they are phallic symbols, and so on.

DIFFERENCES

There are qualitative differences between the outputs of the two hemispheres. Left-dominated connections are, as we would expect, logical, precise, and have a one-to-oneness. I hear your connection and agree that it *is* a likeness. When you state your right-brain connection, it may stimulate quite a different picture in my mind than in yours. In this sense there is ambiguity and a chance that a connection will be misunderstood or will be a "mistake."

POOR INTERNAL COMMUNICATION

My left hemisphere, through selective inattention, develops bad listening habits with right-hemisphere contributions. When I lost my childhood skills at whole-brain thinking, it is almost as if I stopped having a common language in which both hemispheres were fluent. As a result, many of the offerings of my right hemisphere are lost in translation or never understood by my left hemisphere. At the same time the left-hemisphere's requests for help go untranslated by the right hemisphere. The potential for synergy goes unrealized.

THE CULTURAL DIVISION OF SELF

The effects of culture overlap and reinforce the effects of physiology, and it has been useful to me to look at my divided self from both perspectives to appreciate the power of the forces that divide me.

SAFEKEEPING SELF/EXPERIMENTAL SELF

Freud hypothesized three selves: the ego, the id, and the superego. TA also postulates three: the parent, the adult, and the child. I find it useful to think of two selves: the experimental self and the safekeeping self. These are metaphorical conveniences—I really do not have two selves in my one skin.

I first began to examine my two selves in the Cincinnati airport several Christmases ago. The airlines had just started searches to prevent highjacking. I was at the end of a long line, feeling frustrated. An airline pilot passed the line, skipped the search procedure, and, looking trim in his blue suit with the gold stripes, disappeared in the distance. A small, timid voice in my head said, "Hey, George, let's get a blue suit, put some gold stripes on the sleeves, and we can save some time."

"You idiot," said a strong mean voice, "you will get us put in jail." After a long pause the timid voice said, "You know, we tell everyone else to be open-minded about ideas . . . How about you being more open-minded?"

"Oh, all right," replied the mean voice, "It would save us time . . . and it might save us money. We would not need a ticket. Also, it would not be difficult to implement—we already have a blue suit, so all we would need is some gold ribbon (long pause). What if they made us fly the airplane?"

I never did implement that idea, and I continued to think about that inner dialogue. There was such a difference in the way the two voices treated each other that I began to speculate about the personality and

functions that went with each voice. The two types of self are characterized as follows.

Safe-keeping Self	Experimental Self
Censors	Feels
Evaluates	Takes risks
Reassures and supports	Breaks rules
Analyzes	Makes connections
Guides	Recognizes patterns
Is realistic	Plays
Looks at consequences	Speculates
Is logical	Curious
	Sees the fun in things
Alert to possible danger	Likes surprises
Avoids surprises	Open to anything
Avoids wrongness	Makes impossible wishes
Avoids risks	
Makes rules	Does not mind being wrong
Is serious	Does not mind being confused
Cautious	Images
Suspicious	Is intuitive
Fearful	Is impetuous
Punishes mistakes	In touch with unconscious mind
Punishes wrongness	In touch with total experience
	Uses seeming irrelevance
Probably punishes anything my parents disapproved of	Uses dreams

You will immediately see that there is some correspondence between
the safekeeping self and the functions of the left hemisphere, and also a
relationship between right-hemisphere functions and the experimental
self. I prefer not to make any attempt to establish a one-to-one relationship
because I believe that the two concepts are basically different in how
they develop and how they work within me. They tend to reinforce the
dividedness of myself.

In an attempt to better understand how my selves interact I have built
several theoretical models. The one that best captures the ways in which
I think I work with myself is illustrated in Figure 3.

EXPERIMENTAL SELF	EXPERIMENTAL SELF	EXPERIMENTAL SELF	EXPERIMENTAL SELF	
Feels	Feels	Feels	Feels	Probably punishes anything parents disapproved of
Takes risks	Takes risks	Takes risks	Takes risks	
Breaks rules	Breaks rules	Breaks rules	Breaks rules	
Makes connections	Makes connections	Makes connections	Makes connections	Censors
Plays	Plays	Plays	Plays	Is fearful
Recognizes patterns	Recognizes patterns	Recognizes patterns	Punishes mistakes	Punishes mistakes
Speculates	Speculates	Speculates	Avoids risks	Avoids risks
Sees the fun in things	Sees the fun in things	Sees the fun in things	Punishes wrongness	Punishes wrongness
Likes surprises	Likes surprises	Likes surprises	Alert to possible danger	Alert to possible danger
Open to anything	Open to anything	Open to anything	Avoids surprises	Avoids surprises
Curious	Curious	Is logical	Is logical	Is logical
Uses dreams	Uses dreams	Makes rules	Makes rules	Makes rules
Uses seeming irrelevance	Uses seeming irrelevance	Avoids wrongness	Avoids wrongness	Avoids wrongness
In touch with total experience	In touch with total experience	Is serious, cautious, and suspicious	Is serious, cautious, and suspicious	Is serious, cautious, and suspicious
In touch with unconscious mind	In touch with unconscious mind	Is realistic	Is realistic	Is realistic
Is impetuous	Analyzes	Evaluates	Evaluates	Evaluates
Is intuitive	Guides	Analyzes	Analyzes	Analyzes
Imagines	Reassures	Guides	Guides	Guides
Does not mind being confused	Supports	Reassures	Reassures	Reassures
Does not mind being wrong		Supports	Supports	Supports
Makes impossible wishes	SAFEKEEPING SELF	SAFEKEEPING SELF	SAFEKEEPING SELF	SAFEKEEPING SELF

Figure 3 The mind going from total experimental self to total safekeeping self.

Figure 3 illustrates my mind going from total experimental to total safekeeping. I am sure that I operate somewhere in between, with movement either way, depending upon the situation and how I am feeling. For example, if I am working in a high-risk, punishing climate I will tend to be dominated by my safekeeping self. If my surroundings are friendly and supportive, my experimental self will have a chance to emerge.

We know from the observation of groups that a punishing person greatly reduces the quality and quantity of ideas. Punishment, in this sense, interrupts, and is perceived as a put down. Extrapolating from that, I wondered if this might apply to my two selves. If I have a punishing safekeeping self looking over my shoulder and finding fault with what I do, will my experimental self tend to shut itself down? There is no question, that we can increase the productivity of a group by reducing punishment to a minimum. If I reduce self-punishment to a minimum, will I enjoy an increase in personal productivity?

For the past three years I have worked with myself and about 400 volunteers in groups, not all at once. I experimented with ways of learning to reduce self-punishment. The general procedure is to identify a specific instance of self-punishment. I assume the punishment was at one time functional and I ask what are the possible benefits that I am getting from this specific example. I then explore the possible damage of that punishment and then problem-solve to get all the benefits without the actual punishment.

There is evidence that self-punishment is an important factor in keeping my experimental self leashed. There is also evidence that with some effort self-punishment can be greatly reduced. About a third of the subjects in the punishment reduction experiment reported significant changes in their friendliness within themselves and their willingness to take risks and be experimental.

To conclude this section the phenomenon of the divided self is a serious handicap when it comes to using more of my potential. There are two ways of reducing or overcoming this handicap. In the case of my physiological division I can relearn to communicate fluently between right and left. Many of the idea-getting strategies of Synectics® are aimed at this. Much of our research effort is aimed at developing more direct and effective ways of doing this.

The second way is to learn to share time between the safekeeping and experimental selves. If you will image my self model, when I need an idea, safekeeping, by agreement, turns off. Experimental is free to develop any wildness imaginable. As soon as there is a glimmering, safekeeping turns on full blast—except for punishing—and lends its considerable talents to guidance and help. When new modifications are needed, safekeeping turns 'off. This rapid oscillation allows me to use much more of myself than if I have a relatively fixed proportion of each self available.

IMPLICATIONS FOR GROUP WORK

Obviously the group will benefit from every improvement in individual effectiveness. In addition, appreciation of the dividedness within a single person can lead to a better understanding of the importance of interaction between separate persons. The ideal group member will develop ways of listening to and watching other members that allow them to reach and respond to both hemispheres and both selves.

CRITERIA

My criteria or standards determine how I evaluate the effectiveness of my thinking. They are developed, paradoxically, without much conscious thought. Until I reexamined my criteria, I was unaware that there are fundamental differences between routine and speculative thinking. These differences are not so much in the thinking operations as they are in the characteristics of the two kinds of thinking. My criteria or standards tend to lead me to a kind of thinking with the characteristics of routine and away from the kind of thinking with the characteristics of speculation.

Before bringing my criteria or standards into my awareness, I was unconsciously taking an absurd position: New thoughts and ideas delight; the process of *getting* them (speculation) has many characteristics that are abhorrent. For example, in a traditional meeting we have agreed to speculate about how we might develop a more effective room service for a hotel building client. A member suggests that we examine how bees and birds handle *their* problems of fast feeding. Such an excursion from the subject would probably be criticized as a waste of time. "Hey, we are talking about hotel room service for people, not bees and birds."

That person's criteria for effective thinking leads him away from the kind of thinking that is most apt to bring some newness to the problem. In order to clarify and modify my criteria so that I can encourage myself and others to speculate when it is appropriate, it is useful to examine the characteristics of routine and speculative thinking.

I have asked many groups to give me their thoughts about this. The following is a composite list for the two types of thinking: It is meant to be suggestive rather than exhaustive.

Some Characteristics of Routine Thinking

1. Logical.
2. Empirical.
3. Few mistakes are tolerable.
4. Focus is on completing the task.
5. There are specific guidelines.
6. Boundaries.
7. Predicatable.
8. Comfortable.

9. Familiar.
10. Low risk.
11. Socially acceptable.
12. Supported.
13. You know where you are going and there are roadmarks along the way.

Some Characteristics of Speculative/Creative Thinking

1. You do not know where you are going.
2. You do not know whether you are going to get there.
3. Focus is on the process as well as getting there.
4. Many mistakes are necessary.
5. Much confusion.
6. Much uncertainty.
7. High risk.
8. Not provable in advance (and sometimes not after the fact).
9. Makes you anxious.
10. Unpredictable.
11. *Appears* inefficient and wasteful.
12. Easy to reject as impractical or impossible.

This helps explain why some of the process of getting new ideas is abhorrent. Confusion, uncertainty, and wrongness are states of mind that many of us are taught *not* to seek. It is bad practice to pursue a line of thought when I do not know where it is going. Yet if I am to speculate, I must dwell in these states and pursue lines of thought without knowing where I am going. If I deny confusion and uncertainty, I am feigning a routine situation and will do the kinds of thinking that go with it.

IMPLICATIONS FOR THE GROUP AND THE INDIVIDUAL

Innovative thinking—the kind of thinking that produces new approaches and new products—requires *both* routine and speculative thinking. Productivity will be increased if each individual helps the group develop explicit new criteria for good thinking together. Such criteria will recognize that the standards for good logical, realistic thinking will be quite different from the standards for good confused, mistake-ridden thinking. Each learns not to apply the criteria for one type of thinking to the other.

HABITS

The next area of opportunity is habits of thought. Most of these habits are not bad in the sense of being destructive to others or myself. In many ways they are good and necessary to protect me from offending others and making mistakes that can hurt me. But unless I can learn to deal with them, to use them appropriately, they limit my ability to use the speculative/creative areas of myself and this reduces the use of my potential. Each of us, without much conscious thought, has developed many of these semiautomatic, limiting habits. I want to discuss only three. As you encounter other habits you can invent ways of delimiting them when appropriate.

SELF-CENSORING

This habit is established when young. As a child I say anything that comes to mind. I may ask a guest of my parents, "Why are you so fat?"

I learn that this is unacceptable. I find that sometimes my parents laugh at what I say in a friendly way. At other times their laughter hurts. I begin to develop a mechanism to stop doing something before I do it—a censor. In the process of learning to be comfortable with people I find that there are a great many impulses I must cut off at the pass before they get me in trouble. My censor gets more and more powerful and inhibiting. Like any efficient, growing operation, it automates some of its activities. Some of my censor stays under my conscious control. *I* still make the decisions about whether to express the thought. The other part goes underground into my unconscious. It makes decisions without consulting me. Based on criteria of danger, it may refuse to release associations, beginning ideas, impulses, and intuitions into my conscious mind. When I need a beginning idea and I "draw a blank," I know my underground self-censorship is being too strict.

Some of the consequences of self-censorship are obvious. If I work with a group in which there is a punishing climate, I tighten my conscious censorship. I express fewer ideas. This has repercussions in my unconscious. It gets the signal "This is a dangerous environment, batten down the hatches." Fewer beginnings are allowed to see the light of my conscious mind. I literally do not have the flow of ideas I normally would. I reduce the use of my potential.

Conversely, in a cooperative, supportive climate the opposite happens. I loosen the specifications, I express more ideas. My unconscious gets the message, and the flow of ideas into my awareness increases. This is another phenomenon we can demonstrate but not prove. Once a person has experienced this, he or she needs no proof.

Each of us has to some extent devised ways to outwit our censors. A few strategies are to ask other people—they say things outside of my thinking pattern and a censored idea can connect with that; to doodle on a piece of paper and after a while look for a pattern or image and relate it to the problem; to listen to music and sing about the problem, making up the words, and there are many others. One that everyone uses because they have found that it works is to turn away from the problem—do something else, sleep on it, go get some physical excercise, and so on. We observed that this practice was so effective that we made it a part of our procedures as an idea-getting strategy.

We call it an Excursion. I will examine it in detail later, but an example here will make the strategy clear. Suppose, earlier, when you were working on the thermos closure problem that you had drawn a blank. Realizing that your censor was being too strict, you might say to yourself, ''Forget all these specifications for a moment. Think of examples of built-in closures on your body (or in your house, or in nature, etc.). Chances are good that you would think of your eyelid, various sphincters, your throat, and others. Any one of these is a beginning idea that you can then modify (by comparing, retrieving, and transforming). An excursion, by changing the specifications of acceptability, confuses and relaxes my censor.

The major consequence of self-censorship is that I do not avail myself of my vast storehouse of experience. To get some idea of the richness of that storehouse, I built on some calculations of Carl Sagan[11] to estimate that stored in my brain is the equivalent of 7000 volumes of the *Encyclopaedia Brittanica*. My self-censor restricts me to using only a tiny fraction of these for each ten years of my life.

ELIMINATING POSSIBLE CONNECTIONS

This is a habit so established and reinforced that I am hardly aware of it. With the huge amount of data coming at me from outside and inside myself, I have to ignore a lot of it or I will go crazy. What I tend to do is to be automatically inattentive to that which is or seems to be irrelevant to the task at hand. Intruding thoughts and associations are ruled distractions and banished. Like the other habits, this one is essential and important in routine situations and at certain times when speculating. If I permit it to operate across the board, I defeat speculation. This brings me to another hypothesis. Retrievals, associations, distractions, intuitions, connections, and observations will fall somewhere on the spectrum shown in Figure 4. In a routine problem my retrieval will be a precise fit to the problem. In a problem in which I do not know the answer my retrievals will be approximate or, many times, irrelevant. When I am working on a problem in which speculation is necessary, the thought that comes to mind will, at best, be approximate. It is quite difficult to be certain whether such a thought is going to be useful or not. My programming pressures me to consider it irrelevant if it does not instantly have

Precise   Approximate   Irrelevant

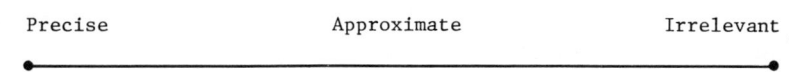

Figure 4 Spectrum of degrees of fit to problem area.

some precise fit to my problem. This effectively keeps me in a routine thinking mode.

This programming starts early. I recall an experience with one of my sons when he was about three. He saw his first horse. "Daddy!" he said, clutching my hand, "There is a big cat." I laughed and said, "No, that is a horse."

In our culture we punish such approximate thinking and pattern recognition in the name of teaching our children to be precisely right. Actually, skill in guessing and connection making, leading to pattern recognition, are the heart of speculation (and learning). As a result of thousands of mild punishments like the one I have described, most children learn to leave their speculation motor safely in neutral and ask questions. They seem to be saying, "OK, if you do not want to practice making connections and inventing, I will stop. It was fun. It kept proving to me that I can think and problem-solve, but if it bothers you, I will give it up."

The speculation muscle gets pretty flabby. There is not much exercise in asking questions.

Today I hope I would deal with my son's connection quite differently by saying, "That was a good approximation. It does have four legs like a cat. What else is like a cat?" and then "this is a horse. What is different from a cat?"

He would thus learn to differentiate, and I would avoid bringing about a damaging change in his learning procedures. The precise–irrelevant hypothesis had for me another usefulness. It clarified a thinking characteristic shared by the really great thinkers and discoverers: their capacity to repeatedly use seemingly irrelevant observations and data to give them breakthrough connections about the problem they were working on. Archimedes used his overflowing bath to lead him to the theory of displacement. Fleming used a spoiled culture to lead him to penicillin. Pasteur used a patch of grass that seemed too green to lead him to the cause of endemic anthrax.

In the history of science these are referred to as happy accidents. I do not believe they are accidents at all. With an outstanding thinker it is a habit of thought: When in the searching, speculative mode anything that captures his attention is relevant until he cannot force a connection out of it. This is the opposite of eliminating possible connections. When he shifts to the rationalizing mode of discovering the logic, such thinkers are noted for their intense concentration. Here again we see the capacity to oscillate: The willingness to be wholeheartedly speculative and approximate, coupled with the ability to be routine, logical, and precise when it is appropriate.

A further important usefulness of this precise–irrelevant hypothesis is that if I can bring myself to appreciate the value of approximate thinking, I can instantly make approximately relevant and helpful many thousands more of the experiences stored in my head. For most of us this storehouse is so rich and varied that we have *some* experience that will illuminate, with approximate understanding, nearly *any* problem or learning task we face. I have experimented with many individuals and groups to test this point. When a participant is willing to use approximately relevant experiences, he or she readily understands complex, strange to his expertise problems in his or her own terms and is able to contribute to their solution. This approximate understanding does not usually lead to precise ideas for solving the problem; it *does* often lead to approximate ideas that expertise can transform.

LISTENING TO EVALUATE

As an inhibitor of speculative thinking, this is the most serious habit of all. And when I am able to modify it, it becomes one of the most powerful speculative and building tools in my whole arsenal. It is rooted in a lesson I learned early that is continually reinforced: When I make a mistake, I will get punished. This is not literally true. I am certain many, perhaps even most, mistakes are not punished. I am *emotionally* convinced otherwise. The fear of making mistakes becomes so deeply ingrained that I not only want to avoid making mistakes myself but I do not want to be associated with the mistakes of others. I begin to live by Mahr's law of limited involvement: *Don't get any on you.* I listen to every idea and proposal with my ear tuned for flaws. If there is anything mistaken about the idea, I discard it. I also develop a strong need to point out the flaws to the owner; I justify this by explaining that I do not want him to make a mistake. A much more powerful motivator is that I want to make it clear that I am *not* supporting or associating myself with this erroneous line of thought. I don't want to get any on me!

Pointing out flaws has two consequences. Beginning ideas tend to be dropped rather than transformed into possible solutions. Just as important, idea getters learn to be so cautious that to avoid the punishment of having their flaws pointed out many promising beginning ideas never see the light of day.

I can overcome both these drawbacks by considering listening for flaws as the first step in a three-phase procedure. Step one, I listen with an open mind, that is, I hear both the useful implications of the idea and also the flaws (I will discuss later how to keep your mind open in spite of flaws). I focus my attention on the flaws. Step two, I invent ways to overcome the flaws and am ready to build them into the idea when the owner is finished. The third step, I use the idea, however flawed, as a stimulus to retrieve a beginning idea without those particular flaws.

This is a form of mental and emotional jiujitsu. I use the very force that would eliminate the idea, the flaws, to stimulate thinking that will make that idea stronger and to produce a second idea potentially stronger than the first.

IMPLICATIONS OF THE HABITS FOR GROUP WORK

The importance of learning to modify these habits is clear for the individual. The implications for group work may be less evident. My posture as a traditional group member, even assuming good will, might be characterized as *judgmental/safekeeping/helpful*, in that order. When a member ventures to offer an idea, I listen to evaluate or judge it. If I decide that it is safe to support the idea, I then help him with his idea. When I modify my habits as suggested above, it reframes my posture. It now becomes *responsibility for success*. I have an obligation to use myself and my team. This means that if I have a glimmer of a thought, I tell them in case they can use it. When one of them has such a glimmer, I use it like Archimedes used his overflowing bath—relevant unless I cannot make it so. When an idea is offered, I go into my three-step procedure to wring out of it every possibility. This, of course, is the stuff that synergy is made of.

Such a posture has a critical impact on the climate of a meeting and within a group. And paradoxically, this posture appears to be practical only if the climate supports it.

CLIMATE

I am here using the word *climate* to represent all the elements that affect a person. What is going on inside me—whether I feel good, healthy, and ready to take a few risks or whether I feel unwell and vulnerable—will have a lot to do with how I operate. What is going on around me also affects how I operate. I want first to look at external climate.

This is governed by three communication channels: words, vocals, and nonverbals. We are all familiar with words. Vocals refer to things done with the voice: They include tone, hesitation, and emphasis. Nonverbals are all stimuli that do not make noise and are not communicated as words, for example, the way in which a person dresses, the way in which a room is decorated, unwritten rules and customs, gestures, exressions, muscle tension, and so on.

Albert Mehrabian did some experiments to establish the relative impact to these channels in face-to-face communication.[12] If total impact is 100%, he asserts that words account for only 7%. Vocals convey 38% of my meaning, and nonverbals 55%. I believe what is important in these figures is that a heavy burden is carried by vocals and nonverbals. It does not matter whether words are 7 or 50%; two channels that I do not know much about are far more important than I thought.

Yet it makes sense. Consider how many ways you can say the words "that is a good idea." Using vocals and nonverbals you can make them into anything from a great compliment to a stinging insult.

We have discovered that there is, practically speaking, no such thing as a neutral action in communication. Everything makes a difference. Either the action helps create a climate in which it is safe to speculate, or it hurts the climate. This becomes critical when we see that climate is the basic determinant of the level of speculation/creativity that will be possible in any given group.

At one end of the climate spectrum are actions that are punishing and competitive, at the other are acts that are supportive. The capacity for problem-solving and creativity is such that even when the climate is punishing, people are productive. As the climate shifts from punishing to supportive, we know from thousands of observations and experiments that there is a significant increase in creative productivity. Figure 5 shows some of the actions referred to that make or break climate.

The reaction of many people to these outlines is first the one I described earlier: "What do you mean a challenge is destructive!"—a general denial that these actions *are* that destructive. The second reaction which comes after there is some acceptance (usually after seeing a few videotapes of themselves in action) is, "If I pay the necessary attention to these, I will not have time for anything else."

This protest has my sympathy. I shared this impatience as I was doing the research. In the course of our development we first identified actions that worked. For example, a group would offer an analogy and the group would come alive and make progress. In one case the group was stuck for ideas on how to persuade and convince a potential alcoholic that he

ACTIONS THAT <u>ENCOURAGE</u> SPECULATION/CREATIVITY

Listen	Build on	See the value in
Paraphrase	Speculate along with	Focus on what is going
Stay loose until	Share the risk	for the idea
rigor counts	Set up win/wins	Assume valuable impli-
Protect vulnerable	Make it "no lose"	cations
beginnings	Support confusion/	Take responsibility
Take on faith	uncertainty	for understanding
Temporarily suspend	Acknowledge	Waste no energy eval-
disbelief	Credit	uating early
Assume it can be	Value learning from	Jump to favorable con-
done	mistakes	clusions
Share the burden	Be attentive	Use ambiguity
of proof	Be interested	Give up all rights to
Connect with	Show approval	punish or disci-
Accept	Give early support	pline
Be open to	Eliminate status/rank	
Join	Be optimistic	

ACTIONS THAT <u>DISCOURAGE</u> SPECULATION/CREATIVITY

Be pessimistic	Nitpick	Correct
Preach/moralize	Interrupt	Name call
Be judgmental	Be bored	Blame
Assume no value	Misunderstand	Set up win/lose
Make no connections	Be inattentive	Be competitive
Put the burden of	Act distant	Make fun of
proof on other	Pull rank	Be dominant
person	Get angry	Command
Take ball away from	Disagree	Order
Ask questions	Argue	Direct
Cross examine	Challenge	Threaten/warn
Give no feedback	React negatively	Demand
Be noncommittal	Discount/put down	Do not listen
Put on a stone-face	Be cynical/skeptical	Do not join
Be critical	Insist on early	Use silence against
Disapprove	precision	Scare
Be impatient	Point out flaws	

Figure 5 Actions that encourage speculation/creativity and actions that discourage speculation/creativity.

was in danger. Nothing new was developing and the group's energy was low. A member said, "You know, a religious conversion must be an example of total persuasion."

There was a charge of energy I could feel. Another member exclaimed, "Jesus! (laughter) if you will excuse the expression. My image of religious conversion is that the convert suddenly sees his future if he continues without God. How could we do that to a 'potential,' except only make him see what will happen if he continues with alcohol?"

"How about using a fortune teller?" said another member.

"Right! have the fortune teller tell the guy some truths about his past and then make some predictions."

The group was off and running. When we recognized that analogies did not *necessarily* work, I undertook to find out why. By studying tapes of instances in which analogies did not "work," I observed that quite often, as the analogy was offered, a member would take some action that might seem innocent but that, in effect, killed the effectiveness of the analogy. He might ask a question about it, or simply disagree that it *was* an analogy, or make fun of it.

When I began examining analogies that worked, a definite contrast was apparent. The working analogy elicited early support—an exclamation, a quick connection.

This led us to the protection of analogies. We would intervene to prevent or turn away or counteract those actions we thought might be damaging. This increased the likelihood that the analogy would work. There are still times, even when there are no destructive actions that can be seen, when analogies and other idea-getting strategies do not work. We are dealing with probabilities.

It was through the study of strategies that usually work that I learned about actions that cause them not to work. I was dismayed at the number of these actions, subtle and crude, which in the hands of a competitive participant could keep a group from being productive. No single action can do this, but if there are many, that becomes the style of the group. It seems to be catching.

I have gone into some detail about this, because it is a basic issue in group work. Cleaning up the transmissions to reduce the destructive actions to a minimum takes time, understanding, and effort. The payoff is not instantly visible. Most group members will have operated in the midst of nearly all the destructive actions and still accomplished much. But unless group members are willing to *invest* in climate, it will not be possible to experience the synergy that is one blessing of a cooperative/ supportive climate.

INTERNAL CLIMATE

I have become persuaded that many of the actions in Figure 5 apply to the way in which I treat myself. For this to make sense and be manageable I use my concept of safekeeping self and experimental self. I experience my safekeeping self as dominant, reacting negatively, devaluing, and correcting—in fact, using most of the discouraging actions on my experimental self. I have been able to modify this by having role-playing dialogues between these two selves. I assume that everything my safekeeping self does is with good intent, even though often outspokenly negative and critical. Safekeeping's mission is to keep me safe, not to do good thinking. I have my two selves problem-solve as though they were two different people, to invent a way of working together that takes into account *both* of their missions. This is one reason the metaphor of two selves can be useful. It allows me to manipulate and improve my ways of using my inner resources by imitating, within me, the processes of a well-trained group.

SUMMARY

In the first part of this chapter I have attempted to show some of the problems and opportunities in both individuals and groups. I will spend the rest of the chapter discussing the methods Synectics® has developed to capitalize on some of these opportunities.

SYNECTICS METHODS

In our research our first steps were toward finding procedures that helped us get new ideas. When we saw something that did this, we incorporated it experimentally. We kept studying it and modifying it until it worked quite well, or we dropped it. We did not spend a lot of energy in discovering *why* a procedure worked. This tended to make us dogmatic. "This works most of the time, so do it this way."

As we understood *why* some of our procedures worked (i.e., increased the probability of success), we realized that there were many other ways to get the results we were getting. We became more relaxed.

In 1962 we were teaching a group of chemists an excursion. It is one we now call The Example Excursion. "Give me an example of a built-in closure from the world of your body." (I have changed the words to those used in a formal excursion). One of the participants challenged me, "I find it more effective to just choose some object around me, almost at random, and use that to give me ideas.''

I remember feeling defensive and slightly angry. ''OK, show us.'' He selected the banister of a staircase in one corner of the room: ''Let me see . . . those spokes, or whatever, form a border around the air . . . I will have to make it hold better than air . . . I will stretch a rubber sheet between the spokes . . '. so my thermos closure will be a foamed rubber sheet, with a lip or flange all around. One side is fastened to the bottle, the other is free to be lifted.''

I attempted to equate this with an Example Excursion. Today, realizing that what he was doing was exactly the kind of thinking we want, I would welcome his verson and perhaps invite the group to experiment with selecting an object at random. The point I want to emphasize is that the Synectics® methods I will introduce are *one* way to solve some of the problems outlined in the first part of this chapter.

ROLES

In every meeting there are three basic roles that are played. These can be most easily defined in a meeting called to solve a problem. First, there is the person who owns the problem. Then there is the person who is running the meeting. Finally, there is the participant. If we had a role-play experiment and asked three people to play purely one role, the problem owner would state his or her problem, the process person would limit himself to process. After the statement of the problem, he might say, ''Now it is time for ideas. Any ideas?'' The participants would then supply ideas. The problem owner would absorb these. The process person might after a while say, ''Now it is time to evaluate the idea you like best, problem owner.''

Meetings do not work this way. Role responsibilities are not explicitly assigned. Each of us feels perfectly free to assume any of the three roles. For instance, the chairman might say, ''It is time for ideas.'' I would feel free to interject, ''Before we go to ideas, I need a little more background.''

Or, when evaluation time was at hand, even though I am simply a participant, I evaluate and give my opinion of the problem owner's decisions. This free-flowing model has the advantage of maximum involvement for all members and in fact, *when all members thoroughly understand and respect the basic role responsibilities*, it makes for an excellent meeting. When group members are at that point, they need only agree about how they are going to operate in this meeting. For example, at one meeting they may agree that everyone will be responsible for process. At another, in which they expect a lot of emotional involvement, they may agree that one of them will be facilitator and have charge of process.

STRUCTURE AND INTERACTIONS*

Every meeting is conducted on two levels. One level is the structure. It is needed and important, and if that is all that is taken into account, the meeting will be inefficient. The quality of interactions between members determines the climate, and climate determines (the quality and quantity of) ideas. An example will clarify this point.

This group is working on the problem of automobile safety. The process facilitator has asked the team to develop a variety of ways of approaching the problem, a step in our structure called goal/wishing.

Jim: I wish for a seat belt that fastens itself.

Joe: That doesn't really solve the problem. The basic problem is safety and that is only one aspect of it.

Joe has just issued an invitation to a win–lose situation. If this were allowed to pursue its course, Jim and Joe would each spend some energy justifying his position. This would make no contribution to the problem-solving and would end with one or the other feeling as though he had lost the discussion. The loser would therefore stop cooperating with the winner and use more energy in setting up a second win–lose situation that he would win.

The process facilitator recognizes Joe's invitation.

Facilitator: Joe, it sounds as though you have a goal/wish with a different objective than Jim's. Please frame it as a wish while I write up Joe's. Then I will get yours.

The facilitator is not dealing with structure here; he is handling an interaction so that instead of getting a win–lose situation, the team gets a win–win situation.

Throughout my discussion of the Synectics® methods I will be concentrating more on facilitating interactions than on structure.

CLIENT

We call the problem owner a client. We want to make explicit the fact that the team is working for this person. We will really win only if we give the client what he or she needs.

---

* For a more detailed discussion see "The Practice of Creativity."[13]

The client's overall responsibility is to get from the team as much of what he or she needs as is possible. All of his or her actions are aimed at that. His or her interactions with team members will be designed to increase their involvement. I will be more specific about this as we go along. Now, the responsibilities are the following.

1. *Give the Team Enough Information About the Problem or Opportunity So They Can Begin to Work.*   We have found that a brief period (about five to seven minutes) to cover all answers to the following questions will generally serve: (1) What is the background of the problem? (b) Why is it a problem? (c) What are some of the things you have thought of or tried? and (d) What do you wish to get from the group in the time available? In addition, it is important to make clear to the group what your personal stake in the problem solution is.

2. *Contribute to the Meeting as Though a Participant.*   It can be particularly useful if the client offers a far-out, wishful goal/wish to demonstrate openness to such goal/wishing on the part of others.

3. *Be Alert for Opportunities to Show Appreciation for Good Thinking.*   When a participant makes a pleasing contribution, use words, vocals, and nonverbals to let him or her know about your pleasure. Do not attempt to fake this.

4. *Resist the Temptation to Comment on the Contributions When They Do Not Please.*   You have complete control over the direction in which the team will go. When making choices, simply ignore those that do not aim where you want to go.

5. *Listen Approximately.*   Many of the suggestions made will not be precise fits to your problem. Exercise your skill in three-step listening. Listen for flaws, listen to overcome flaws, and listen to get an idea without flaws.

6. *Model the Ways You Wish Participants to Act.*   For example, practice crediting: When someone says something that stimulates a thought in your mind, let her or him know. "Sally, when you said that, it gave me the idea that. . . ." Basically, if you check the outline of actions that encourage speculative thinking and do whatever you can to conform to that in your operations, you will fulfill this responsibility.

7. *Prove That You Are There to Find Ideas That Will Work for You.*   For example, in the goal/wish phase the facilitator will be getting from the team a mixture of ideas, wishes, beginning ideas, and directions. During this period, if a new idea that appears feasible is offered, it will be written up by the facilitator. Even though you have not been com-

menting on the offerings as they happen, you can say something like, "Hey!, there is a new one I could experiment with next week." You may not choose to do so, but you have let your team know that you are alert for ideas that will be useful to you.

8. *Indicate the Sort of Direction That Interests You.* Here, again, you will not be reacting to every goal/wish, but when a goal/wish suggests a direction you particularly like, let the team know.

9. *Be Decisive.* When the facilitator asks you for directions, give him what he needs. If you need time to consult, take all you want, but when through, be definite.

10. *When You Are Not Getting What You Want, Let the Facilitator Know.* If the team is way off base, ask for a break and discuss your feelings with the facilitator privately. Often you can express what you want in a goal/wish, for example, "I wish I could get more ideas on how to get distribution."

11. *When Evaluating an Idea Use Itemized Response.* This is a procedure that gives the members of the team a balanced evaluation that educates them about the kind of thing you are looking for and are concerned with. If you choose to proceed with developing ideas, they know just what needs to be worked on. In the itemized response you articulate three or four advantages or positive aspects of the idea and then you shift your attention to the concerns or gaps in the idea. These are aspects that need further inventing. If possible, you phrase your concerns as *how to's*.

For example, in a car safety problem, suppose you are the client and the idea to be evaluated is a padded mechanical bar that lowers to embrace the occupants of both seats, much like the bar on a ski lift.

Jane: I like the relative simplicity of that, and another plus is that it would be easier for the user to pull it down and put it in place than the seat belt. It would be easier to automate than a seat belt. Another thing I like—and this may be sneaky—we could make it so that in the up position it would get in the way of driving. The user would almost *have* to put it on.

My concerns are how can we make it comfortable, like a belt and how can we prevent sideways slipping. I guess my only worry is how can we make it simple—not a big mechanical deal.

In summary, the client is the reason for the meeting. Satisfying him or her is the clear objective of each member. How the client interacts with the individuals and ideas will have a profound effect on the productivity of the group.

FACILITATOR/LEADER

The overall responsibility of the facilitator is to manage the process of the group in order to get for the client what he or she needs. To this end the facilitator guides the team, *mindful* of structure, but not bound by it. He or she uses his or her judgment about what he is after. The facilitator's primary job is to create and protect the climate. Here are some specifics about how this can be done:

1. *Make Yourself Thoroughly Familiar With the Actions Outlined in Figures 5 and 6—Those That Discourage Speculation and Those That Encourage It.*   I will be suggesting some ways of intervening or avoiding the discouraging actions, but the best way to facilitate is to improvise in your own style to create and protect the climate.

2. *Listen to Team Members.*   This is the foundation upon which nearly *every* encouraging action is built. Permit the speaker to paint any picture he wishes; your aim is to understand from his point of view. If in doubt, or if you think that the team member may be in doubt, paraphrase to be sure that you understand to his satisfaction. This sounds easy, but it is not. You will catch yourself making judgements, tuning out, listening to your own thoughts, and otherwise failing to really comprehend what the speaker is saying.

The importance of listening cannot be overemphasized. Skill in good listening has a pervasive effect on the team's productivity because it directly affects climate.

You will also on occasion have a member who tries to dominate the meeting. He will have immediate responses and go into endless detail if you permit. These people are usually bright and valuable, but they can ruin a meeting. You will need to control such a person without alienating him (he may be your boss). Here are three ways of dealing with this— there are others and you can invent your own:

> When you believe you understand the point, interrupt to say, "Thank you, I've got it," and move quickly to someone else.
>
> Avoid the compulsive talker's eyes when asking for a response.
>
> More drastically, when you ask for a response, look at someone else and hold your hand to the talker in a casual stop sign.

3. *Keep the Energy Level High.*   This may seem an impossible assignment, but it is not if you use the tools available to you. There are a number of things that affect the energy in a group, including some that are beyond our control, such as a member's hangover. But there is a lot

you *can* do. Here are some suggestions:

> Your interest, alertness, and intensity are contagious, so when you take over as facilitator, give it your best. Use your natural body language: Move around, move close to a speaker, use your hands—anything that is comfortable for you.

> Use excursions lavishly when the group is tired. It is often like an actual vacation from the problem and members return refreshed and with renewed material banks.

> Keep the pace fast but not hurried. Do not linger on any one step too long. When group members give signals of boredom, do something different.

> Humor can be invaluable. If amusing associations occur to you, bring them out. When a member jokes, show that you enjoy it too—if you really do. You are probably not a stand-up comic and so do not push yourself to become one. Just be yourself, encourage humor, but do not let the meeting degenerate into a joke-telling session.

> Surprise the group. After running a few excursions that are alike, run an excursion they do not expect. I will give you some examples later in the chapter, but you can make up excursions of your own.

> Have a plan to shake things up for the session right after lunch, and for later in the afternoon. These are low-energy times.

4. *Keep Your Eye on the Client.* When members are giving ideas, watch your client with great care. If he or she shows interest, check to see if he or she would like to pursue that line of thought. When in doubt about what kind of content your client wants, ask him or her: "Client, we have 20 goal/wishes. Would you like to select one to pursue or would you like some more goal/wishes?"

5. *Rotate the Facilitator Role.* Like the Pony Express changed horses, it is wise to change facilitators to keep up the pace. It has other benefits: This is a demanding role and until everyone has tried it, they will not appreciate the importance of their cooperation as a participant.

6. *Do Not Pussyfoot.* Because climate is such a critical element, facilitators often believe that being very gentle and hesitant establishes a climate that encourages actions that foster speculation. There is nothing the matter with gentleness or hesitance if you are stuck for the right word, but you can be crisp and definite in intervention to protect the climate. It is your responsibility and you have the authority to carry it out. You will demonstrate that you are in charge of process.

In summary, the facilitator/leader role is complicated and demanding. You are like the conductor of the orchestra. You do not play any instrument and you are responsible for making beautiful music.

PARTICIPANTS

Participants are the heart of any meeting. All the skills of the facilitator and the constructive responses of the client are designed to help each participant make his unique contribution. To emphasize the true relationships in a meeting we view the leader as a servant to the group. The group is, of course, servant to the problem. The client is the problem's representative and except in matters of behavior his opinions are honored. Differences with him are welcome too. They are aired, written, and the decision of how to use them is left to the client.

The basic responsibility of a participant is to use his or her wits to help the client get what he or she needs. The best way to do this is to dig as deeply into your personal potential as you can. Here are some specifics:

1. *Pay Intense Attention to Yourself and to Your Impulses.* You think at the rate of about 900 words a minute. People talk at the rate of 125 words per minute. Use only a small part of your energy in attending to what is being said. Use most of your energy in following the thoughts stimulated by the speaker. Even when your images and thoughts seem irrelevant to the problem, note them and attempt to connect them.

2. *Use Your Pad.* When you are "out listening," that is, out of the meeting, listening to yourself, keep notes on your pad. This way you need not interrupt, yet when the time comes you will have your ideas ready. When you get an idea, note it then rejoin. Do not depend on your remembering.

3. *Do Not Censor Something That Feels Important Even If It Does Not Make Sense.* Let the group hear it. They will listen to it as a stimulus and may be able to make a connection that you were not making.

4. *Make Three-Step Listening Your Modus Operandi.* (a) Listen for flaws. (b) Listen to overcome flaws. (c) Go for an idea without the flaws.

5. *Practice Open-Mindedness.* This means that when you are listening to an idea, you pay attention to the positive implications of the ideas as well as to the flaws. If you find yourself unable to find any positive implications, you know that you are closed to that idea.

6. *Know the Actions That Discourage Speculation and Police Yourself So You Do Not Slip into Any of Them.* Know the actions that encourage speculation and use them at every opportunity.

7. *Cooperate With Your Facilitator.* Even when you do not understand exactly what he or she is asking, cooperate as best you can. Guess, and do it. After the session you can ask questions.

In summary, as a participant, your responsibility is to bring to the meeting your whole self and use as much of it as possible.

STRUCTURE

Structure in a meeting furnishes guideposts for the facilitator so that he or she can lead the group from the beginning of the problem through to a solution, without missing any essential steps. On the other hand, knowing the structure permits the facilitator to ignore it. He can pay less attention to formal structure and focus on the activity that will most help the client at any given moment. For example, the group may be doing goal/wishes. When the second one is offered, the client says, "Wow! I never thought of that and I would like to explore it."

The facilitator skips several steps in the structure and goes to an immediate evaluation, first assuring the members that he will come back and get the goal/wishes they have developed.

GENERAL PURPOSE

The first step is to develop an understanding of the problem. We have found that this phase is most effective when we get both an analytical understanding and a speculative understanding. As you have seen, we ask the client to give a brief explanation of the problem. As that is going on, we ask the participants to be translating the explanation into goal/wishes. At this stage we want the thinking to be unfiltered by reality. If the client says, "Potential alcoholics refuse to believe that they are," we want participants to be thinking of goal/wishes as wishful and unreasonable as, "I wish we could give a potential some truth serum and have him talk to himself," "How to inoculate a potential to make him immune to alcohol," "I wish a potential would turn green one day a week to persuade himself."

We are "exploring" the problem to give the client as many different ways of seeing it as we can. We are not interested in precision at this stage.

The next step is focusing on one direction. The client selects one or two of the goal/wishes that appeal to him or her, and the team explores *that*. For example, an adventurous client might choose, "How to make a potential turn green.'' The team then develops more goal/wishes aimed toward the specific, "How to give him a litmus paper that tells him the truth," "How to make a potential instantly recognizable to everyone," "I wish that whenever a potential looks in a mirror, he is reminded of the truth.''

The third step is to focus again on a specific. The client might now pick the mirror goal/wish. The team develops ideas that might make something like the mirror goal/wish practical. For example, in this case the team developed the idea of giving the potential access to a lie detector. He answers a series of questions. Whenever he denies his addiction to alcohol, the machine tells him he is lying.

DEALING WITH IDEAS

The most profitable way to use an idea is as a stimulant. The least profitable is to evaluate. During the session we ask participants to practice three-step listening until it is time to evaluate. When that time comes, the client itemizes the advantages implicit in the idea and then voices his concerns as how to's. The client may ask the team to join him in his itemized response.

Then, if the client wants the help of the team, he selects one of his concerns and the team gives him ideas on how to overcome it by adding to or modifying the original idea. When the client decides that his concerns are overcome, we go to the next phase.

NEXT STEPS

The final step in the process is to record whatever next steps the client wishes to take to proceed with the implementation of the possible solution.

IDEA-GETTING STRATEGIES

The excursions are specifically designed to help participants originate beginning ideas. While this is their primary objective, excursions can do far more than that for the team. Over the years I have asked groups to tell me what excursions have done for them. Below is a list of some of their responses:

1. Gets me around a mental block.
2. Facilitates the nurturing of other's ideas as well as my own.
3. Changes my synapses.

4. It gives me permission to be irrelevant.
5. Broadens my tolerence for approximations.
6. Recharges my enthusiasm/fun batteries.
7. Makes me feel good.
8. Makes nothing off limits.
9. Builds confidence and trust in the group.
10. It helps me appreciate the thinking of others in my group.
11. Makes me less nervous and anxious.
12. Sponsors an "I can" frame of reference.
13. Helps me see that it really is possible to look at this problem in a new way.
14. Helps me to mine my right hemisphere.
15. It is a model of nonpunishment for any kind of a contribution.

An excursion, unless it interrupts a promising line of thought, is nearly always a good investment in climate. The signals that tell me an excursion is a must are (1) the team does not have *any* beginning idea, (2) the team is recycling ideas that have no newness, (3) the team is sending signals that they are bored or restless, (4) people are finding a lot wrong with what is going on, (5) people are being very precise, and (6) my team is not having fun.

SOME SAMPLE EXCURSIONS

The team is working on the problem of convincing potential alcoholics that they really are alcoholics. (Some personality profiles and people have a much higher than average chance of becoming alcoholics. They are skillful at rationalizing their behavior to keep themselves believing that they are simply social drinkers.)

Example Excursion

Facilitator:	Team, give me an example of persuasion from the world of rocks.
Sam:	A landslide.
Facilitator:	Tell us a little more, Sam.
Sam:	A landslide is overwhelmingly persuasive. It sweeps everything in its path, or it buries it.
Janet:	A volcano. The molten rock gradually persuades the mountain that it has to give, and suddenly BOOM!

Jim:	A diamond.
Facilitator:	How so, Jim?
Jim:	You know what they say . . . a diamond is a girl's best friend? (Laughter) Well, that makes a diamond pretty persuasive.
Facilitator:	OK, team, let's examine landslide. What comes to mind when you image landslide?
Jim:	Tremendous leverage . . . it can take only a sneeze to start the thing.
Sally:	It is a huge force but it is made up of small pieces . . . pebbles, dirt, even boulders are small compared to the overall force.
Sam:	It gives the impression of moving slowly, almost slow motion. Actually it is racing along very fast . . . I'd guess it is 40 miles an hour.
Facilitator:	Now, team, let's take these examination thoughts and go to our problem of persuading potential alcoholics that that is exactly what they are. Use this material the way Archimedes used his overflowing bath. Make it give you new beginning ideas.

He waits. . . .

Sally:	Overwhelm him with the facts, but do it a little at a time.
Jim:	I have a build on that . . . make the facts not general ones about what happens to alcoholics, he can "outrun" those. Somehow make these facts specific to *him*, personally.
Kitty (the client):	You know it would be possible to *do* something like that . . . the part about facts. We know with quite a lot of detail and precision what is going to happen to the problem drinker.
Sam:	Get his wife or a friend to give you background on him so you know where in the cycle of alcoholism he is. Then you begin to tell him what is going to happen to him next. Giving us his name is the sneeze that starts us on the way to overwhelming him.
Facilitator:	Kitty, are you ready to paraphrase the idea that is coming out?

Kitty summarizes the idea, making it fit her precise knowledge of what her treatment center can actually do. Briefly, the program starts with a call or visit from a spouse or concerned friend. A case history is taken, and 10 predictions are developed from the case history and center's knowledge. Once a month the potential alcoholic gets a note predicting some specific event that will happen to him or her in the coming month. For example, the note might say, "In the month to come you will find it increasingly difficult to remember what you had for dinner the night before. This is the pattern followed by a person who is oversensitivie to alcohol. When you decide you might be ready to explore ways of normalizing your life, call: (and number is given).

Imaging Excursion

The wish the team is working on is, "How to have a voice in his head keep telling him that he is a potential alcoholic."

Facilitator: Now, team, I want to go on an excursion using the word "voice." Put the problem out of your mind and focus on the word (he writes it on one of the large pads in the front of the room). Let it stimulate an image in your mind. Scan that image until you see something in it that intrigues you. Give me one word for the intriguing something. . . . Sally?

Sally: Dog.

Facilitator: (writes *dog* under *voice*, then covers *voice* with his arm) Sam, do the same for dog . . . and if you pull a blank, just say "Pass."

Sam: Attack.

Facilitator: (writes *attack* under *dog* and covers *dog* with his arm.)

As with all excursions, this one invites right-brain activity. There is a tendency for a team to slip into associations rather than imaging. This will not ruin the excursion, but it is not as effective. The idea is for a participant to look at his or her image and be sensitive to the feeling of being intrigued (I believe this is exercise for intuition). When she gives her word, the team images with her and attempts to make a connection to the previous word—good right-brain activity. If you get associations, ask that the participants give you a "once removed" association. For example, looking at *dog* my impulse will be to say, "Cat"; I resist this and think of a second, more distant association like "days."

The facilitator collects a word from each member.

Facilitator: Now, select your favorite word from this list. Write it at the top of a clean sheet of paper. Let that word stimulate an image in your mind's eye. Turn on the movie in your mind— all that means is that you animate your image—let things happen and make notes on your pad about what happens.

The Facilitator allows two or three minutes.

Facilitator: OK, team, let's go back to the problem. How to have a voice in his head keep telling him he is a potential alcoholic—take the material you have written that seems totally irrelevant and use it like Archimedes used his overflowing bath. Make it give you a beginning idea. Remember that a beginning idea does not need to work, but if it did, it would help us.

The Facilitator allows two or three minutes.

Jim: The idea is to get a skillful neurolinguistic programmer to make a tape that will program the listener to seek help at one of your centers.

Facilitator: Great! Where did you get that, Jim?

Jim: I picked the word *attack*. My movie was about the Trojan war and some Greeks were pushing one of those big towers toward the Trojan fort. The tower was high and overlooked everything. I wished I could have something in the potential alcoholic that would "overlook" him all the time. I thought of a conscience to direct him to your centers and I thought of Milton Erickson. He was a psychiatrist who was very skillful at transmitting directions understood by the right brain or unconscious while having an ordinary conversation with the left brain. This is what neuroliguistic programming is.

Kitty   What a marvelous thought! We know that the potential al-
(the client):   coholic turns to alcohol for comfort. We can have the tape program him to come to our centers for comfort instead of to the bottle. And perhaps we can train our therapists to reinforce what the tape promised.

OTHER EXCURSIONS

There is no limit to the number of different excursions. Chapter 6 in *The Practice of Creativity*[14] introduces some in more detail, you will find others in the *Journal of Creative Behavior*,[15] and perhaps the best source of new excursions is yourself. You can design them to suit you. The important elements are (1) participants must put the problem out of their minds, (2) your instructions lead them to use some right-brain functions (imaging, connection making, pattern recognition) apparently unconnected with the problem, and (3) participants use the seemingly irrelevant material back on the problem.

OTHER KEY SYNECTICS® STRATEGIES

GOAL/WISHES

I have mentioned this earlier and it is such an important tool that I want to discuss it further. History: in early problem-solving sessions it seemed that each participant, before attacking the problem in earnest, felt the need to restate or redefine the problem in his or her own terms. We made this part of the process. After the problem statement and analysis we asked each participant to restate the problem as he or she understood it. The facilitator wrote these restatements as part of the notes.

We then discovered that after everyone had a restatement, some wished another chance to restate. Since each of these restatements was somewhat different from the others, we welcomed as many as we could take time for. It became apparent that these varied points of view were helpful to the client (sometimes) in understanding his problem differently. It happened almost as a joke that people offered restatements that were absurd, impossible, or very wishful. The effect of these wishful Goals was much like that of an analogy—they stirred up unusual energy and stimulated a different kind of thinking. We began to encourage them and called this element goal/wishes.

Many people have trouble with wishing. I asked a business friend why he found wishing difficult and distasteful. "I have spent my adult life doing my best to be realistic and deal with situations the way they *really* are, not the way I *wish* they were," he said.

"If you don't wish about a situation, how do you know how it ought to be?" I asked.

"You have a point, but I do not call that wishing. I call that having a goal or objective—it is not a wish, it is something possible to achieve. Wishing, by my definition, is hoping for something to happen that you *know* cannot happen," he replied.

It is understandable that practical people have trouble tolerating wishfulness. However, wishing may be seen as an additional form of exploratory thinking, of goal setting. Because it is not concerned with reality, it has the capacity for opening one's eyes to new possibilities. If one is constantly realistic and precise in wanting (goal setting), one automatically rules out exploring many lines of thought that might be profitable.

When we look at the history of various developments, we see that some-one must have once indulged in the following unrealistic and impractical wishing:

> I wish I had a carriage that would propel itself.
>
> I wish I could make instant drawings of happenings I want to remember.
>
> I wish I did not have to wait to see the photograph I take. (Dr. Edwin Land can document this wish. It was made by his daughter and led to the Polaroid camera and film.)
>
> I wish I had a magic drain in my sink so that garbage, bones, and waste would go down it and disappear.

You get the idea.

ITEMIZED RESPONSE

Sometimes called an open-minded evaluation, this procedure is one of the most valuable in Synectics.® Although I briefly described this earlier, I would like to add that itemized response is based on the assumption that I am seldom dealing with fools. If someone offers an idea, it must have seemed useful to the offerer or she or he would not have presented it. It is not only polite to deal with the idea courteously, it is prudent and economical. When I am looking for ideas, it does not make sense to throw out any without examination. *After* examination, if appropriate, I can discard.

This procedure in itemized response is simple. When I understand the idea (having paraphrased it if there is any doubt) I smother my natural inclination to point out the flaws. I will not ignore or forget the flaws, I will deal with them, but *not first*. First, I focus on the advantages of the idea, both explicit and implied. When I have spoken of these, I turn my attention to the flaws or my concerns about the idea. I express these as how to's, thus making it clear that these are problems that we could overcome if we chose to pursue this idea.

When the itemized response is completed, I make a decision about whether to go further with it or to put it aside.

This procedure recognizes that most ideas, even the great ones, begin with many weaknesses. If I let my natural safekeeping inclinations make my decisions for me, I will not pursue anything but the sure winners—and when an idea is a sure winner of that sort, it will have been around for a while, it will not be new. Itemized response presses me into a behavior that recognizes that strong ideas are built by inventing ways to overcome the flaws in beginning ideas.

FORCE FIT

Force fit is a term we coined to describe the moment of truth in speculation: When I take the seemingly irrelevant observation or material and I use it as Archimedes used his overflowing bath, I make it give me a connection. It is the most difficult step in the process. To make it work I must be truly free to move from one end of the thinking spectrum to the other, free to oscillate from irrelevant to precise, open to confused images and impulses from my right brain as well as logical, realistic inputs from my left brain, in short, to be good at force fitting I need my whole self operating at its cooperative best.

When Jim explained where his neurolinguistic programming idea came from, he was describing his force fit. The important elements are the following: (1) He developed his Trojan War attack scene without consciously thinking of its possible application to the problem; (2) he shifted gears back to the problem and went into a delicate, tolerant, tentative mode—totally open, and not in any hurry to nail down a practical idea, he considered his material to see what specific appealed to him or caught his attention; (3) when the tower seemed interesting, he took time to image the tower and how it seemed—very tall and kind of rickety . . . higher than anything else in his image . . . it overlooked everything; (4) he shifted his attention back to the potential alocholic and asked, "What is this tower trying to tell me about this person?" and the overlooking idea became someone looking over the shoulder of the potential alcoholic; (5) He found the beginning idea to be almost in place. He asks himself how he can make it more practical? He thinks of conscience as something to work with. He asks himself how he can make conscience work on the problem? He thought of hypnotism . . . the right direction, and he did not see how to make it practical. He remembered reading of the effectiveness of a neurolinguistic recording. That was the link he needed, and the idea was ready to give to the group.

If the tower had not resulted in an idea, Jim would have returned to his image and picked another aspect that interested him and repeated his force fit thinking. Each such cycle actually takes less than a minute to explore for promise. Developing the promise into an idea may take a little longer.

I found that as I became more skillful at using myself, I would shift into force fit thinking whenever anyone was explaining an idea. I am often able to develop builds and even new ideas this way.

SUMMARY

From observation and experiment with several thousand problem-solvers, Synectics® believes that nearly all of us have enormous potential for good thinking. Because most of us are in a constant competition with each other, we do not cooperate with and support each other as much as we could. We must spend a good deal of our energy in protecting ourselves and in win–lose contests. Synergy in groups, if it happens at all, is at a low level.

We are convinced that each of us, by systematically manipulating ourselves and the climate around us, can markedly increase the amount of that good thinking potential that we use. In addition, we can profit together from the deliberate synergy we create. Not the least of the blessings that flow from wholehearted cooperation and support is that they make meetings more exciting and more fun.

REFERENCES

1. W. J. J. Gordon, *Operational Creativity*, Harper and Row, New York, 1957.
2. W. J. J. Gordon, *Synectics*, Harper and Row, New York, 1961.
3. George M. Prince, *The Practice of Creativity*, Harper and Row, New York, 1970.
4. Eric Berne, *Games People Play*, Grove Press, New York, 1964.
5. George M. Prince, *Mindspring!*, in press.
6. Warren McCullough, *Embodiments of Mind*, MIT Press, Cambridge, Mass., 1965.
7. Thomas R. Blakeslee, *The Right Brain*, Anchor Press/Doubleday, New York, 1980.
8. George M. Prince, "Putting the Other Half of the Brain to Work" *Training*, Vol. 15, No. 11, November 1978, pp. 57–58, 60–61.
9. Harry Stark Sullivan, *The Interpersonal Theory of Psychiatry*, W. W. Norton, New York, 1953, pp. 170, 233–234.
10. *Mindspring*, p. 112ff.
11. Carl Sagan, *Dragon in Eden*, Random House, New York, 1977, p. 44.
12. Albert Mehrabian, "Communication Without Words", *Psychology Today*, Vol. 2, No. 4, September 1968, p. 52.
13. George M. Prince, *The Practice of Creativity*, Harper and Row, New York, 1970.
14. Ibid.
15. George M. Prince, "The Mindspring Theory: New From Synectics Research," *Journal of Creative Behavior*, Vol. 9, No. 3, 1976, p. 175.

CHAPTER 12

VALUE ENGINEERING

THOMAS SNODGRASS
SANDRA GILL

Value engineering is a special purpose-technique for creating functionally equivalent or improved designs for products or services while simultaneously reducing overall production or delivery costs. It has been used primarily to solve problems in which the end result has been a product. However, it can also be used to solve social issue problems in which the end result is a plan or policy. In this presentation of value engineering the basic concepts and procedures will be described first. These will be followed by a case history in which value engineering has been successfully applied to a human services problem.

The basic techniques of value engineering (VE) were developed in 1949 by Lawrence D. Miles of the General Electric Company.[1] During this period of scarce resources Miles focused on one objective, "equivalent performance for lower cost." What resulted was not only a technique, but in his words, "a philosophy implemented by the use of a specific set of techniques, a body of knowledge, and a group of learned skills."[2] This approach subsequently became known as value engineering or value analysis (VA/VE), where "better for less" is provided with dramatic reductions in cost. The essential focus in VA/VE is the identification of product or service *functions*. Costs are subsequently reduced through creative speculation and critical evaluation of new alternatives to the delivery of those functions.

SOURCES OF VALUE

The concept of *value* and barriers to optimal value are key concepts in
VE. The "value standard" in any VA/VE approach is to provide desired
and essential functions at the lowest possible cost without loss of high
user acceptance. In other words, value is the relationship of functional
worth, as determined by various users, to cost of production.[3] VE therefore differs from 'cost reduction programs, in which the focus is on the
substitution of less costly parts in a product or service. In a VA/VE
sequence the form of the product or service may be totally abandoned.
The functions provided through the product or service are the focus of
group attention. Cost reduction occurs through the evaluation and selection of less expensive alternative means by which critical functions are
provided.

One difficulty in VA/VE centers on alternative concepts of value.
Clearly, value derives from use: Unless a product or service has a use,
it typically has no value. However, products and services are used by
different people for a variety of reasons. The item may provide a *practical
use*, such as a tool or an information system, or it may bear some *esteem
value* or prestige. Other items provide *exchange value*, such as money,
coupons, collaterals, and so on. A large part of our mechanical world
focuses on pragmatic use value. In contrast, much of our physical and
professional world places greater significance on esteem and exchange
values, for example, personal possessions such as clothing, automobiles,
homes and neighborhoods, and professional offices. Each of these various
sources of value must be taken into consideration in order to preserve
the integrity of the basic value equation:

$$value = \frac{function + high\ user\ acceptance}{cost}$$

In short, value is always comparative, and a key element in the VA/VE
approach is to evaluate alternatives for important functional delivery.

BARRIERS TO VALUE

The reasons for "poor value" derive from many sources. Often the primary task, basic functions, and supporting functions for which a product
or service is designed are simply not known. They may be the result of
professional or cultural traditions, personal habits, or individual attitudes.
We continue to design and provide for much of our world on the basis
of what has been done in the past. Another reason for poor value is lack

of information. Many decisions must be made for political reasons before all the information is in or prior to an opportunity to test and evaluate what we know. Sometimes our decisions are based on "honest wrong beliefs," that is, our information is incorrect but we do not realize it. Still another reason is our lack of creativity or its counterpart, evaluation; in the absence of comparison between alternatives, evaluation cannot take place. Our motivation to be creative is often dampened by habits and attitudes about "one best way." These same habits and attitudes may also preclude us from incorporating new facts and information that lead to new visions.

The fullest value is obtained when we begin correctly. This requires (1) a careful identification of the goals and objectives we wish to achieve, (2) the identification of a primary task, then basic and supporting functions to be provided, and (3) a carefully designed and monitored period of creative design followed by evaluation of alternatives. Finally, the development and implementation of products or services according to design specifications is essential for full value to be realized. (These conditions will not be met without substantial support of the VA/VE program and will be discussed following the job plan.)

PROCEDURAL STEPS IN VA/VE: THE JOB PLAN

The job plan has become an accepted element in VA/VE programs. It is more than a sequence of work; it is a strategy for concentrated, unbiased thinking and analysis. It is a sequential process in which the output of one step provides data for the next, but this does not preclude a team from returning to an earlier phase for the purpose of gathering more information for a more reliable result.

The job plan generally includes four or five major phases, each of which will be described: the information phase (with four major components); the speculation or search phase; the evaluation phase, and the implementation phase (see Figure 1).

I. THE INFORMATION PHASE

Task 1: Collect Data

There are four major tasks to be completed in the information phase. First, collection of accurate data regarding the product or service is initiated. User satisfaction data, user attitudes and market research regarding competitive items or services, user needs and expectations, pro-

THE JOB PLAN

1. Information Phase	Collect data Identify functions Determine costs
2. Speculation	Create alternatives Record ideas
3. Evaluation/Analysis Phase	Evaluate alternatives
4. Implementation Phase	Develop proposal Present proposal Evaluate implementation

Figure 1   The job plan for VA/VE.

duction costs, specifications, user experience with the product or service, and so on are all potentially critical to the development of feasible alternatives. This information will be particularly valuable in the speculation and evaluation phases. Information reflective of at least three user groups should be included: (1) the attitudes from users who have direct contact with the product or service; (2) the attitudes of management and/or administration regarding the product or service, for example, production costs and difficulties; and (3) the attitudes and experiences of prior providers or producers/competitors. These three groups typically produce very different perspectives, which may be critical for ultimate acceptance. This information helps determine what is required for acceptance, what is tolerated by users, and, perhaps most important, what is not tolerated.

Task 2: Identify Functions

The second major task is to determine what the product or service does, in numerous two-word (verb–noun) phrases. A primary *task* must be identified, that is, that function which is the reason all basic and supporting functions are performed. A hospital, for example, may "provide health care," "conduct surgery," or "monitor illness." An electric motor may "drive equipment," "move parts," and "provide power." Then basic and supporting functions must be identified. The *basic function* is absolutely essential to the performance of task. Without the basic function(s), the product or service cannot be performed. Finally, all supporting functions must be identified. *Supporting functions* are all those that are subordinate to and supportive of basic functions. Supporting functions are often intangible and difficult to identify. They are often added in product or service design to appeal to user attitudes and increase sales potential. They often amount to 50% or more of the total functional cost of a product or service and are usually aimed at the attitudes and emotions of particular user groups. For this reason many products and services have different supporting functions, depending upon the geographical area or market.

The use of the verb–noun phrase for functional analysis is a hallmark in the VA/VE approach. This aspect of the information phase is usually accomplished through brainstorming, until the team feels it has identified all the important functions. Then, in the process of deciding which are basic and supporting, other functions are identified or omitted through discussion. Supporting functions may be further categorized into primary, secondary, and tertiary groupings, depending on the complexity of the product or service (see Figure 2). A complete functional analysis will answer "How?" with each verb–noun phrase moving left to right and will answer "Why?" with each move in the opposite direction.[4]

Task 3: Determine Costs

Having established what a product or service does, the third major task in the information phase is to determine functional costs. A variety of costing procedures are helpful for this detailed and tedious aspect. One example is illustrated in the case history that follows this overview of concepts and procedures. Regardless of the strategy, the purpose of this task is to allocate meaningful costs. Miles has provided many illustrations of how partial cost information can result in less reliable value analyses. At least four sources of cost are essential, that is, material, labor, design, and overhead. In addition, many other factors may be important. Cost equations are often used by experienced VA/VE specialists to determine the many parameters that affect total costs and their interrelationships. Consequently, a variety of recording, calculation, and posting techniques have been developed so that total function costs can be determined and posted.

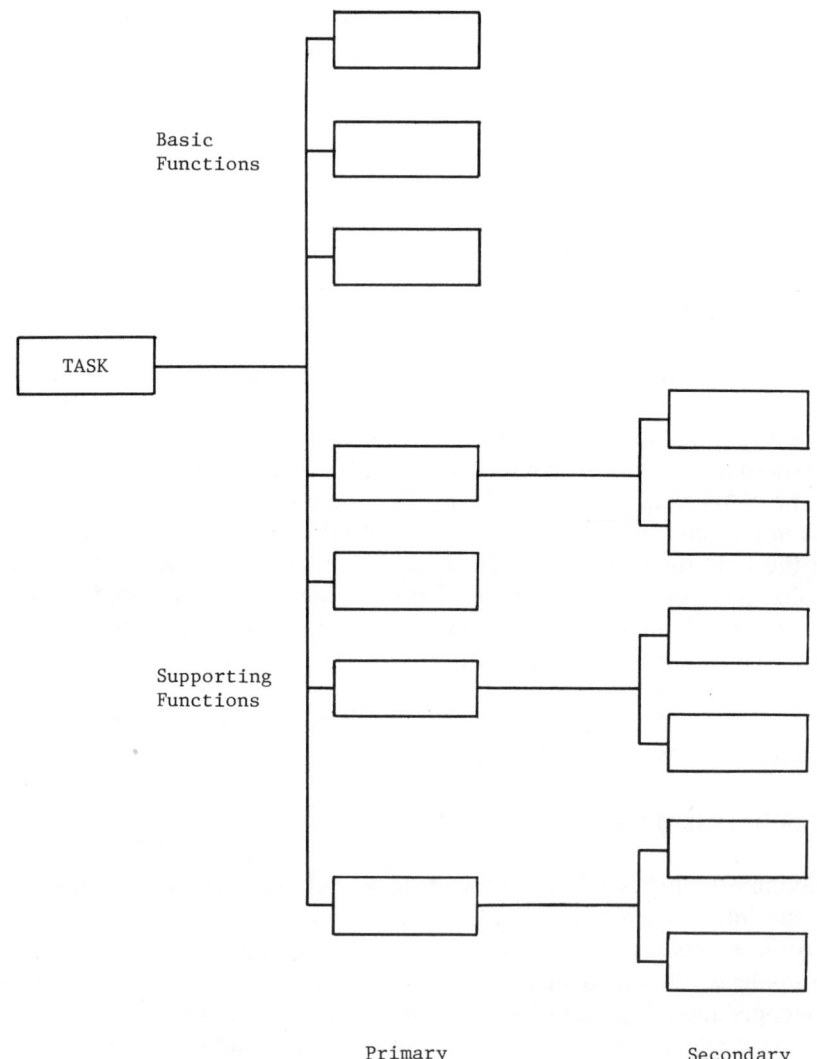

Figure 2   FAST (Function Analysis System Technique) diagramming.

Task 4: Determine Users' Reactions

The fourth task in the information phase is to answer, "What are the users' reactions?" User attitudes can be allocated to functions just as costs can be. When completed, the VA/VE team can determine what functions are most important, what supporting functions are most critical to sales, acceptance, and so on. Additionally, functions to which users are indifferent may be seen as potential areas for cost reduction. This identification of high, low, and indifferent acceptance toward function categories is critical to the speculation and evaluation phases.

II. THE SPECULATION OR SEARCH PHASE

In contrast to the information phase, in which a product or service is "blasted" into its many functions, the purpose of the speculation phase is to "create and refine." The essential output of this phase is to create responses to, "What else could do the job? In what other ways can each function be performed?" A primary maxim in this procedure is that "you cannot be wrong!" During the speculation phase, it is essential that no evaluation of any kind be made, including questions, frowns, comments, or any other manner of criticism.

Since the success of the speculation phase depends on the creative potential of the team, an understanding of the conditions for creativity is included here. In general, groups are more creative when the team is sufficiently diverse to bring wisdom and alternative perspectives to the group. This may be achieved through the presence of different professional backgrounds, for example, marketing, sales, architecture, engineering, and construction. It may also include diversities in chronological age among members and in length of tenure with the organization. Representatives from different organization hierarchical levels is another factor to consider.

Where individuals are diverse, however, there must be a structured group process for generating creativity so that evaluation and criticism are indeed ruled out. To enforce this norm VA/VE specialists often provide a cuing system to participants, for example, bells, whistles, and so on for members to "sound" when participants habitually begin to evaluate one another. Team members must be sufficiently committed to the VA/VE approach to discipline themselves to this required nonevaluative behavior until the creativity phase is completed.

Another requirement for full output in the speculation phase is the recording of creative ideas. Two forms of recording may be especially valuable during this phase. First, individual recordings of their own ideas on paper or $3'' \times 5''$ cards will increase their retention of new ideas. Individuals tend to forget their own ideas during brainstorming unless they are written before idea verbalization begins. Second, numerous ideas from all group members must be recorded for the entire team to see. This is not only essential for the next phase, evaluation, but also allows hitchhiking, that is, building on or expanding upon another member's idea.

$3'' \times 5''$ or $5'' \times 7''$ cards, flip charts, blackboards, or transparencies are useful media for this process.

III. EVALUATION AND ANALYSIS

This phase follows the high energy of the speculation stage. Here all ideas and solutions are critically evaluated in terms of cost, feasibility, economy, and so on. Most group members experience high satisfaction in this process if the principle of "the better idea" is followed, that is, current ideas are eliminated only by the development of better ones. A variety of formats have been developed by VA/VE specialists for the evaluation phase. Regardless of the particular technique, the importance of this step is to maintain a constructively critical approach. It is at this point that much of the data and information from the information phase becomes critical, since it keeps the team aware of alternative needs expressed by a variety of users. Finally, it is important to focus on alternatives that meet "high acceptance" standards as well as on those areas in which high costs have been identified.

IV. THE IMPLEMENTATION PHASE

The outcome of the previous phases is culminated in this final step. The previous efforts have produced new facts and examined previous information in new ways. The presentation of this material is an extremely important step. Brief et al.'s research[5] on proposal adoption provides important clues for successful proposal presentations and implementation planning.

First, successful proposals are seen as feasible, particularly by management and administration. This means that technical, managerial, and administrative concerns are important if not critical highlights in addition to overall cost savings. Time, budget, scheduling, and production constraints are important factors to be developed and rehearsed. New professional/technical skills or talents are important to consider. Evaluation, monitoring, and feedback to determine product or service results in both the short and long range after a VA/VE design is approved are important components for validation and future support of the VA/VE approach.

Further, information is best presented by an individual who is capable, credible, and competent to discuss the product or service in a clear, concise fashion. Lengthy summary documents or proposal formats may be required by the organization, but simple one- or two-page formats are suggested for the verbal presentation. Simple, clear, but colorful visual aides are also recommended for illustration purposes. Emphasis is best applied to highlights; details may be discussed during questions and answers following the basic presentation.

In short, successful VA/VE proposals match necessary functions with available resources—human, financial, material, and time—as much as feasible. They are evaluated in terms of overall cost reduction, cost avoidance, product or service improvement, or increased sales/user acceptance. Usually the last three standards provide the largest return on investment in product-oriented VA/VE programs.

IMPLEMENTATION OF VA/VE PROGRAMS

The recurring, systematic use of VA/VE strategies may be quite difficult in many organizations. Because most VA/VE teams are multidisciplinary, skills in team dynamics are critical to the success of the effort.[6]

Multidisciplinary VA/VE teams require a skilled leader who is experienced in VA/VE as well as in group leadership skills. He or she must be competent, credible with peers, and have appropriate credentials in the technical or professional area to be examined. The most successful teams are further motivated by top management support and active interest in program progress. Well-supported, active VA/VE teams and departments potentially enhance the return on investment and problem-solving capacities of the entire organization. But they demand commitment to a process that requires interpersonal collaboration and discipline to a specific problem-solving procedure.

CASE HISTORY

VE is most effective when team members are self-motivated to problem-solve and when the problem-solving process can be addressed in a systematic fashion.

Such was the case in the fall of 1971 in School District #67, Lake Forest, Illinois. Two major factors were the common motivators. First, there was a strong indication that the district had a serious problem of not completely understanding the public. The school board was faced with a situation common to many school districts, that of increasing enrollments and higher costs. It took the accepted approach of sponsoring a referendum requesting higher educational taxes. For the first time in the history of the Lake Forest school system the referendum experienced a resounding defeat. Out of a community of 10,000 voters, the "no" vote won two to one. Such a situation definitely and seriously affected the school administration, the teachers, and, most significantly, the parents of children in the school district. These parents were ably represented by a very active organization, the Association of Parents and Teachers (APT). The second factor required systematic planning. The Superintendent of Public Instruction for the State of Illinois required that each local school district develop a plan that would serve as the basic document that the Office of the Superintendent of Public Instruction would use to evaluate and supervise the local school district program. The plan made it very clear that the voters should be involved, and this presented a real dilemma to the school board and administration.

One of the APT officers worked for the consulting firm Value Standards, Inc. She requested that the firm propose a VE study to the school board and administration as a means of attacking the problems created by the two situations and coming up with workable solutions that could be implemented.

THE PROPOSAL

A meeting was held in early January 1972, with Dr. Albert Poole, School Superintendent, to discuss the proposal and determine the means of preparing for carrying out the requirements of the proposal. The proposal consisted of four requirements.

The first requirement was to assemble problem-solving teams. Four teams with five individuals per team, for a total of 20 individuals, would be needed. Each team was to be made up of a member of the school board, a member of the school administration, officers and/or members of the APT, and teachers. The team members must be willing to work on Saturdays over a period of approximately three or more months in carrying out the team responsibilities.

The second part of the proposal was the requirement for budget data to be assembled in a form that could be entered into a computer program which would provide a very important manipulation of figures during the course of the study.

The third requirement was the need for a community attitude study using a statistically significant sample of voters. The necessary preparation of field documents, the filling out of questionnaires, and a tabulation had to be done.

The fourth requirement was the application of the VE job plan, or as Value Standards called it, the VITAL action plan. VITAL, they were told, stands for Value Information Techniques using Analytical Language. The whole system was designed to assure that the end recommendations of the teams would assure a better value school system as considered by the voters of the school district. The VE job plan, or VITAL action plan, consisted of three phases—information, creativity, and evaluation—to be carried out by the teams. Two phases—planning and implementation— were to be carried out by the administration and school board. Finally, it was pointed out that it was essential that such a VITAL study, or VE study, be carried out on a planned scheduled basis so that everyone would understand his or her commitment and that the end result would be available at a predictable time. The schedule upon which the administration and school board agreed is shown in Figure 3. The schedule extended over an approximately five-to-six month period.

The proposals were accepted and the collection of the cost data was initiated. This involved the collection of all the salaries of the administration, the teachers, and the various other employees of the school district, along with materials and other related expenses. As previously mentioned, this budget data was entered into a computer program that was the property of Value Standards, Inc., and would be used in the information phase.

The schedule was used to spell out the responsibilities. The school system was responsible for collecting the cost or budget data during the first two weeks of the study. Value Standards would put it into a structured form that could be entered into the computer and have it ready for the teams by the fifth week. You will also note that an item, FAST diagramming (Function Analysis Systems Technique), was indicated in the second week for both the school system and Value Standards, and this really initiated the first team meeting as a part of the information phase.

THE INFORMATION PHASE

The information phase is the unique phase in the VE or VITAL system, and it is absolutely essential that the teams involved be directed by a trained and experienced value specialist. Thomas J. Snodgrass of Value Standards, Inc., started out on a Saturday morning, reviewing the background and overall elements of VE.

One of the first things that was done was to define value in terms meaningful to the teams. This relationship was described as being made up of two sides of a seesaw that must be in balance. On one side the voter has certain expectations of a service, such as a school system at a price, represented by his taxes. When the system meets or exceeds his expectations, at this price he feels that it is a very effective system. On the other side the school board, or government agency, must operate within certain budgets that equal the taxes provided. This budget is made up of salaries, services and supplies, buildings, and so on, and it must have a

TIME SCHEDULE FOR ATTITUDE AND COST STUDY/SCHOOL SYSTEM

WEEKS	1	2	3	4	5	6	7	8	9	10	11	12	13	14	15	16	17	18	19	20
VALUE STANDARDS																				
Attitudes																				
Planning & Organization	▨	▨																		
FAST Diagramming		▨	▨																	
Documents			▨																	
Task Force Training and Supervision		▨		▨	▨		▨	▨			▨	▨	▨							
Cost (Budget) Analysis																				
Structured Budget			▨	▨																
Costed Bill of Material					▨															
Allocation of Costs						▨														
Function Cost Reports									▨	▨										
Evaluation																				
Analysis														▨	▨					
Creative Session																	▨			
Report Writing																			▨	
SCHOOL SYSTEM																				
Attitudes																				
Planning	▨																			
FAST Diagramming		▨																		
Name List					▨															
Document Approval & Pilot																				
Document Delivery & Follow-up									▨	▨										
Tabulate											▨									
Allocate														▨						
Posting																				
Cost (Budget) Analysis		▨																		
Prepare Cost Data																				
Training						▨														
Allocation of CBM																				
Evaluation																				
Analysis															▨					
Creative Session																▨				
Report Writing																		▨		
Distribution of Report to Voters																				▨

Figure 2. A time schedule for an attitude and cost study.

sufficient amount in each of these categories to provide the necessary services expected by the voter. Therefore throughout the study it was emphasized that the objective was to achieve a good value relationship, that is, high voter acceptance and sufficient funds to meet the budget requirements for this high acceptance.

It should be pointed out at this time that it was initially difficult for the school board and administration to recognize that the voters were really their "customers." They felt that their primary customer was the student; but it was the voter who really established the acceptance criteria. This became more and more obvious as we proceeded into the course.

It was pointed out that the information phase accomplishes three things. First, it identifies what it is that the service does, and to do that the teams had to carry out the function analysis. Secondly, it was necessary to transpose the budget costs of the district into the various functions to obtain a function cost. Finally, once the voter attitudes had been obtained, it would be necessary to assign the various attitude comments to these same functions in determining what the voters' attitudes were as they related to the various functions.

The first session, however, tackled the problem of determining what the school district did to meet and exceed expectations of the voters. The system called FAST was used. it was explained that the system is not fast in the sense of quick, but it did communicate and it defined the relative importance of the various functions, which all had to be defined with a verb and a noun.

It did not take the 20 participants long to determine that the verb–noun that describes the overall need which the school system met was *educate children*—this was the TASK. It was then pointed out that the TASK, *educate children*, was performed by two types of functions. One type consists of basic functions, which were the essential functions stripped down to their bare necessities. Without them the school system could not exist.

The second type of functions consists of supporting functions. These were functions that were added to build voter and student acceptance. It is the judicious selection and emphasis of the supporting functions that achieve the good value relationships. The teams were instructed that basic functions vary from service to service, and product to product, but there were four primary functions that are constant regardless of product or service. These were described as assure convenience, assure dependability, satisfy users, and attract users. Each of the teams was assigned to one or more of the supporting functions and one team was assigned the basic functions.

A very specific and logical procedure is used to develop the FAST diagram and Figure 4 is an example of how the team working on basic functions evolved the various functions. They started out with the overall need that they called the TASK, *educate children*, and drew a line to the right hand side of the TASK. Using 2″ × 3″ cards, they started printing the various functions that were necessary, in their opinion, to educate children. One of these functions was *develop individuals* (Figure 4). They then asked the question, "How do you educate children?" They were satisfied that *develop individuals* was one of the functions that occur directly to the right of this scope line. It was one of the basic functions. When they asked the question, "Why do you develop individuals?" they were satisfied that *educate children* was the answer. They then looked at the second level by asking, "How do you develop individuals?" and one of the responses was *instill basics*. When they asked, "Why do you instill basics?" they were satisfied that *develop individuals* answered that. The team then asked, "How does our school district instill basics?" The answers they came up with were *teach reading, teach communications, teach math*. The "why" met their requirements. This approach was repeated for each of the nine secondary functions and in turn these were expanded to the third level. It was felt at this point that when they asked the "how," they would really be describing specific techniques and materials and so on, so they stopped.

The same procedure was used by the team assigned the responsibility of expanding the primary supporting function, *satisfy user* (Figure 5). User, in this case, would be considered both student and citizen voter. It should be kept in mind that when the "why" question is applied to *satisfy user*, the answer implies high acceptance by the voter to the TASK, *educate children*. When the "how" question was applied to the primary supporting function, *satisfy user*, the team came up with four answers— *provide benefits, furnish enrichments, furnish public relations*, and *assure quality education*; but when they asked, "How did the system *furnish public relations*?" they came up with three more, one of which was *provide communication*. Asking the question, "How to *provide communication*?" they indicated the various individuals or groups to which the communication was necessary, mainly parent/teacher, parent/administration, and teacher/student. During this session frustrations ran high during the difficulty of finding the exact verb and noun to describe what a system or product does; there were many false trys, but with the encouragement from the project leader and with the knowledge available in the team, an organized and logical function analysis gradually developed. The complete function analysis is shown in Table 1. It has been put into outline form so that costs and attitudes can be referenced directly to the individual functions.

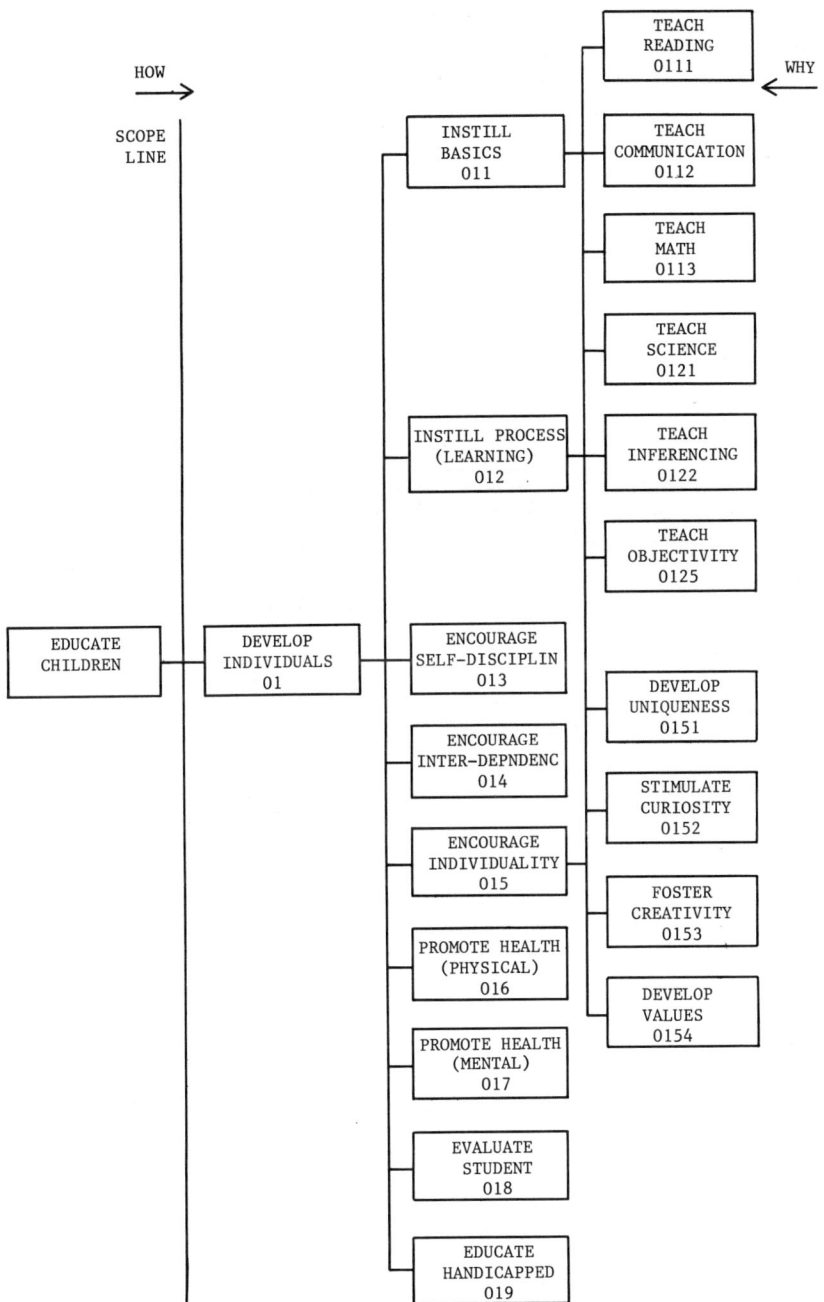

Figure 4 Developing FAST diagram basic function.

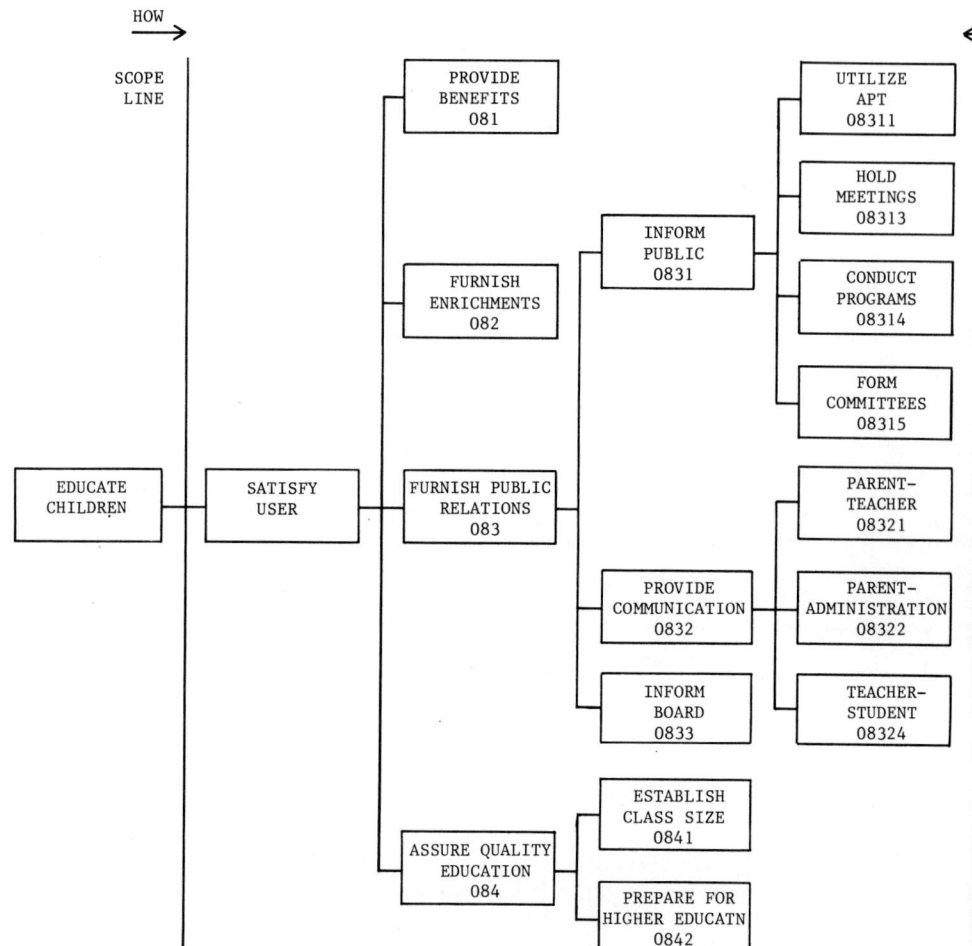

Figure 5 Developing FAST diagram supporting function.

Once the function analysis was prepared, it was then possible to prepare the necessary field documents and obtain the voter attitudes required in the third part of the information phase. The drop and mail technique was used. One of the members of the APT delivered a questionnaire to a voter's residence that had been randomly selected from the voter files. The selected voter filled out the questionnaire and mailed it. This maintained confidentiality.

The voter files were biased in terms of the population. One of the major surprises to many of the participants was that only one-third of the voters actually represented parents with children in the school system.

Another 17% had children in private schools; finally, 50% were called, with some degree of disdain, "empty nesters." Obviously, this attitude changed drastically as the study progressed. The questionnaires were tabulated using the APT personnel and were ready for the scheduled session.

The second and very important part of the information phase was to

---

FUNCTION ANALYSIS FOR LAKE FOREST SCHOOL DISTRICT –
DISTRICT 67

TASK: Educate Children

BASIC FUNCTIONS

01	Develop Individuals
011	Instill Basics
0111	Teach Reading
0112	Teach Communication
0113	Teach Math
012	Instill Learning Processes
0121	Teach Science
0122	Teach Inferencing
0123	Teach Relationships
0124	Teach Study Habits
0125	Teach Objectivity
013	Encourage Self-Discipline
0131	Foster Independence
0132	Develop Self-worth
0133	Promote Motivation
014	Encourage Inter-Dependence
0141	Provide Group Experiences
0142	Recognize Individuality
0143	Respect Individuals
01431	Foster Citizenship
015	Encourage Individuality
0151	Develop Uniqueness
0152	Stimulate Curiosity
0153	Foster Creativity
0154	Develop Values
016	Promote Physical Health
0161	Teach Positive Habits
0162	Improve Fitness
017	Promote Mental Health
018	Other
0181	Evaluate Student
01811	Grade Student
01812	Group Student
019	Educate Handicapped

Table 1 Function Analysis

```
02              Administer System

021             Set Objectives
  0211            Determine Needs
    02111           Study Laws
    02112           Consult Staff
    02113           Consult Public
022             Establish Priorities
023             Administer Personnel
  0231            Pattern Organization
    02311           Delegate Authority
    02312           Establish Responsibility
  0232            Employ Staff
    02321           Develop Job Descriptions
    02322           Train Employees
    02323           Establish Rules
    02324           Other (Travel)
  0233            Evaluate Staff
    02331           Measure Performance
  0234            Support Staff
    02341           Teach Basics
024             Administer Finance
  0241            Purchase Resources
  0242            Pay Bills
    02421           Pay Employees
025             Manage Curriculum
  0251            Develop Curriculum
    02511           Provide Balance
    02512           Explore Innovative Methods
      025121          Options
    02513           Promote Creativity in Teaching
    02514           Improve Language Arts
026             Administer Services
027             Administer Facilities
  0271            Maintain Buildings
    02711           Clean Buildings
    02712           Repair/Maintain Buildings
    02713           Alter Buildings
  0272            Schedule Use
028             Discipline Student
  0281            Teach Respect (Authority)
  0282            Establish Dress Code
```

Table 1 Cont.

```
03              Obtain Funds

031             Create Budget
 0311               Establish Priorities
 0312               Analyze Needs
  03121                 Consult Public
  03122                 Consult Staff
032             Assess Resources
 0321               Review Balance
 0322               Analyze Income
  03221                 Project Federal Aid
  03222                 Project State Aid
  03223                 Project Local Aid
  03224                 Project Other Aid
  03225                 Project Taxes
033             Adjust Budget
034             Establish Levy
035             Request Funds
036             Receive Funds


04              Provide Facilities

041             Determine Needs
 0411               Project Population (Where)
 0412               Evaluate Facilities (Present)
  04121                 Determine Safety
  04122                 Project Usefulness (Time)
  04123                 Consult Regulations
  04124                 Consult Professionals
042             Buy Land
 0421               Examine Sites
 0422               Determine Cost
 0423               Select Location
 0424               Meet Regulations
043             Build Schools
 0431               Determine Needs
  04311                 Consult Staff
  04312                 Consult Community
  04313                 Compare Facilities
  04314                 Develop Educational Specifications
  04315                 Establish Priorities
   043151                   Stress Practicality
 0432               Select Architect
  04321                 Interview Applicants
  04322                 Evaluate Applicants
 0433               Invite Bids
 0434               Employ Contractor
 0435               Erect Building
 0436               Develop Site
```

Table 1 Cont.

```
05              Provide Resources (People, Time, Space, Materials)

051             Examine Curriculum
052             Assess Needs
 0521               Diagnose Individuals
  05211                 Consult Staff
  05212                 Consult Specialists
  05213                 Select Tests
 0522               Provide Up-to-date Materials
 0523               Evaluate Instructional Materials
053             Set Priorities
 0531               Assess Funds
054             Orient Teachers
 0541               Provide Training (In-service)
 0542               Provide Delivery
  05421                 Develop Learning Centers
  05422                 Equip Library
  05423                 Equip Class
 0543               Design Experiences
```

SUPPORTING FUNCTIONS

```
06              Assure Convenience

061             Provide Transportation
 0611               Select Method
 0612               Establish Routes
 0613               Arrange Schedules
 0614               Operate Transportation
062             Feed Pupils
 0621               Assess Needs
  06211                 Choose Service
  06212                 Provide Facilities
  06213                 Hire Personnel
  06214                 Assure Diet
063             Use Facilities
 0631               Schedule Use
 0632               Provide Maintenance
 0633               Collect Fees
064             Provide Auxiliary Help
 0641               Employ Teacher Aids
 0642               Utilize Volunteers
065             Pay Fees
 0651               Mail Information
 0652               Receive Payments
 0653               Distribute Materials/Supplies
066             Other
```

Table 1   Cont.

```
07              Assure Dependability

071             Establish Fiscal Policy
 0711               Establish Financial Controls
  07111                 Hire Accountants
  07112                 Hire Auditors
 0712               Invest Funds
 0713               Plan Purchasing
 0714               Guarantee Pay
072             Protect Students
 0721               Involve Community Agencies
  07211                 Police
  07212                 Recreation Department
 0722               Discipline Students
  07221                 Fire Drills
  07222                 Hall Safety
  07223                 Preserve Facilities
 0723               Assess Health
  07231                 Employ Nurse
   072311                   Test Vision
   072312                   Test Hearing
   072313                   Teach Health
   072314                   Keep Records
   072315                   Administer First Aid
   072316                   Report Information
 0724               Centralize Supplies


08              Satisfy Users

081             Provide Fringe Benefits
082             Provide Enrichment
 0821               Encourage Culture
  08211                 Establish Clubs
  08212                 Establish Libraries
  08213                 Provide Athletics
  08214                 Teach Music
  08215                 Teach Art
  08216                 APT Cultural Activities
  08217                 Other
083             Provide Public Relations
 0831               Inform Public
  08311                 Utilize APT
  08312                 Issue Reports
  08313                 Hold Meetings
```

Table 1   Cont.

08	Satisfy Users (Continued)
08314	Conduct Programs
08315	Form Committees
0832	Provide Communications
08321	Between Parent-Teacher
083211	Reporting
08322	Parent - Administration
08323	Parent - Board of Education
08324	Teacher - Student
0833	Inform Board of Education
084	Assure Quality Education
0841	Establish Class Size
0842	Prepare for Higher Education
085	Other

09	Attract Users
091	Remain Compatible
0911	Plan Design
0912	Choose Materials
092	Create Atmosphere
0921	Create Decor
09211	Choose Colors
09212	Use Texture
09213	Provide Accessories
0922	Employ Light
0923	Employ Space
093	Design Landscape
0931	Hire Architect

Table 1   Cont.

view each of the elements of cost and assign it to one or more of the functions that had been previously developed. This technique built on the knowledge developed in the function analysis. Not only were they being forced to understand all of the various things that the service does, but they then had to look at a set of costs and determine what these costs accomplish.

An example is shown in Figure 6, where a number was assigned to the pupil contact hours of the English teachers. These pupil contact hours had been determined to total an annual cost of $27,582. It was further determined that 575.35 hours were involved in pupil contact hours. They then looked at each of the various functions and determined how much, in terms of hours, should be assigned to that particular function. Figure 7 shows part of a team made up of a teacher, an administrator, and a parent struggling with such an allocation.

3013 - SS	RELATED ARTS	1	54692.	0.	0.	54692.
3014 - SS	PUPIL PERSONNEL	1	85200.	0.	0.	85200.
3015 - PP	LIBRARIANS	1	37700.	0.	0.	37700.
3016 - SS	ADMINISTRATION	1	213550.	0.	0.	213550.
3017 - SS	CLASSIFIED STAFF	1	224190.	0.	0.	224190.
4035 - PP	PUPIL CONTACT HRS	1	27582.	0.	0.	27582.

(0111, 275, 575.35) (012, 172, 575. 35) (0121, 6, 575.35)
(01811, 50, 575.35) (0281, 20, 575. 35) (08216, 3, 575.35)
(0133, 6, 575.35) (0133, 6, 575.35) (0141, 6, 575.35)
(08324, 25, 575.35) (08324, 25, 575. 35) (0152, 8)

4036 - PP	HOMEROOM	1	5910.	0.	0.	5193.
4037 - PP	STUDY PERIOD	1	4363.	0.	0.	4363.
4038 - PP	NON CONTACT HRS	1	30795.	0.	0.	30795.

Figure 6   Costs assigned to functions in function analysis.

Figure 7   A teacher, administrator, and parent in the process of allocating pupil contact hours to each function.

They determined that 275 out of 575.35 hours should be assigned to the function 0111, which is *teach reading*, one of the secondary basic functions. They also determined that 25 hours were expended in the supporting function 08324, which is *satisfy user, provide communication* (between *teacher and student*). The allocating of cost was initially viewed as a tedious exercise by some team members, but when completed, it was recognized as probably as important an exercise as any in which they were involved. The cost allocation again communicated among the teams information that was either not previously available or that was certainly not understood in common terms as they had been forced to establish in this exercise.

This third part of the information phase, dealing with function attitudes, was carried out at the end of the twelfth week of the study, when the questionnaires had been returned, tabulated, and were ready for the group to review. The hand tabulation was used for each of the questions and was reviewed in a form shown in Figure 8. This represented the public school list and represented the question, "What is important to you in the Lake Forest Elementary School?" The responses were collected and tabulated in the usual form and represented as percentages. The same teams sat down and reviewed these questionnaires and assigned the function numbers to the various comments. As an example, a comment, *motivate child*, was made by 11% of the respondents and assigned the function number 0133. The function, *promote motivation*, headed *develop individual, encourage discipline*. The teams found this very stimulating and again very informative.

In the next period of time the material cost and attitudes were processed. The attitudes were all collected by function, posted, and totaled to determine the degree of acceptance and nonacceptance for the various functions and the cost data processed by the computer program to accumulate the amount of cost and types of service involved for each of the functions. They in turn totaled up to the next highest category, and finally there was a total for basic functions and a total for supporting functions. The four teams were now ready for the next step, which is a small portion of the evaluation phase, namely, analysis, where the teams reviewed the data to determine the high-cost functions and those with low acceptance. This was a relatively short activity, taking approximately one to two hours of team time. They were now ready to move into the next very important phase—creativity.

CREATIVITY

The teams received a short lecture on some of the techniques and challenges of creativity, but the primary message was that for the rest of the

QUESTIONNAIRE TABULATION				Question Number ___3___

Page ___1___ See Page ___2___

Type of Respondent ___MASTER PUBLIC___ Project Number ___180___

Location _____ Make _____

QUESTION NUMBER AND WORDING OR VERBATIM RESPONSES TO QUESTION	Accumulate Responses	Cumul. No.	% of 1	Function Alloc.
3. WHAT IS IMPORTANT TO YOU IN LF ELE SCHOOL				
Motivating child		14	11	0133
Treat child as individual		23	18	0142/015
Creativity in teaching		15	12	02513
Personal attention to both slow & fast child		7	5	0151/0521
Good basic training		19	15	0110
High academic standards		18	14	084
Self confidence		7	5	0132
Develop skills to make choices		4	3	0121
Good study habits		9	7	0124
Smaller class size		11	9	0841
Graded assignments		3	2	01811
Good all around education		11	9	084
Recognize learning problem early		1	1	0521
Quality of teachers		24	19	02331
Good administration		3	2	08322
Adequate facilities		3	2	0271/0431
Recreational opportunities		1	1'	08213
Child should enjoy learning		9	7	0133
Continue the arts		3	2	08215
Discipline		3	2	028
Close working relationship between the school, parent and child		5	4	08321/2/3
Teach basics		11	9	011
Keep abreast with trends		1	1	02512
THIS QUESTION		127		
DON'T KNOW		2		
NO ANSWER		3		
TOTAL RESPONDENTS		132		

Figure 8 Questionnaire tabulation.

day they could not be viewed as wrong. No evaluation of any of the suggestions or ideas made by the other team members would be permitted. To enforce this and to put a little "spirit" in the sessions, a small bell was provided for each of the teams. If anyone questioned another member's idea, the team member was permitted to hit the bell to bring things back into the creative session. It was an animated session, with bells ringing and ideas popping and exploding in all areas. The important thing was that the ideas were being generated from the questions, "In what other ways can we teach math?" "In what other ways can we provide communication—parents/teachers?" and so on.

In each case there were 40–50 ideas generated for the various functions. This completed the creativity phase in which each of the teams participated.

EVALUATION

The teams were now ready to review and determine what was possible. This was described as the objective of evaluation. They first of all had to look at the low-acceptance areas. The teams were furnished with summary data that showed a relationship of the function to the way the various respondents rated that function's adequacy. They found that there were a number, such as *encourage self discipline, provide communication—parent/teachers, discipline students*, and so on, that fell below 75%, which they considered to be a minimum percentage of adequacy (Figure 9). They also looked at the high budget cost areas and found a number that represented costs higher than the various teams felt represented the good value relationship.

They then went to their creative lists and began to determine the ideas and approaches that would accomplish the two objectives to increase the acceptance for the various voter groups and at the same time reduce costs so that the budget would cover these required functions.

All of the preliminary team work now paid off in big dividends. The team members had a common understanding of the various things that the school district did, which ones were considered most important by the voters, which ones required improvements, where high costs existed, and what was needed to reduce these costs. They were able to come up with specific recommendations to achieve the objectives of increasing the acceptance among the various types of voters and reducing the costs so that they would not have to go back again and ask for a higher tax level to increase the budget. In their opinion both of these objectives were achieved. "The function analysis was an education in itself," said June Seaman, a school board member. "It forced everyone to sit down and

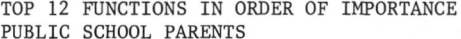

TOP 12 FUNCTIONS IN ORDER OF IMPORTANCE
PUBLIC SCHOOL PARENTS

1. Instill learning   012
   process
2. Encourage self-   013
   discipline
3. Promote motivation   0133
4. Foster citizenship   01431
5. Develop uniqueness   0151
6. Discipline students   0280
7. Grade students   01811
8. Teach communications   0112
9. Teach math   0113
10. Provide communication   08321
    (parent-teacher)
11. Prepare for higher   0842
    education
12. Teach science   0121

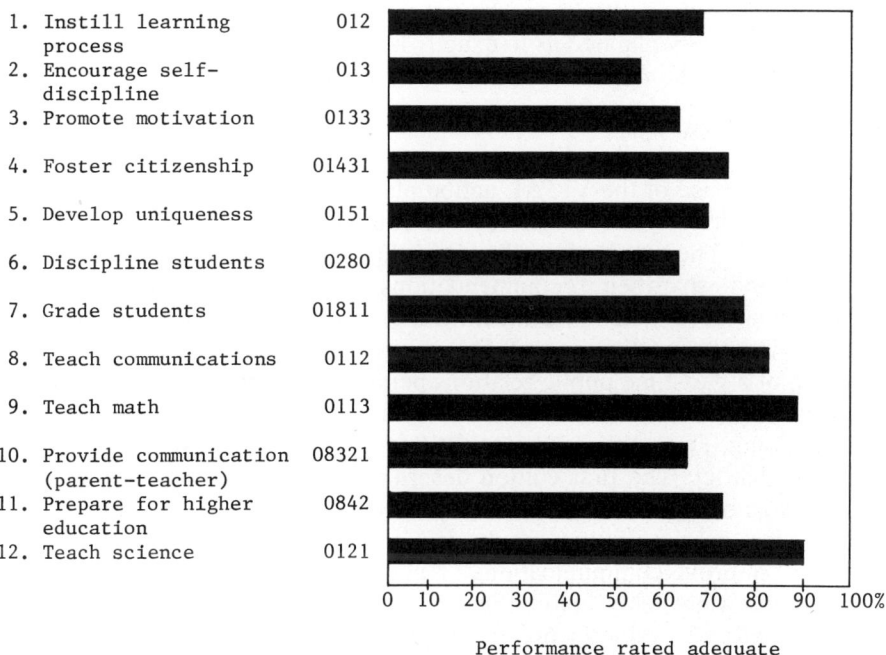

Performance rated adequate

Figure 9 Top 12 functions in order of importance to public school parents.

place on paper in specific and exact terms the purpose of our educational system," added Fred Nicola, a principal.[7]

PLANNING

The planning was primarily the responsibility of the administration and, most specifically, the school superintendents. Two were involved: Dr. Albert Poole was in on the initiation of the study and in the early phases did an excellent job of organization and data collection; Dr. Alan Klingenberg took over and was involved in the final planning and very much in the implementation phase. The initial planning has been described and included the preparation of the costs, the gathering of the attitudes, the organizing of the teams, and a certain amount of public relations. An article was placed in the *APT News*, winter edition.[8] An article was printed in the suburb's local paper, *The Lake Forester*, on February 17, 1972, titled, "School Plan Bare Bones Self Study."[9] The article presented the reasons for the study and how it was to be carried out. The follow-up was carefully planned as to exactly how the various recommendations were going to be implemented.

IMPLEMENTATION

The final phase of the VITAL action plan first involved the presentation of the results. This was done by the combined groups and various aids were used. The groups described how they had gone through the various stages and why their recommendations would meet the objectives of improving acceptance and meeting the requirements of the voters/students with the existing budget. One of the recommendations involved the hiring of a part-time public relations person. Another was to issue a new publication to acquaint the citizens and parents with what the schools were doing. This publication was entitled "Spirit of 67," using the number of the district. The first edition described the results of the report and what was being done.[10] It further spelled out some of the areas in which they were going to put an emphasis.

To "improve communications" a bimonthly newsletter, a speaker bureau, and a biweekly column in the *Lake Forester* had been instituted. A district "67" bulletin board would be displayed in the Lake Forest Railway Station. Learning centers had been developed in various schools and teacher courses had focused on individual instruction. The article stated that a cost analysis had been initiated using a program planning and budgeting system so that they could "better insure cost control and ascertain existing programs." "The Office of the Superintendent of Public Instruction requirement for voter input information in each district planning and goal setting has been very adequately fulfilled using the information phase from this study."[11]

Another important realization was that the study had taken one year from the time that the initial realization for its need was expressed until all of the information was available for decisions and future planning. However, it was felt that the information collected would be useful in varying degrees for the next five years, particularly considering the relatively stable nature of this community.

SUMMARY

The success of VA/VE programs depend on commitment to the job plan by team members: information, speculation or search, evaluation, and implementation. Careful attention to each of these elements is required for the VA/VE effort to achieve its full potential. The basic ingredient, of course, is the expertise of the VA/VE manager/team leader. Beyond that prerequisite meeting management and group leadership skills are additional resources for the successful VA/VE team leader. A variety of tools and techniques such as forms, data analysis methods, and computer programs can be used in structuring the most efficient efforts of the team. Consultations with VA/VE specialists may be very useful in the life cycle of the VA/VE program. The organization's willingness to be open to the VA/VE approach, to support it throughout its program development, and to sponsor it in its multidisciplinary format are important conditions for full success.

REFERENCES

1. Lawrence D. Miles, *Techniques of Value Analysis and Engineering*, McGraw-Hill, New York, 1961, p. 1.

2. John H. Fasal, *Practical Value Analysis Methods*, Hayden Book Company, New York, 1972, p. 6.

3. Ibid.

4. Thomas J. Snodgrass, Value engineering course materials, University of Wisconsin-Extension, Department of Engineering and Applied Sciences, Madison, Wisc.

5. Arthur P. Brief, André L. Delbecq, Alan C. Filley, and George P. Huber, "Elite Structure and Attitudes: An Empirical Analysis of Adoption Behavior," *Administration and Society*, Vol. 8, No. 2, August 1976, pp. 227–248.

6. Sandra L. Gill, "Guidelines for the Management of Temporary Task Force Teams," in Jesse F. McClure (Ed.), *Managing Human Services*, International Dialogue Press, Davis, Calif., 1979, pp. 86–99.

7. Charles H. Harrison, "Community Survey Inspires Innovation," *Nation's Schools*, 1973.

8. Bev McRae and Joyce Fox, "Survey Planned to Aid Community School Communications," *APT News*, Lake Forest Association of Parents and Teachers, winter 1972.

9. The School Scene, "School Plan Bare Bones Self Study," *Lake Forester*, February 17, 1972.

10. "APT Board Survey Results Show How Residents Rate Our Schools," *Spirit of 67*, February–March 1973.

11. Alan J. Klingenberg and Thomas J. Snodgrass, "Evaluation, Supervision, Recognition and Market Research," *Illinois Journal of Education*, Vol. 64, No. 2, 1973, pp. 67–71.

METHOD ABSTRACTS

BRAINSTORMING

1. DESCRIPTION

   "Brainstorming is a group process in which the members, usually from different backgrounds, respond to a central question or theme. Emphasis is placed on generating a large number of ideas while deferring criticism and evaluation. Brainstorming is especially useful for attacking new problems or for identifying new ways of looking at old problems" [Delp et al., p. 3].

2. BRIEF HISTORY

   Brainstorming was developed "and first used by Alex F. Osborn in 1939 in the advertising agency, Batten, Barton, Durstine and Osborn, which he then headed" [Taylor et al., p. 24]. "Osborn devised the . . . procedure to create a free and uninhibited atmosphere which would increase the creativity of group members" [Collaros and Anderson, p. 159]. "Osborn's earliest brainstorming concentrated on commercial and educational problems. Later the technique was to be tested, and modifications developed in a variety of circumstances on problems of a technical, military, financial, and aesthetic nature by groups throughout the world" [Rickards, p. 60]. Brainstorming has become widely known through Osborn's book, *Applied Imagination*, first published in 1953.

3. BASIC PREMISE OR ASSUMPTION

   Based on the principle of deferred judgment, which allows for "a given period of time for listing all the ideas that come to one's mind regarding a problem . . . One should forget about the quality of the ideas entirely. Quality only is stressed" [Parnes, 1963, p. 284]. Evaluation comes later, *after* the free flow is allowed under deferred judgment.

4. PURPOSE/GOAL

The goal is to produce a large quantity of ideas during the ideation phase of problem-solving in order to increase the probability of novel and creative solutions to a problem. In recent years it has been applied in the divergent phases of *each* problem-solving step.

5. MENTAL OPERATIONS SUPPORTED

Cognition

Memory

Divergent production

6. BENEFITS

Emphasizes postponement of judgment, which helps to overcome mental blocks and relieve stress that may inhibit participants [Parnes, 1963, p. 284].

7. LIMITATIONS

"Brainstorming may be unproductive if the group members are meeting each other for the first time during the session. The effectiveness of a brainstorming session is greatly enhanced when the members know each other before the session and when they are motivated to solve the problem under consideration" [Bouchard, 1971].

"Superior–subordinate relationships outside the session may affect the free interchange of ideas within the session" [Delp et al., p. 3].

"Discussion may be dominated by one or two members, and may stifle the participation of other members" [Delp et al., p. 3].

8. PARTICIPANTS

Quantity

"Five or six people, recorded round robin; six to eight, Osborn (classical), the larger size being suited to less experienced groups; larger groups can be handled in trigger sessions (up to about 15 people), but still larger groups should be split into subgroups" [Rickards, p. 66].

Qualifications

"Good and constructive attitudes in normal meetings" [Rickards, p. 66].

"Wide spread of backgrounds" [Rickards, p. 66].

Roles of Group Members

Facilitator or leader

Idea contributors (four to eight) (One of these should be an expert on the topic being brainstormed.)

Secretary

9. INPUT

"A statement of the problem, usually in the form of a question," which serves to focus participants' ideas [Delp et al., p. 4].

10. OUTPUT

"The principal result of a brainstorming session is a large number of ideas which may serve as possible answers to the question. These answers are not qualified or evaluated. Other techniques are empolyed for this, such as Decision Trees, Cost-Benefit Analysis, [or Value Engineering]" [Delp et al., p. 4].

11. BASIC ELEMENTS

Facts in fact finding.

Problems in problem finding.

Ideas in idea finding.

Criteria in solution finding.

Implementation ideas in acceptance finding.

12. RULES FOR USE/SYNTAX

"Criticism is ruled out" [Osborn, p. 84].
"Free-wheeling is welcomed" [Osborn, p. 84].
"Quantity [of ideas] is wanted" [Osborn, p. 84].

Combination and improvement of ideas are encouraged [Osborn, p. 84].

Novel ideas may be created through the use of such devices as the following:

1. Magnification [Parnes, 1977, p. 126]

2. Minification [Parnes, 1977, p. 126]

3. Rearrangement [Parnes, 1977, p. 126]

4. Forced relationships [Parnes, 1977, p. 127]

5. Attribute listing [Parnes, 1977, pp. 127, 128]

6. Morphological approach (relating attribute listing to a forced relationship in a matrix) [Parnes, 1977, pp. 128, 129]

7. Personal analogy (identification with a task or object) [Bouchard, p. 192]

13. OPERATIONAL PROCEDURES

The following activities may be found in the idea-finding stage. They constitute just one example of the procedures that may vary greatly from situation to situation and from stage to stage.

Presession Activities

1. Organizer/facilitator reviews problem and does fact finding and problem finding.

2. Organizer/facilitator identifies and selects group participants, keeping in mind the nature of the problem.

3. Facilitator circulates two days in advance of session a "background memo" explaining the problem area.

Session Activities

1. Facilitator explains "the four guiding principles of brainstorming."

2. Facilitator states "the question or problem under consideration and discusses it briefly (with the participants) in order to clear up (possible) misunderstandings" [Delp et al., p. 5].

3. Group spontaneously provides ideas in response to question.

4. "If necessary, facilitator reminds members about the rules of brainstorming during the session. Sometimes members need a new direction, or they may be tactfully asked to curtail extraneous discussion. An atmosphere of cordiality and free expression must be maintained at all times" [Delp et al., p. 5].

5. "Sometimes it is useful to present a mock problem to the group to familiarize members with the method. An exercise in which members generate all possible uses of a machine or technique is helpful. A problem example might be 'list conventional and unconventional uses for a bicycle.' Such an exercise might relax the members and make the main session more effective" [Delp et al., p. 5].

6. "Have all ideas recorded as they are generated. A tape recorder may be used to provide a record, but this method doesn't allow

for immediate reference to previous responses" [Delp et al., p. 5].

Postsession Activities

1. "A member of the brainstorming group, usually the secretary, checks with each participant the day after the session in case a person has had some afterthoughts. As one might expect, ideas that have been slept on often turn out to be very valuable" [Potter, p. 55].
2. "The [group] secretary prepares a triple-spaced, typewritten list of all ideas suggested both during the session and afterwards" [Osborn, p. 250].
3. "The [facilitator] then edits the list, making sure that each idea is succinctly but properly stated. At the same time, he/she classifies the ideas within logical categories" [Osborn, p. 250].
4. "The [organizer] or individual concerned with the problem that gave rise to the brainstorming session screens the ideas in collaboration with selected associates" [Potter, p. 55].
5. "The screened ideas are passed on to the individual or group that has responsibility for implementing the accepted ideas" [Potter, p. 55].
6. "The action of the implementing group is reported . . . to the original brainstormers, thus completing the cycle of communication" [Potter, p. 55].

14. TIME REQUIRED

    10 minutes to one hour or longer, depending on nature of the problem. Often continuation sessions are scheduled after incubation.

15. MEETING ROOM/ENVIRONMENT

    The more informal the better.

16. SUPPLIES

    Blackboard and chalk or flip chart and marking pens are used to record all suggested ideas [Delp et al., p. 4].

    Writing materials for each member [Delp et al., p. 4].

    A large placard placed on wall in front of group stating the four basic rules of brainstorming [Osborn, p. 241].

    A tape recorder to record the session (optional) [Osborn, p. 242].

17. TECHNOLOGY

    Not computer supported.

18. FORMER USERS

U.S. Treasury. How to sell more U.S. Savings Bonds? [Osborn, p. 90]

U.S. Air Force. "What can Headquarters Civilian Personnel Division do to further the growth of Executive Development Programs throughout the Air Force?" [Osborn, p. 91]

The Scott Paper Company. How to improve methods of purchasing? [Osborn, p. 91]

RCA, Advanced Development Section. How to improve TV receivers? [Osborn, p. 92]

Chesapeake and Ohio Railway "Better Service Conference." How to improve conference's program? [Osborn, p. 94]

University of Southern California, Delinquency Control Institute. How to improve the public relations of police departments? [Osborn, p. 94]

Community Chest. How to improve money-raising techniques? [Osborn, p. 94]

19. BIBLIOGRAPHY

Biondi, Angelo M., Ed. *The Creative Process,* DOK Publishers, Buffalo, N.Y., 1972.

Bouchard, Thomas J. "Whatever Happened to Brainstorming?," *Journal of Creative Behavior,* 5 (1971), pp. 182–189.

Collaros, Panayiota, and Lynn Anderson. "Effect of Perceived Expertness Upon Creativity of Members of Brainstorming Groups," *Journal of Applied Psychology,* 53, 2 (1969), pp. 159–163.

Delp, Peter, Arne Thesen, Juzar Motiwalla, and Neelakantan Seshadri. *Systems Tools for Project Planning,* International Development Institute, Indiana University, Bloomington, Ind., 1977.

Noller, Ruth B. *Scratching the Surface of Creative Problem-Solving,* DOK Publishers, Buffalo, N.Y., 1972.

Noller, Ruth B., Sidney J. Parnes, and Angelo M. Biondi. *Creative Actionbook,* Charles Scribner's Sons, New York, 1976.

Osborn, Alex F. *Applied Imagination,* Charles Scribners Sons, New York, 1963.

Parnes, Sidney J. "The Deferment-of-Judgment Principle: A Clarification of the Literature," *Psychological Reports,* 12 (1963), pp. 521–522.

Parnes, Sidney J., Ruth B. Noller, and Angelo M. Biondi. *Guide to Creative Action,* Charles Scribner's Sons, New York, 1977.

Potter, David. *Discussion in Small Groups: A Guide to Effective Practice,* Wadsworth Publishing, Belmont, Calif. 1976.

Rickards, Tudor. *Problem-Solving Through Creative Analysis,* Gower Press, Essex, 1974.

Taylor, Donald, Paul C. Berry, and Clifford Block. "Does Group Participation When Using Brainstorming Facilitate or Inhibit Creative Thinking?," *Administrative Science Quarterly,* 3 (1958–1959), pp. 23–47.

DELPHI METHOD

1. DESCRIPTION

   The Delphi Method is a survey technique for achieving consensus among isolated anonymous participants with "controlled opinion feedback" [Linstone and Turoff, p. 10].

2. BRIEF HISTORY

   Delphi was first applied by the Rand Corporation in a U.S. Air Force sponsored study in the 1950s. "The subject of this study was the application of 'expert opinion to the selection, from the point of view of a Soviet strategic planner, of an optimal U.S. industrial target system and to the estimation of the number of A-bombs required to reduce the munitions output by a prescribed amount'" [Linstone and Turoff, p. 10]. It was a later effort that brought Delphi into the science and business worlds—a Rand paper, "Report on a Long-Range Forecasting Study," by Olaf Helmer and T. J. Gordon in 1964. They concentrated on long-range trends in science and technology (technological forecasting). Today, its use is widespread as the need grows to incorporate subjective information into evaluation models dealing with more complex societal problems within such areas as the environment, health, education, and transportation.

3. BASIC PREMISE OR ASSUMPTION

   That a structured and iterative questionnaire, a kind of remote conferencing procedure, can serve as an effective means to draw on expert opinion to develop useful forecasts.

4. PURPOSE/GOAL

   "To determine or develop a range of possible alternatives."

   "To explore or expose underlying assumptions or information leading to different judgments."

   "To seek out information which may generate a consensus on the part of the respondent group."

   "To correlate informed judgments on a topic spanning a wide range of disciplines."

   "To educate the respondent group as to the diverse and interrelated aspects of the topic."

   [Source: Delbecq et al., pp. 10–11]

5. MENTAL OPERATIONS SUPPORTED

Memory

Divergent production

Convergent production

6. BENEFITS

"The process of writing responses to the questions forces respondents to think through the complexity of the problem, and to submit specific, high-quality ideas."

"The anonymity and isolation of respondents provides freedom from conformity pressures."

"Simple pooling of independent ideas and judgments facilitates equality of participants."

"The Delphi process tends to conclude with a moderate perceived sense of closure and accomplishment."

"The technique is valuable for obtaining judgments from experts geographically isolated."

[Source: Delbecq et al., pp. 34, 35]

"Prevents domination by certain individuals" [Delbecq et al., p. 83].

7. LIMITATIONS

"The lack of opportunity for social-emotional rewards in problem-solving leads to a feeling of detachment from the problem-solving effort."

"The lack of opportunity for verbal clarification or comment on the feedback report creates communication and interpretation difficulties among respondents."

"Conflicting or incompatible ideas on the feedback report are handled by simply pooling and adding the votes of group respondents. Thus, while this majority rule procedure identifies group priorities, conflicts are not resolved."

[Source: Delbecq et al., p. 35]

8. PARTICIPANTS

Quantity

Number varies with the problem.

Qualifications

Vary with the problem.

Roles of Group Members

Decision-maker(s). "The individual or individuals expecting some sort of product from the exercise which is used for their purposes."

A Staff Group. "The group which designs the initial questionnaire, summarizes the returns, and redesigns the follow-up questionnaires."

A Respondent Group. "The group whose judgments are being sought and who are asked to respond to the questionnaires."

[Source: Delbecq et al., p. 10]

9. INPUT

Issue to be addressed.

Panel of experts.

Questions pertaining to issue.

[Judd, p. 30]

10. OUTPUT

Forecasts.

Plans.

Problem definition.

Guidelines.

Program or policy.

Summary of current knowledge.

Selection from alternatives.

11. BASIC ELEMENTS

Topics.

Questions.

Issues.

Opinions.

12. RULES FOR USE/SYNTAX

Anonymity.

Iteration with controlled feedback.

Statistical group response.

[Martino, p. 20]

13. OPERATIONAL PROCEDURES

Estimated Time

		Estimated Time
1.	Develop Delphi question(s).	$\frac{1}{2}$ day
2.	Select and contact respondents.	2 days
3.	Select sample size.	$\frac{1}{2}$ day
4.	Develop questionnaire #1 and test.	1 day
5.	Type and send out.	1 day
6.	Response time.	5 days
7.	Dunning time (if used).	3 days
8.	Analyze questionnaire #1.	$\frac{1}{2}$ day
9.	Develop questionnaire #2 and test.	2 days
10.	Type and send out.	1 day
11.	Response time.	5 days
12.	Dunning time (if used).	3 days
13.	Analyze questionnaire #2.	1 day
14.	Develop questionnaire #3 and test.	2 days
15.	Type and send out.	1 day
16.	Response time.	5 days
17.	Dunning time (if used).	3 days
18.	Analyze questionnaire #3.	1 day
19.	Prepare final report.	4 days
20.	Type report and send out.	1 day
21.	Prepare respondents.	1 day
22.	Type report and sent out.	1 day
	Minimum time	$44\frac{1}{2}$ days

[Source: Delbecq et al., p. 87]

14. TIME REQUIRED

See 13, Operational Procedures.

15. MEETING ROOM/ENVIRONMENT

See 16, Supplies.

16. SUPPLIES

Analysis of Questionnaire #1

Needed by each member of the work group consisting of decision-makers and staff:

1. Copies of each variable or item from every Delphi questionnaire, placed on a $3'' \times 5''$ card.
2. Pad of paper.
3. Tape.
4. Pencils.
5. Scissors.

Needed by total group:

1. Flip chart.
2. Felt pens.
3. Role of masking tape.

Analysis of Questionnaire #2

Needed by each member of the work group:

1. Copies of each response to Questionnaire #2.
2. Pad of paper.
3. Tape.
4. Pencils.
5. Scissors.
6. Vote tally sheet.

Analysis of Questionnaire #3

Needed by each member of the work group:

1. Copies of each response to Questionnaire #3.
2–6. Same as for Questionnaire #2.

[Source: Delbecq, p. 94]

17. TECHNOLOGY

Hardware

For "Delphi Conference" or *real-time Delphi* as opposed to conventional Delphi [Linstone and Turoff, p. 5].
Each conferee needs access to the following:

1. A remote terminal (a keyboard with letters, numbers, and symbols linked to the control computer, which is programmed to sort, store, and transmit each conferee's message) (Price, p. 498).
2. A cathode ray tube (CRT) display device.
3. A printer.

18. FORMER USERS

University of Wisconsin, School of Nursing. Development of a structure for a conference on nursing role realignment [Delbecq et al., p. 106]

U.S. Navy. A forecast of computer technology [Martino, p. 44].

Smith Kline and French (S K & F) Laboratories. The future of medical care [Martino, p. 44].

Michigan Sea Grant Program. Resource management in the Great Lakes Region [Ludlow, pp. 102–123]

National Institute of Drug Abuse and National Coordinating Council of Drug Education. National drug abuse policy formulation [Jillson, pp. 124–159]

U.S. Air Force Laboratories. Priorities in system concept options [Jones, pp. 160–167]

Bell-Canada, Business Planning Group. Identification of corporate opportunities (or threats) that will arise through changes in society and/or technology in the next decade or two [Day, p. 172]

Parsons & Williams, Inc. (international consulting firm). "Forecast 1968–2000 of Computer Developments and Applications" sponsored by the International Federation of Information Processing Societies (an examination of future computer applications in business, the home, government, and institutions; and the projection of the future of specific computing and technological developments) [Day, p. 169]

Social Engineering Technology (SET), Inc. "Delphi Panel on the Future of Leisure and Recreation," conducted for a group of companies interested in future market opportunities in the recreational area that could develop through the impact of cultural change [Day, p. 169]

Institute for the Future (IFF). The Future of the Telephone Industry, sponsored by the American Telephone and Telegraph Co., New York [Day, pp. 169–170]

19. BIBLIOGRAPHY

Day, Lawrence H. "Delphi Research in the Corporate Environment," in Harold A. Linstone and Murray Turoff, Eds., *The Delphi Method*, Addison-Wesley Publishing, Reading, Mass., 1975, pp. 168–193.

Delbecq, André L., Andrew Van de Ven, and David Gustafson. *Group Techniques for Program Planning*, Scott, Foresman, Glenview, Ill., 1975.

Jillson, Irene Anne. "The National Drug-Abuse Policy Delphi: Progress Report and Findings to Date," in Harold A. Linstone and Murray Turoff, Eds., *The Delphi Method*, Addison-Wesley Publishing, Reading, Mass., 1975, pp. 124–159.

Jones, Chester G. "A Delphi Evaluation of Agreement," in Harold A. Linstone and Murray Turoff, Eds., *The Delphi Method*, Addison-Wesley Publishing, Reading, Mass., 1975, pp. 160–167.

Judd, Robert C. "Delphi Method: Computerized Oracle Accelerates Consensus Formation," *College and University Business,* 49 (Sept. 1970), pp. 30–34.

Linstone, Harold A., and Murray Turoff, Eds. *The Delphi Method*, Addison-Wesley Publishing, Reading, Mass., 1975.

Ludlow, John. "Delphi Inquiries and Knowledge Utilization," in Harold A. Linstone and Murray Turoff, Eds., *The Delphi Method*, Addison-Wesley Publishing, Reading, Mass., 1975, pp. 102–123.

Martino, Joseph P. *Technological Forecasting for Decisionmaking*, American Elsevier, New York, 1972.

Price, Charlton R. "Conferencing via Computer: Cost Effective Communication for the Era of Forced Choice," in Harold Linstone and Murray Turoff, Eds., *The Delphi Method*, Addison-Wesley Publishing, Reading, Mass., 1975, pp. 497–516.

INTERPRETIVE STRUCTURAL MODELING (ISM)

1. DESCRIPTION

   Interpretive structural modeling is a computer-assisted, interactive, learning process, whereby complex issues or problems may be organized.

2. BRIEF HISTORY

   ISM is founded on various branches of mathematics—graph theory (Euler, 1736), logic (Boole and de Morgan, 1847), matrices (Cayley, 1858), relation theory (de Morgan; Pierce, 1892), and lattice theory (Birkoff, 1948).

3. BASIC PREMISE OR ASSUMPTION

   That computer algorithms, combining the communication tools of words, graphics, and mathematics are useful in organizing and structuring the large quantities of quantitative and qualitative data connected with complex societal problems.

4. PURPOSE/GOAL

   To extend capacity to define complex systems and enhance interdisciplinary efforts to communicate about system improvement [Warfield, p. 205].

5. MENTAL OPERATIONS SUPPORTED

   Memory

   Convergent production

   Evaluation

6. BENEFITS

   Makes use of graphical modes of communication to illuminate complex issues, systems, or concepts [Warfield, p. 199].

   Provides for a means of organizing and giving structure to large quantities of information [Warfield, p. 198].

   Facilitates learning, comprehension, and communication [Warfield, p. 194].

   Provides for the development of a rationale that will support a decision [Warfield, p. 198].

7. LIMITATIONS

   The size of the element set that can be accommodated in a given time is limited by the speed and memory capacity of computing equipment.

The process is tiring, so that the time of a session cannot normally exceed about four hours.

8. PARTICIPANTS

Quantity
A group of no more than eight members.

Qualifications
Broker. Should understand context requiring study, learning, and organization.

Facilitator. Should be skilled in helping groups work together and be familiar with ISM process.

Technician. Should understand equipment needed for ISM process and know how to use ISM software.

Participants. Should be knowledgeable about problem area, capable of contributing to implementation of ISM results, aware of sources of information; politically sensitive, capable of representing constituency in articulate nondogmatic manner, and capable of engaging in focused dialogue.

Observers.

Roles of Group Members
Broker. Identifies participants and encourages them to take part in ISM process, takes care of financing, and selects facilitator and technician.

Facilitator. Oversees metaprocessess (those processes necessary to maintain group stability and help the group reach its desired goals).

Technician. Arranges for installation, implementation and maintenance of computer software and hardware.

Participants.

Observers.

9. INPUT
Set of elements and variety of relations [Lendaris, pp. 346–347].

10. OUTPUT
Interpretive structural model [Warfield, p. 204].

11. BASIC ELEMENTS

Set of elements germane to the problem, for example, variables, subsystems, objectives, or goals [Warfield, p. 71].

Contextual relation [Warfield, p. 349].

12. RULES FOR USE/SYNTAX

Transitive embedding [Warfield, p. 349].

Weighted embedding [Warfield, p. 354].

Scanning method [Warfield, p. 254].

Coupling method [Warfield, p. 354].

13. OPERATIONAL PROCEDURES

    1. Theme is selected.
    2. Developer is identified.
    3. Elements and contextual relation are identified.
    4. Leader is identified.
    5. ISM program is entered in computer.
    6. Adequate computer time is allocated.
    7. Facilities are ready.
    8. Session plan is complete.
    9. Computer contains elements and contextual relation.
    10. Session can begin.
    11. Element set is edited.
    12. Reachability matrix is complete.
    13. Total structure is available.
    14. Amendments are complete.
    15. Final structures are satisfactory.

    [Source: Warfield, p. 347]

14. TIME REQUIRED

Each session, not more than three hours. A rough approximation can be stated as follows:

$$T \text{ (hours)} = \frac{1}{600} e^2 p^{0.5}$$

where e is the number of elements in the element set and p is the number of particpants engaged in the model development process [Warfield, p. 353].

15. MEETING ROOM/ENVIRONMENT

Television display screens accessed remotely through a telephone line connected to a computer terminal (kept in low profile).

Tables and comfortable chairs.

Blackboard.

[Source: Warfield, pp. 348–349]

16. SUPPLIES

Dictionary.

Paper and pencils.

Chalk.

17. TECHNOLOGY

Software

Computer program.

Hardware

Computer terminal with telephone access to a computer containing ISM software. Display units consisting either of television screens or a large projection-type display controlled by the computer terminal. Automatic drafting equipment to construct the maps (when the software becomes available.)

18. FORMER USERS.

Wide variety of applications in industry, education, and government.

19. BIBLIOGRAPHY

Bryce, H. J., Ed. *Managing Fiscal Retrenchment in Cities,* Academy for Contemporary Problems, Columbus, Ohio, 1980, Chapter 6.

Crim, K. O. "Use of ISM in Environmental Studies at the Senior High Level," University of Dayton Report, UDR-TR-79-27, Dayton, Ohio, April 1979.

Fitz, R. W., D. M. Gier, and J. Troha. "A Methodology for Project Planning Using ISM," *Proceedings of the International Conference on Cybernetics and Society,* IEEE, New York, 1977, pp. 297–302.

House, Robert W. "Application of ISM in Brazil's Alcohol Fuel Program," *Proceedings of the International Conference on Cybernetics and Society,* IEEE, New York, 1978, pp. 1008–1012.

Lendaris, George G. "On the Human Aspects in Structural Modeling," *Technological Forecasting and Social Change,* 14, 4 (1979), pp. 329–351.

Malone, D. W. "Strategic Planning: Applications of ISM and Related Techniques," *Proceedings of the International Conference on Cybernetics and Society,* IEEE, New York, 1978, pp. 995–1000.

Mizoguchi, F., K. Tahara, and M. Saito. "Use of ISM to an Analysis of Expert Role in Simulation Modeling Process," *Proceedings of International Conference on Cybernetics and Society,* IEEE, New York, 1978, pp. 983–988.

Warfield, J. N. *Societal Systems,* Wiley, New York, 1976.

ISSUE BASED INFORMATION SYSTEM (IBIS)

1. DESCRIPTION

   "An IBIS is designed essentially to develop a *discourse* relative to some initially unstructured problem area or *topic* (e.g., mass transit in Los Angeles, 'stagflation' in the U.S. economy, the energy crises internationally, fair employment practices in a firm). *Issues* are raised and disputed from various positions and points of view, and *arguments* are advanced for and against in an attempt to provide decision makers with an appreciation of the problem situation" [Swanson, p. 258].

2. BRIEF HISTORY

   "The IBIS, or Issue-Based Information System, was developed by Kunz and Rittel as an argumentative planning model. The characteristics of the IBIS encourage thoroughness in problem coverage and aid in making decision bases explicit and retraceable. The IBIS can be used to structure arguments for and against proposed plans either by groups or by an individual. The IBIS serves to construct an explicit account of the way in which decisions are reached" [Grant, p. 185].

3. BASIC PREMISE OR ASSUMPTION

   "Rests on a model of problem-solving by cooperatives as an argumentative process. . . . in the realm of values, there are no experts; there exists a 'symmetry of ignorance' (Rittel) with regard to values between the planner and his clients" [Dehlinger and Protzen, p. 38].

4. PURPOSE/GOAL

   ". . . to organize and structure the processes by which decisions and choices . . . are reached" [Dehlinger and Protzen, p. 38]. ". . . to support coordination and planning of political decision processes. IBIS guides the identification, structuring, and settling of issues raised by problem-solving groups and provides information pertinent to the discourse" [Kunz and Rittel, p. 13].

5. MENTAL OPERATIONS SUPPORTED

   Memory
   Divergent production
   Convergent production
   Evaluation

6. BENEFITS

   ". . . encourages thoroughness, exposes underlying values during the course of debate, and makes it possible to communicate, record and argue about decision-bases in situations in which there are multiple clients with conflicting values and interest" [Grant, p. 185].

7. LIMITATIONS

". . . easy to smother arguments under mountains of paper work; that the prospect of being frank and accountable about decision bases can be uncomfortable and even dangerous, and may prevent many people from participating in any argumentative planning procedure; and that the logical, monethematic nature of the technique is uncomfortable and/or unacceptable to many people" [Grant, p. 185].

8. PARTICIPANTS

Quantity

Varies with each problem.

Qualifications

Vary with each problem.

Roles of Group Members

Participants

Experts

Judges/critics

Documentarians

9. INPUT

A "trigger phrase" that denotes the problem area [Kunz and Rittel, p. 13].

10. OUTPUT

A plan, policy, or solution to a given problem [Grant, p. 189].

11. BASIC ELEMENTS

Core Components

Issues (questions) of which there are four major types:

Factual	Is x the case?
Deontic	Shall x become the case?
Explanatory	Is x the reason for y?
Instrumental	Is x the appropriate means to accomplish y in this situation?

Positions
Arguments
References

Secondary Set of Components

Questions of fact or consensus

Answers

Supporting evidence

Third Set for Recording and Interrelating the Core Components and Secondary Components

Topics that group several issues and questions

Forms

Tables

Matrices

Trees

12. RULES FOR USE/SYNTAX

Debate. "IBIS is expected to encourage, even to generate as many conflicts as possible" [Dehlinger and Protzen, p. 38].

Raising questions and issues.

Contributing answers and arguments

Proposing solutions and plans.

Deciding.

Evaluating.

Setting priorities.
[Source: Mann, p. 163]

13. OPERATIONAL PROCEDURES

1. "The session begins, after an explanation of the rules and procedure, with the introduction of the topic or problem. This can be done either by presenting a problem (discrepancy) description, one or two controversial solution proposals, or a set of issues to get the debate started" [Mann, p. 163].

2. "Questions, issues, and solution proposals are submitted freely by all participants at any time, by filling in the appropriate form for each category and 'posting' these in a place visible and accessible to all" [Mann, p. 163].

3. "Answers and arguments to the questions raised and posted are submitted freely at any time by all participants, by entering or attaching the contribution (as short and concise as possible) to the appropriate sheet for each question" [Mann, p. 163].

4. "Group decisions are achieved in special 'decision sessions':"

    4.1. "Whenever a player wants to obtain a group decision on an issue, the item is entered as a 'candidate' for the next decision session, to be scheduled periodically or whenever a number of candidates has accumulated'' [Mann, p. 163].

4.2. ''The decision session starts by the group ordering the agenda by deciding which questions it will discuss, and setting priorities among these'' [Mann, p. 163].

4.3. ''The items on the agenda are taken up in the established order of priority. They may be: decided offhand (by acclamation), rejected (dropped from the agenda), reworded, substituted, postponed, or decided by vote or formal evaluation procedure. Any of these actions are effected upon appropriate motion and debate, which consists of reviewing arguments submitted thus far, and adding new ones (in writing). Standard rules of procedure may be followed for the discussion'' [Mann, pp. 163, 164].

4.4. ''After debate on a question has ended, the agenda may be modified by motion and vote'' [Mann, p. 164].

5. ''Group decisions are recorded (decision file) and become the basis for the subsequent work. Revision of decisions once taken may be considered only if there is new information added to the previous arguments; the question will then be reentered as a candidate'' [Mann, p. 164].

6. ''The session ends when the predetermined time limit is reached, with a formal evaluation of the final solution (or several alternatives). If no solution has been decided upon, the decision file will serve as its description'' [Mann, p. 164].

14. TIME REQUIRED

Several hours to several weeks or months.

15. MEETING ROOM/ENVIRONMENT

''. . . appropriate space and possibilities for adequate display of 'posted' issues is essential. A large wall where sheets of paper can be easily fastened and removed, visible and accessible to all players would be best; alternatively, a long table might serve the same purpose. The participants must be able to write conveniently at their places. Prepared forms for the various types of questions, arguments, evaluation, etc. may be helpful. Larger size displays for the game rules, the current state of the discourse (issues raised; candidates, current agenda) and the decision file should be provided to help the participants keep the overview of the game proceedings. Issue 'maps' or graphs of the networks of various relations among issues may also be useful for this purpose. They may be prepared by the 'bookkeepers' or the players themselves'' [Mann, p. 164].

16. SUPPLIES.

    See 15, Meeting Room/Environment.

17. TECHNOLOGY

    Software

    Issue Bank. A file of living, settled or abandoned, and latent issues.

    Evidence Bank. File of questions and their answers (answered and open).

    Handbook. Collection of model problems

    Topic list

    Issue map. Representation of the various relations between issues, questions, and so on by graphic display of the state of the argument.

    Documentation system. Search and analysis in view of living or latent issues and positions, descriptor index and thesaurus construction, regular scanning in view of the topic list.

18. FORMER USERS

    University Planning System HKP (Hochschule-Kapazitäts-Planning). [Dehlinger and Protzen, p. 39]

    Martin Luther King Development Corporation Planning Information System. [Dehlinger and Protzen, p. 39].

    Environmental Planning Information System. [Dehlinger and Protzen, p. 40]

19. BIBLIOGRAPHY

Dehlinger, Hans, and Jean-Pierre Protzen. "Some Considerations for the Design of Issue Based Information Systems (IBIS)," *DMG-DRS Journal: Design Research and Methods,* 6, 2 (April–June 1972), pp. 38–45.

Grant, Donald P. "How to Use the IBIS as a Procedure for Deliberation and Argument in Environmental Design and Planning," *Design Methods and Theories,* 11, 4 (Oct–Dec. 1977), pp. 185–220.

Kunz, Werner, and Horst W. J. Rittel. "Issues as Elements of Information Systems," *The DMG 5th Anniversary Report,* Jan. 1972, pp. 13–15.

Mann, Thorbjoern. "Play IBIS. A Design Game Based on the Argumentative Model of Design,"*DMG–DRS Journal: Design Research and Methods,* 6, 4 (Oct–Dec. 1972), pp. 161–165.

Swanson, E. Burton. "A Methodology for IBIS Data Gathering and Development," *Design Methods and Theories,* 11, 4 (Oct–Dec. 1977), pp. 256–261.

KANE SIMULTATION (KSIM)

1. DESCRIPTION

   KSIM is a language utilizing two modes of communication—the verbal, written, and spoken; and the visual, symbolic and graphic. Cross-impact analysis is used as a means of exploring the structural dynamics of a system and geometry of linkages. It accommodates both qualitative and quantitative data, subjective and objective communications. [Kane, p. 129]

2. BRIEF HISTORY

   Developed by Dr. Julius Kane in the early 1970s. [Vallee, p. 395] Originally invented as a tutorial tool for interdisciplinary workshops at the University of British Columbia.

3. BASIC PREMISE OR ASSUMPTION

   "The structure of the system (the nature of its interactions) is far more important than the state of the system" [Kane, pp. 140–141.].

4. PURPOSE/GOAL

   To provide a language whereby members of a policy making, planning, or design team, who may not be knowledgeable about the technical aspects of computer programming, computer simulation, and higher mathematics, can analyze the structural dynamics of a system through cross-impact analysis to project future states of a system.

5. MENTAL OPERATIONS SUPPORTED

   Cognition

   Memory

   Divergent production

   Convergent production

   Evaluation

6. BENEFITS

   Allows for disagreement and multiple criteria.
   Promotes debate and interaction.
   Once learned, easy to use.
   Flexible.

7. LIMITATIONS

   None; not meaningful in this case.

8. PARTICIPANTS

   Quantity
   > May be used by individuals and groups. The number of members in a group found to function most effectively is five.

   Qualifications
   > Requires emotional involvement and element of risk.
   > Leader must be experienced in group processes and KSIM.

   Roles of Group Members
   > Leader
   >
   > Panel or workshop members

9. INPUT

   Statement of the problem

   Relevant variables

10. OUTPUT

    Projections about future states of a system that may be used to plan, make policy, or solve problems.

11. BASIC ELEMENTS
    Variables (12 or less).

12. RULES FOR USE/SYNTAX

    Cross-impact analysis

    Debate

    KSIM, simulation language to construct model

13. OPERATIONAL PROCEDURES
    1. Define variables.
    2. Establish cross-impact matrix.
    3. Compute.
    4. Review.
    5. Go back to step 1 and repeat process until problem solved.

14. TIME REQUIRED
    Several hours to several weeks [Linstone et al., p. 9].

15. MEETING ROOM/ENVIRONMENT
    Sessions should be held away from "home base" of all of the participants, that is, in neutral territory.

16. SUPPLIES

Blackboard, chalk and/or pencil, and paper for constructing inter-action matrix

Computer scope for computer graphics (optional)

17. TECHNOLOGY

Computer not required; however, computer graphics can greatly magnify the impact of KSIM.

18. FORMER USERS

U.S. Army Corps of Engineers.

NASA Headquarters and Johnson Space Center.

The Canadian Government—"Make or Buy" Research Policy.

The OECD, Paris.

19. BIBLIOGRAPHY

Kane, Julius. "A Primer for a New Cross-Impact Language—KSIM," *Technological Forecasting and Social Change*, 4 (1972), pp. 129–142.

Linstone, Harold A., George Lendaris, Steven Rogers, Wayne Wakeland, and Mark Williams. "The Use of Structural Modeling for Technology Assessment," *Technological Forecasting and Social Change*, 14, 4 (1979).

Vallee, Jacques. "Modeling as a Communication Process; Computer Conferencing Offers New Perspectives," *Technological Forecasting and Social Change*, 10 (1977), pp. 391–400.

NOMINAL GROUP TECHNIQUE (NGT)

1. DESCRIPTION

   Nominal group technique (NGT) is a special-purpose (single issue) method of structuring communication in decision groups faced with an unstructured problem situation [Delp et al., p. 14]. Individual judgments are combined to arrive at decisions for problem-solving or idea generation strategies [Delp et al., p. 14].

2. BRIEF HISTORY

   NGT was developed by André Delbecq and Andrew Van de Ven in 1968 from "social-psychological studies of decision conferences, studies of industrial engineering problems of program design in the NASA aerospace field, and social work studies of citizen participation in program planning" [Van de Ven, p. 2].

3. BASIC PREMISE OR ASSUMPTION

   That the process of silent generation of ideas in writing by individual members of the team that are later analyzed and evaluated by the group, serves to overcome problems typical of interacting groups, these being the following:

   1. Unbalanced participation among members.
   2. Dominance by high-status, aggressive, or articulate members of the group.
   3. Premature evaluation and criticism of ideas.

4. PURPOSE/GOAL

   To identify elements of a problem.

   To identify and rank goals or priorities.

   To identify experts whose experience or skills may be useful in solution development.

   To involved decision-makers from multiple levels and proposal review to determine strengths and modifications that would improve preliminary designs and to promote the acceptability of the final decision.

5. MENTAL OPERATIONS SUPPORTED

   Memory

   Divergent production

   Convergent production

6. BENEFITS

"Dominance by high-status, aggressive, or articulate members is reduced since each has an equal opportunity to participate."

The group remains problem-conscious; and premature evaluation, criticism or focusing on ideas is avoided."

"The silent generation of ideas minimizes the interruptions in each person's thought process."

'A written record increases the group's ability to deal with a large number of ideas. It also avoids the loss of ideas."

"Discussion only to clarify items helps eliminate misunderstanding, without reducing the group's efficiency."

"Enhances the conditions for creativity when generating information on a problem. It avoids rambling discourse and other deficiences found in group processes."

[Delp et al., pp. 14, 15]

7. LIMITATIONS

NGT is a single agenda process and a highly focused meeting.

NGT does not necessarily promote group cohesion.

Bringing group members together may be cost prohibitive in certain situations.

8. PARTICIPANTS

Quantity

Five to nine members.

Qualifications

Group make-up may be homogeneous or heterogeneous. "Studies have shown that heterogeneous groups exhibit more creativity" [Delbecq, p. 80]. "But interpersonal differences and communication problems may increase for such groups" [Delp et al., p. 15].

Roles of Members of the Group

Leader

Participants

9. INPUT

The NGT question that provides the basis for generating the ideas [Delp et al., p. 15].

10. OUTPUT

    A list of ideas rank ordered according to importance. Overlapping ideas may be combined under a common heading (by the leader) [Delp et al., p. 15].

11. BASIC ELEMENTS

    Nominal question

    Ideas

12. RULES FOR USE/SYNTAX

    Round robin (a process for serially recording of an idea that each participant provides in turn) [Delp et al., p. 15].

    Discussion and evaluation of ideas.

    Rank ordering of ideas.

    Independent voting.

13. OPERATIONAL PROCEDURES

    1. "Silent generation of ideas in writing" [Van de Ven, p. 2].
    2. "Round robin feedback from group members. Each idea is recorded in a short phrase on a blackboard or flip chart" [Van de Ven, p. 2].
    3. "Discussion of each recorded idea for clarification and evaluation" [Delbecq, p. 8].
    4. "Silent individual voting on priorities" [Van de Ven, p. 2].

14. TIME REQUIRED

    NGT requires approximately four hours to prepare, analyze, and summarize one group session; the session itself usually takes a minimum of 90 minutes.

15. MEETING ROOM/ENVIRONMENT

    The seating arrangement must allow all members to easily focus on the ideas listed on the flip chart or blackboard. Some means of displaying the completed chart pages is necessary (e.g., tacks or masking tape).

16. SUPPLIES/EQUIPMENT

    A blackboard or flip chart

    Chalk or marking pens

    Index cards for each participant

    Paper and pencils for each participant

17. TECHNOLOGY.

    Not computer supported.

18. FORMER USERS

Health planning groups, hospital administrators and staff, boards and trustees, land-use planners, engineers, architects, draft persons, fine arts administrators and consumer groups.

19. BIBLIOGRAPHY

Delbecq, André L., Andrew H. Van de Ven, and David H. Gustafson. *Group Techniques for Program Planning: A Guide to Nominal Group and Delphi Processes*, Scott, Foresman, Palo Alto, Calif. 1975.

Delp, Peter, Arne Thesen, Juzar Motiwalla, and Neelakantan Seshadri. *Systems Tools for Project Planning*, International Development Institute, Bloomington, Ind., 1977, pp. 14–18.

Huber, George P., and André L. Delbecq. "Guidelines for Combining the Judgments of Individual Group Members in Decision Conferences," *Academy of Management Journal*, 15, 2 (June 1972), pp. 161–174.

Van de Ven, Andrew H. *Group Decision Making and Effectiveness*, Comparative Administrative Research Institute of the Center for Business and Economic Research, Graduate School of Business Administration, Kent State University, Bowling Green, Ohio, 1974.

PROGRAM PLANNING METHOD (PPM)

1. DESCRIPTION

   The Program Planning Method is a systematic and structured planning strategy through which clients or consumers, experts, and decision-makers are explicitly involved. It is "a method in which internal exchange across organizational units and extraorganizational interfaces can be sequenced, and offers an explicit process for structuring the character of participation within each phase of planning" [Delbecq and Van de Ven, p. 469].

2. BRIEF HISTORY

   PPM was developed by André Delbecq and Andrew Van de Ven in 1971 at the University of Wisconsin. This model was originally developed from social-psychological studies or decision sequencing and distilled from aerospace planning studies. It was then tested in social planning contexts.

3. BASIC PREMISE OR ASSUMPTION

   "Program Planning Method assumes that problem identification is necessary before planning a program. The best qualified people to identify the problem are the groups affected by potential programs or current inadequate programs. These people are brought together with planners and program personnel to identify problems and to rank them" [Delp et al., p. 228]. Experts and decision-makers are also strategically involved to facilitate the organizational change process for optimal program success.

4. PURPOSE/GOAL

   To strategically identify problems.

   To develop appropriate and innovative programs to solve them [Delbecq and Van de Ven, p. 467].

5. MENTAL OPERATIONS SUPPORTED

   Divergent production

   Convergent production

   Evaluation

6. BENEFITS

   "Organizes client, consumer, decision maker, or community participation for problem-solving."

   "Increases the legitimacy of the program in view of the participants' involvement."

"Decreases potential resistance to the implementation of the program."

"Increases the program's effectiveness because all concerned parties participate in the design."

"Facilitates proper problem identification and reduces the chances of solving the wrong problem."

"Facilitates the use of outside experts in the planning process."

"Incorporates the advantages of the Nominal Group Technique."

[Source: Delp et al., p. 227]

7. LIMITATIONS

"The participation of consumer and client groups may be unfeasible either financially or logistically."

"The role of the group leader is more important than in single applications of the Nominal Group Technique. The same leader is often in charge of three different groups and thus can control the success of the exercise."

"Client/citizen participation may have a negative effect on the planning process, e.g., premature expectations may arise."

[Source: Delp et al., pp. 227–228]

8. PARTICIPANTS

Quantity, Qualifications, and Roles

Client group. Participants usually include user groups (clients or consumers), experts, and decision-makers; they may be of all ages, abilities, and geographic locations. This selection process and the number of people involved in PPM will vary according to the substantive and political complexity of the planning issue.

Planner/Leader/Facilitator. This person or the planning staff will need to be knowledgeable about PPM, Nominal Group Technique, meeting management, and group decision-making strategies. Tasks include the following:

Analyzing the planning process and making each phase operational.

Identifying and involving appropriate decision-makers in each phase.

Facilitating effective group performance.

9. INPUT

    Problem recognition.

    Knowledge of problem area.

    Knowledge of target area, for example, a region, community, or ethnic group [Delp et al., p. 228].

    Nominal Group Technique and related group decision-making strategies.

10. OUTPUT

    List of problems ordered to show priorities of clients or consumers.

    Problem definition.

    Identification of key solution elements or criteria for problem-solving.

    Program for solving the problem.

    [Source: Delp et al., p. 228]

11. BASIC ELEMENTS

    Nominal question(s).

    Ideas.

    Facts and additional evidence as may be available from other sources and similar problems.

12. RULES FOR USE/SYNTAX

    Applicable when NGT is used with participants for problem identification and/or solution exploration:

    1. Round robin (a process for serially recording of an idea that each participant provides in turn) [Delp et al., p. 15].
    2. Discussion and evaluation of ideas.
    3. Rank ordering of ideas.
    4. Independent voting.

13. OPERATIONAL PROCEDURES

    1. Identify the various clients, experts, and decision-makers.
    2. Identify problems through various groups processes.
    3. Generate problem-solving ideas from participants.
    4. Develop specific problem-solving alternatives and proposals.
    5. Conduct proposal evaluations.
    6. Implement and evaluate program and program transfer.

14. TIME REQUIRED

    Depends entirely on the complexity of the problem.

15. MEETING ROOM/ENVIRONMENT

When NGT is used, a meeting room in which groups can cluster around a table is needed [Delp et al., p. 228]. The seating arrangement must allow all members to easily focus on the ideas listed on the flip chart or blackboard.

16. SUPPLIES/EQUIPMENT

When NGT is used, the following is needed:

A blackboard or flip chart

Chalk or marking pens

Index cards for each participant

Paper for each participant

Pencils or pens for each participant

Tacks or masking tape for displaying completed chart pages.

17. TECHNOLOGY

Not computer supported, although data may be stored for analysis and design uses.

18. FORMER USERS

Since PPM is a method published in the public domain, its use is more widespread than documented. There is a wide variety of uses, including the following:

Governor's Task Force for Planning Health Services, 1972, State of Wisconsin. [Delp et al., p. 229]

Health planning agency, Texas 1973. [Delp et al. p. 229]

19. BIBLIOGRAPHY

Bryson, John M., and André L. Delbecq. "A Contingent Approach to Strategy and Tactics in Project Planning," *Journal of the American Planners Association*, April 1979, pp. 167–179.

Delbecq, André L., and Andrew H. Van de Ven. "A Group Process Model for Problem Identification and Program Planning," *Journal of Applied Behavioral Science*, 7, 4 (1971), pp. 466–492.

Delbecq, André L., Andrew Van de Ven, and David Gustafson. *Group Techniques for Program Planning*, Scott, Foresman, Glenview, Ill., 1975.

Delp, Peter, Arne Thesen, Juzar Motiwalla, and Neelakantan Seshadri. *Systems Tools for Project Planning*, International Development Institute, Bloomington, Ind., 1977, pp. 227–230.

Van de Ven, Andrew H. "Problem Solving, Planning and Innovation, Part I: Test of The Program Planning Method; Part II: Speculations for Theory and Pratice," *Human Relations Journal*, 33 (Nov.–Dec. 1980), pp. 10–11.

A ROLE-ORIENTED APPROACH TO PROBLEM-SOLVING

1. DESCRIPTION

   A role-oriented approach to problem-solving is "a procedure for solving problems by structuring the efforts of a group according to roles which control particular types of information. . . . This procedure is augmented in practice by operational methods for stimulating creativity and principles drawn from the study of group dynamics" [Burnette et al., p. 481].

2. BRIEF HISTORY

   "The Role-Defined method has been developed over the past several years by Charles Burnette and received its first major public exposure at the 1971 Conference of Environmental Design Educators in Key Biscayne, Florida" [Burnette et al., p. 482].

3. BASIC PREMISE OR ASSUMPTION

   "The approach is based on a theoretical structure for a comprehensive information-processing system conceived to complement thought and expression." [Burnette et al., p. 482].

4. PURPOSE/GOAL

   ". . . for structuring the efforts of problem-solving teams, . . . stimulating creativity, and facilitating the interpersonal dynamics of task-oriented groups" [Burnette et al., p. 482].

5. MENTAL OPERATIONS SUPPORTED

   Cognition
   Divergent production
   Convergent production
   Evaluation

6. BENEFITS

   Role categorizations facilitate descriptive analysis of communication.

   Roles provide the structure for studying interpersonal dynamics and creative behavior within groups.

   Method as a whole can be used for studying and developing personal aptitudes for various roles, as well as for the group as a whole.

   [Source: Burnette et al., p. 409]

7. LIMITATIONS
   None.

8. PARTICIPANTS

   Quantity
   Seven.

   Qualifications
   Vary with each problem.

   Roles of Group Members
   Problem designator/client.
   Resource specifier.
   Resource organizer.
   Form giver.
   Actualizer-doer.
   User.
   Evaluator.
   Facilitator.

9. INPUT
   Brief statement and analysis of the problem.

10. OUTPUT
    Policy, plan, or problem solution.

11. BASIC ELEMENTS
    Facts.
    Opinions.
    Ideas.

12. RULES FOR USE SYNTAX
    Roles.
    Constructive listening policy.
    Positive selections and structured feedback policy.
    Synthesis policy.

    [Source: Burnette et al., p. 483]

13. OPERATIONAL PROCEDURES

    Training
    1. Group dynamics policies.
    2. Informational roles and their use.
    3. Creative idea-generation techniques.

Problem Session

1. Problem statement by problem designator or user(s).
2. Role-oriented goals or objectives from all participants.
3. Selection of initial goals by problem designator or user(s).
4. Sequential introduction of information from all roles.
5. Idea-generation trips (as necessary).
6. Development of ideas and information in all roles.
7. Feedback on ideas and possible solutions from problem designator and user(s).
8. Procedural feedback from evaluator.
9. Refining of ideas toward possible solutions.
10. Recycling through generation of more goals.

14. TIME REQUIRED

Several hours to several weeks or months.

15. MEETING ROOM/ENVIRONMENT

Large, circular table next to a blackboard. Organizer and form giver sit nearest blackboard. Clockwise, from the form giver, are the actualizer-doer, user(s), evaluator, problem designator, and resource specifier. "The Facilitator roves around outside the group . . ." [Burnette et al., p. 485].

16. SUPPLIES

None.

17. TECHNOLOGY

Not computer supported.

18. FORMER USERS

Model Cities project in Miami.

How to involve the outside world in the Affairs of the Martin Luther King Boulevard Project?

A project in landscape architecture in Seattle.

How to specify the population and density of development in a site to assure ecologically sound development?

Department of Architecture, Yale University.

How to renovate an abandoned railroad station to serve as a community center for several user groups?

19. BIBLIOGRAPHY

Burnette, Charles H., Gary T. Moore, and Lynn Simek. "A Role Oriented Approach to Problem-Solving by Groups," in Wolfgang Preiser, Ed., *Environmental Design Research,* Vol. 1, selected papers, 4th International EDRA Conference, Dowden, Hutchinson, and Ross, Stroudsburg, Pa., 1973.

SYNECTICS

1. DESCRIPTION

   Synectics is the name of a company and it is also used to represent a body of knowledge about the processes of creative thought, the dynamics of group problem-solving, and theories about individual difficulties with speculative, as opposed to routine thinking. The word *synectics* was coined in an attempt to capture the essence of an intention—the joining together of diverse elements. It applies equally well to group work, where, to get synergy, one must get diverse individuals to join and cooperate in the accomplishment of the enterprise at hand.

2. BRIEF HISTORY

   "Synectics (stands for) a set of procedures introduced and developed primarily by Synectics, Inc., an American consultancy organization. The name was coined by W. J. J. Gordon, co-founder of the company, to describe a process leading to new insights through bringing together elements that are normally unrelated" [Rickards, p. 78]. "In 1961, . . . Gordon summarized the results (of 10 years) of research into creative individuals and creative process (in his book, *Synectics*)." Following the publication of this book, additional contributions were made to creative problem-solving by groups, by George Prince who recognized forces that operated in groups which neutralized some procedures for spurring creativity. The focus of research was then expanded to investigate ways in which groups of people could work together in more efficient and productive ways [Rickards, p. 79]. "The elements of Synectics today can be traced back to the ideas of Gordon and Prince" [Rickards, p. 79].

3. BASIC PREMISE OR ASSUMPTION

   Few people use more than a fraction of their potential for creative thinking. By making explicit the thinking operations and the various forces that keep us thinking routinely, both as individuals and in groups, we can develop skills to use more of our latent capacities.

4. PURPOSE/GOAL

   To generate insightful and inventive solutions to a problem.

   To increase the probability of success by engaging more of the latent potential of individuals and groups.

   To obtain commitment to a line of action where members of the group hold adversary positions.

5. MENTAL OPERATIONS SUPPORTED

   Cognition

   Divergent production

   Convergent production

6. BENEFITS

   The organization and procedures get a group cooperating and supporting quickly.

   The protected climate of the meeting encourages the participants to use more freely their abilities to speculate.

   The idea-getting strategies help group members explore possibilities that tend to be left untouched.

   Beginning ideas of promise are not discarded because they have flaws. The flaws are systematically attacked.

   The power of a synergistic group is far greater than that of a person working on his or her own. As a result, Synectics groups tend to produce more and more novel solutions.

   Untrained people can quickly (within 5 or 10 minutes) learn to participate.

   Participation in meetings tends to increase use of potential outside of meetings.

7. LIMITATIONS

   Some of the idea-getting strategies of Synectics that stimulate right-hemisphere imaging may make very logical, left hemisphere-dominated people uncomfortable.

   Superior–subordinate relationships must be discarded in the sessions and some superiors resist this.

   Synectics sessions are not always successful, and if expectations are high, the group may abandon the method without giving it a chance to help them.

   A successful session may take from 15 minutes to two days, depending upon the importance and complexity of the problem. It is difficult to get managers to commit sufficient time for problems even if they are important.

8. PARTICIPANTS

   Quantity

   The number of members should be determined by the task. If a high degree of innovation is needed, the group should be small, that is, five to seven, including the facilitator. If what is wanted is opinion and input from a large number of people, the meeting may be as large as needed. Creative efficiency declines slightly when there are more than five people. It begins to decline sharply with over seven.

Qualifications

It has been found, with experience, that nearly anyone can be a valuable participant if he or she is willing to cooperate with the facilitator and the other members of the group. It *is* necessary to have the required expert knowledge in the group.

Roles of Group Members

Leader/facilitator

Client (problem owner)

Participants (these will include whatever experts the problem owner may feel will be relevant)

9. INPUT

A concise statement of the problem and a brief explanation.

10. OUTPUT

A number of speculative restatements of the problem (goal/wishes). A number of possible solutions or options.

11. BASIC ELEMENTS

Ideas.

12. RULES FOR USE/SYNTAX

Constructive Interpersonal Behavior

Open-minded listening.
Building.
Crediting.
Supporting.

Constructive Interpersonal Behavior

Flow Chart.
Itemized responses (evaluation).
Interacting with pads.

Creativity Spurring Processes

Nonpunishing, appreciative climate.
Goal/wishes.

Strategies to encourage right brain thinking:
1. Metaphor and analogy
2. Imaging and analogy
3. Doodling

13. OPERATIONAL PROCEDURES

Presession Activities
1. Planning meeting between client and facilitator to discuss and clarify the objectives of the meeting and to select appropriate participants (people with expert knowledge or influence who may be needed to implement the possible solutions produced). One to three hours.
2. Acquaint the group with the process (two hours). (If the group is to be trained in the process; participant training takes three days, facilitator training to entry level takes five days; advanced level, another 5 days.)

Session Activities
1. Client presents the problem and a brief explanation (five minutes).
2. At the same time the group is listening for goal/wish clues and writing these on their pads.
3. Team generates more goal/wishes as these are recorded on wall pads by facilitator (30 minutes).
4. Client selects goal/wish that appeals to him/her.
5. As client explains why this goal/wish is appealing, team develops immediate ideas and these are recorded by the facilitator (15 minutes).
6. Facilitator takes team on excursion to develop more speculative beginning ideas. Force fitting is done within the excursion at the end.
7. Client selects the most interesting and promising of the ideas, or a combination thereof, and does an itemized response (evaluation of the idea) (five minutes).
8. Team develops ideas to overcome the concerns of the client (10 minutes).
9. If the idea is tested and turned into a possible solution by the client, the facilitator gets up and the next steps are dictated by the client (five minutes).

10. End of meeting (meeting times range from about an hour to two or three days—during which there are many one-hour meetings starting with different goal/wishes).

Postsession Activities

.1. Client and people of his selection decide which possible solutions to pursue.

2. They then test feasibility.

14. TIME REQUIRED
From one hour to three days.

15. MEETING ROOM/ENVIRONMENT
Any comfortable, quiet room will do that is large enough to hold the group.

16. SUPPLIES

Three or four large easel pads

Large crayons

Masking tape to hang the sheets

Pads of paper for each individual

Pens

Pencils

17. TECHNOLOGY
Not computer supported.

18. FORMER Users
About 50% of the companies listed in *Fortune*'s 500 have been or are clients of Synectics, Inc. The technique has been successfully applied in the following areas:

Administration

Marketing

Personnel management

Production

Research

Cost reduction

New product development

Process improvement

19. BIBLIOGRAPHY

Delp, Peter, Arne Thesen, Juzar Motiwalla, and Neelakantan Seshadri. *Systems Tools for Project Planning,* International Development Institute, Indiana University, Bloomington, Ind., 1977.

Gordon, William J. J. *Synectics,* Collier-Macmillan, London, 1970.

Rickards, Tudor. *Problem-Solving Through Creative Analysis,* Gower Press, Essex, 1974.

Prince, George. *The Practice of Creativity,* Harper and Row, New York, 1970.

VALUE ENGINEERING/VALUE ANALYSIS

1. DESCRIPTION

   Value engineering/value analysis (VE/VA) is a special purpose technique for creating functionally equivalent or improved designs for products or services while simultaneously reducing overall production or delivery costs. VE/VA requires a multidisciplinary design group to follow a four-step job plan to create and evaluate alternatives to provide the delivery of key functions desired in a product or service.

2. BRIEF HISTORY

   Value engineering was developed by Lawrence D. Miles of the General Electric Company in 1949 and has been applied in product engineering, architectural, civil, and chemical engineering, and human services since that time.

3. BASIC PREMISE OR ASSUMPTION

   That products and services derive value by providing a means to desired functions, which can be identified in two-word (verb–noun) phrases, and that existing methods of product design or service delivering can be usually improved through creative brainstorming and critical evaluation.

4. PURPOSE/GOAL

   To identify functions a product or service provides to users.

   To diagram these functions into basic and supporting categories that support a primary task for which the product or service is intended.

   To creatively identify other ways in which key selected functions may be delivered.

   To evaluate alternative methods for function delivery according to user-based criteria.

   To present a proposal to decision-makers that manifests high user acceptance, significant cost reduction, improved functional design, and feasibility.

5. MENTAL OPERATIONS SUPPORTED

   Cognition

   Divergent production

   Convergent production

   Evaluation

6. BENEFITS

"Better for less," in terms of increased capacity for functional delivery along with usually significant cost reductions.

Identification of high-cost components in product or services design, which subsequently allows for targeted redesign foci.

Identification of user acceptance ratings on product or service components, which subsequently guides retention of areas of high acceptance, and modification or elimination of areas of less user acceptance.

Creation of dozens of new alternatives for provision of important functions.

Structured approach for group analysis, creativity, and evaluation, which remain focused on functions and problem-solving.

Enormous return on investment (ROI) where VE/VA programs are sponsored and led by well-trained value specialists.

7. LIMITATIONS

Multidisciplinary group membership may be difficult to obtain from functionally organized, hierarchical organizations.

Interpersonal collaboration may be difficult to manage unless top management support is strong.

Group commitment to the job plan may wane as the process continues.

8. PARTICIPANTS

Quantity

Five to seven members per team.

Qualifications

Group make-up may be homogeneous but is preferably heterogeneous, for example, engineering, sales, marketing, user, manufacturing, and so on. Heterogeneous groups usually exhibit more creativity, but they may have more difficulty communicating.

Roles of Members of the Group

Leader, preferably a certified value specialist (CVS)

Participants

9. INPUT

Product or facsimile of service to be subjected to analysis

All available data, for example, market, engineering, manufacturing, and so on.

10.   OUTPUT

    FAST diagram

    Cost analysis

    Proposal

11.   BASIC ELEMENTS

    FAST diagramming

    User data.

    Cost data.

    Sample product or service for team observation.

    Ideas.

12.   RULES FOR USE/SYNTAX

    FAST diagramming.

    Brainstorming.

    Cost analysis.

    Discussion and evaluation of alternatives.

    Proposal development and presentation.

13.   OPERATIONAL PROCEDURES

    1.   Collect information.

    2.   Conduct FAST diagramming.

    3.   Conduct cost analysis.

    4.   Brainstorm alternatives for target functional areas.

14.   TIME REQUIRED

    Depends entirely on availability of data and team members; when use and cost data are available and team members are able to schedule joint session, VE/VA can be done in an abbreviated fashion within two hours; usually one and a half to three days are needed for complex products and services. Completion of individual phases can be staggered over several days if team cannot remain in one physical setting from beginning to end.

15.   MEETING ROOM/ENVIRONMENT

    The work table must be large enough to allow the product or service sample to be displayed, plus accommodate chart pads and other supplies listed below.

16.   SUPPLIES AND EQUIPMENT

    Flip chart

    Marking pens

Index cards for FAST diagramming

Paper and pencils for each participant

Tape for posting index cards onto flip-charted FAST diagram

17. TECHNOLOGY

Computer supported; program for cost analysis developed by Value Standards, Inc., Madison, Wisconsin.

18. FORMER USERS

Lake Forest, Illinois School District.

General Electric.

Emerson Electric Corporation.

Cutler-Hammer Manufacturing.

Numerous production, engineering and manufacturing organizations.

19. BIBLIOGRAPHY

Fasal, John H. *Practical Value Analysis Methods,* Hayden Book, New York, 1972.

Miles, Lawrence D. *Techniques of Value Analysis and Engineering,* McGraw-Hill, New York, 1961.

Snodgrass, Thomas J. Value engineering course materials, University of Wisconsin-Extension, Department of Engineering and Applied Sciences, Madison, Wis.

AUTHOR INDEX

SUBJECT INDEX

Accident, happy, 341
Accountability, 212
Actions, destructive and encouraging, 322
Actualizer-Doer, 310
Adjacency digraph, 173-174
Ambiguity, 41
Amendment software, 164
Analogies: protection of, 346
  use of, 346
Analysis, 285
Anonymity, 106, 123
Answers, 211, 217, 232
Antecedents, 175
Aptitude, 305
Architectural design, 243
Arcs, 167
Argument, 203, 211, 212, 215, 216, 217, 231
Argumentation, 211
Argumentative decision-making procedure, 209
Argumentative model of design, 204
Arithmetic processes, 256
Array, 169, 176
Arthur D. Little, The Invention Design Group, 324
Articulation, 250
Assertions, 180
Assignment costs, 386
Associations, 168
Associating, 55
Associator, 58
Attribute enhancement structures, 183, 188
Attribute lists, 86
Automated indexing, 240
Automated transcription of voices, 242

Bargainer, 237
Bargaining, 211, 237, 238, 239, 242

BASIC, 247
Basic functions, 369
Behavior: competitive, 47
  cooperative, 47
Blocks, 172
Bottom vertex, 175
Boundary loss, 46
Boundary maintenance, 46
Brain division, 330
Brainstorming, 48, 77, 78, 79, 312, 324, 395
  basic elements, 84, 397
  basic premise or assumption, 77, 78-79, 395
  benefits, 396
  bibliography, 400
  case study, 96-101
  description, 77, 395
  former users, 400
  history, 77-78, 395
  input, 83-84, 397
  limitations, 396
  meeting room/environment, 399
  mental operations supported, 60, 396
  operational procedures, 82-96, 398
  output, 79, 82, 94, 397
  participants, 82, 396
  purpose/goal, 77, 396
  references, 101
  rules for use/syntax, 85-93, 397
  supplies, 399
  technology, 399
  time required, 97, 399
Brief vacation, 312
Broker, 159
Building and maintenance functions, 35

California Polytechnic State University, 244
Causal-factual images, 211